THE HOUSING AND ECONOMIC EXPERIENCES OF IMMIGRANTS IN US AND CANADIAN CITIES

Since the 1960s, new and more diverse waves of immigrants have changed the demographic composition and the landscapes of North American cities and their suburbs. *The Housing and Economic Experiences of Immigrants in US and Canadian Cities* is a collection of essays examining how recent immigrants have fared in getting access to jobs and housing in urban centres across the continent.

Using a variety of methodologies, contributors from both countries present original research on a range of issues connected to housing and economic experiences. They offer both a broad overview of these issues and a series of detailed case studies that highlight the experiences of particular communities. This volume demonstrates that, while the United States and Canada have much in common when it comes to urban development, there are important structural and historical differences between the immigrant experiences in these two countries.

CARLOS TEIXEIRA is a professor in the Department of Geography at the University of British Columbia Okanagan, Canada.

WEI LI is a professor in the School of Social Transformation and the School of Geographic Sciences and Urban Planning at Arizona State University, U.S.A.

# The Housing and Economic Experiences of Immigrants in US and Canadian Cities

EDITED BY
CARLOS TEIXEIRA AND WEI LI

UNIVERSITY OF TORONTO PRESS
Toronto Buffalo London

Toronto Buffalo London
www.utppublishing.com

ISBN 978-1-4426-5035-0 (cloth)
ISBN 978-1-4426-2838-0 (paper)

**Library and Archives Canada Cataloguing in Publication**

The housing and economic experiences of immigrants in US and Canadian cities / edited by Carlos Teixeira and Wei Li.

Includes bibliographical references and index.
ISBN 978-1-4426-5035-0 (bound). – ISBN 978-1-4426-2838-0 (pbk.)

1. Immigrants – Housing – United States. 2. Immigrants – Housing – Canada. 3. Immigrants – United States – Economic conditions. 4. Immigrants – Canada – Economic conditions. 5. Urban economics – United States. 6. Urban economics – Canada. I. Teixeira, Carlos, editor II. Wei, Li, 1957–, editor

HD7288.72.U5H69 2015 304.8'73 C2014-906942-1

This book has been published with the help of a grant from the Federation for the Humanities and Social Sciences, through the Awards to Scholarly Publications Program, using funds provided by the Social Sciences and Humanities Research Council of Canada

University of Toronto Press acknowledges the financial assistance to its publishing program of the Canada Council for the Arts and the Ontario Arts Council, an agency of the Government of Ontario.

Canada Council for the Arts
Conseil des Arts du Canada

University of Toronto Press acknowledges the financial support of the Government of Canada through the Canada Book Fund for its publishing activities.

# Contents

*List of Figures and Tables* ix

*Preface* xiii
AUDREY KOBAYASHI

*Acknowledgments* xvii

**Introduction**

1 The Housing and Economic Experiences of Immigrants in Canada and the United States 3
WEI LI AND CARLOS TEIXEIRA

**Part One: The Housing Experiences of Immigrants**

Introduction to Part One: The Housing Experiences of Immigrants 23
CARLOS TEIXEIRA

2 Homeownership among Immigrants in Canada and the United States: Similarities and Differences 43
JOE T. DARDEN

3 Household Formation and Homeownership: A Comparison of Immigrant Racialized Minority Cohorts in Canada and the United States 69
MICHAEL HAAN AND ZHOU YU

4 How Are Sri Lankan Tamils Doing in Toronto's Housing Markets? A Comparative Study of Refugee Claimants and Family Class Migrants 98
SUTAMA GHOSH

5 A Two-Sided Question: The Negative and Positive Impacts of Gentrification on Portuguese Residents in West-Central Toronto 121
ROBERT A. MURDIE AND CARLOS TEIXEIRA

6 The Good, the Bad, and the Suburban: Tracing North American Theoretical Debates about Ethnic Enclaves, Ethnic Suburbs, and Housing Preference 146
VIRPAL KATAURE AND MARGARET WALTON-ROBERTS

7 Housing Experiences and Trajectories among Ethnoburban Chinese in Los Angeles: Achieving Chinese Immigrants' American Dream 176
WAN YU

**Part Two: The Economic Experiences of Immigrants**

Introduction to Part Two: The Economic Experiences of Immigrants in Canada and the United States 207
JOHN R. MIRON

8 The Colour of Money Redux: Immigrant/Ethnic Earnings Disparity in Canada, 1991–2006 227
KRISHNA PENDAKUR AND RAVI PENDAKUR

9 Immigrant Underemployment in the US Urban Labour Markets 261
TETIANA LYSENKO AND QINGFANG WANG

10 The Latino Commercial Landscape and Evolving Hispanic Immigrant Population in Two Midwestern Metropolitan Areas 281
ALEX OBERLE

11 Immigrant Entrepreneurship in the Washington Metropolitan Area: Opportunities and Challenges Facing Ethnic Minorities 302
ELIZABETH CHACKO AND MARIE PRICE

12 Financing Immigrant Small Businesses in Canada and the United States 328
WEI LI AND LUCIA LO

**Conclusion**

13 Immigrant Experiences and Integration Trajectories in North American Cities: An Overview and Commentary on Themes and Concepts 355
JOHN W. FRAZIER

*Contributors* 377

*Index* 383

# Figures and Tables

**Figures**


1.1 The structure of a welcoming society 4
3.1 Headship and homeownership rates by race/ethnicity and immigrant status over the early 2000s 84
3.2 Relative risk ratios by race/ethnicity: Assessing variable household formation 92
4.1 Residential locations of Sri Lankan Tamils in Toronto, created from 2006 census data 108
4.2 The current residential locations of Sri Lankan Tamil respondents in Toronto, 2011 109
5.1 Portuguese ethnic origin by census tracts, Toronto CMA, 2006 124
5.2 Portuguese ethnic origin by census tracts, West Central Toronto, 2006 125
5.3 Gentrification in inner-city Toronto 131
6.1 The Greater Toronto Area, CMA and municipalities 160
7.1 Percentage of Chinese among suburbs in Southern Los Angeles County, 1990 181
7.2 Percentage of Chinese in Southern Los Angeles County, 2010 182
7.3 Percentage of Chinese in Monterey Park by census tract, 2000 183
7.4 Percentage of Chinese in Rowland Heights by census tract, 2000 184
7.5 Chinese household income distribution in Monterey Park, Rowland Heights, and Los Angeles County, 2000 186

7.6 Condominiums in Monterey Park 192
7.7 Single-family homes in Monterey Park 193
7.8 Single-family homes in Rowland Heights 195
7.9 Apartments on Colima Road, Rowland Heights 195
8.1 Earnings differentials between selected groups and Canadian-born white men and women, 1990–2005, Canada 238
8.2 Earning differentials between selected groups and Canadian-born white men and women, 1990–2005, Montreal 238
8.3 Earnings differentials between selected groups and Canadian-born white men and women, 1990–2005, Toronto 239
8.4 Earnings differentials between selected groups and Canadian-born white men and women, 1990–2005, Vancouver 239
8.5 Earnings differentials among women, selected European origins versus Canadian-born British origin, 1990–2005, Canada 247
8.6 Earnings differentials among women, selected non-European origins (1) versus Canadian-born British origin, 1990–2005, Canada 248
8.7 Earnings differentials among women, selected non-European origin (2) versus Canadian-born British origin, 1990–2005, Canada 248
8.8 Earnings differentials among men, selected European origins versus Canadian-born British origin, 1990–2005, Canada 249
8.9 Earnings differentials among men, selected non-European origins (1) versus Canadian-born British origin, 1990–2005, Canada 249
8.10 Earnings differentials among men, selected non-European origin (2) versus Canadian-born British origin, 1990–2005, Canada 250
9.1 Underemployment rate of immigrants in US urban areas by country of origin 268
10.1 Case study area 287
10.2 La Tapatía Tienda Mexicana in Des Moines, Iowa 290
10.3 Birrieria Reyes de Octotlán Restaurant in the Pilsen neighbourhood in Chicago, Illinois 294

10.4 Strip mall illustrating part of the Latino commercial landscape in Carpentersville, Illinois 295
11.1 Distribution of Bolivian population and businesses in the Washington Metropolitan Area 312
11.2 Distribution of Ethiopian population and businesses in the Washington Metropolitan Area 313

**Tables**

1.1 Comparison of immigration in Canada and the United States 11
1.2 US businesses owned by foreign-born owners by gender and race-ethnicity, 2007 14
2.1 Mean homeownership rates by place of birth and ratio of equality with homeownership rate with immigrants born in Europe, Canada, 2006 51
2.2 Homeownership rates and place of birth by metro area in Canada, 2006 53
2.3 Mean homeownership rates by place of birth and ratio of equality with homeownership rate with immigrants born in Europe, United States, 2010 54
2.4 Homeownership rates and place of birth by primary metropolitan area in the United States, 2010 56
2.5 National homeownership rates by place of birth and ratio of equality with homeownership rates with immigrants born in Europe – Comparison of United States in 2010 and Canada in 2006 58
3.1a Descriptive statistics (2001, 2006) and coding information, Canada 75
3.1b Descriptive statistics (2000, 2005) and coding information, US top 20 metros 76
3.2a Per capita homeownership and headship rates by birth cohort in Canadian Census Metropolitan Areas, 2001 and 2006 79
3.2b Per capita homeownership and headship rates by birth cohort in top 20 US metropolitan areas, 2000 and 2005 79
3.3a Per capita homeownership and headship rates in Canadian Census Metropolitan Areas, 2001 and 2006 80
3.3b Per capita homeownership and headship rates in top 20 US metropolitan areas, 2000 and 2005 81

3.4a Per capita homeownership and headship rates in Canadian Census Metropolitan Areas, 2001 and 2006 82
3.4b Per capita homeownership and headship rates in top 20 US metropolitan areas, 2000 and 2005 82
3.5a The relative risk ratios of the determinants of household formation in Canadian Census Metropolitan Areas, 2001 and 2006 86
3.5b The relative risk ratios of the determinants of household formation in top 20 US metropolitan areas, 2000 and 2005 88
4.1 Demographic and socio-economic conditions of respondents at the time of interview 104
5.1 Total population and Portuguese population: Little Portugal, West Central Toronto, and Toronto CMA, 1971, 1981, 1991, 2001, and 2006 128
6.1 Evolution of ethnic urban settlement theories in the United States and Canada 147
7.1 Percentage of Chinese among total population in Monterey Park and Rowland Heights, 1990–2010 185
7.2 Selected demographic and socio-economic characteristics of Chinese in Monterey Park and Rowland Heights, 2010 187
7.3 Selected housing characteristics of Chinese in Monterey Park and Rowland Heights, 2010 189
8.1 Earnings differentials (coefficients) between white Canadian-born and immigrant and visible minorities by sex, Canada, Montreal, Toronto, and Vancouver, 1991–2006 234
8.2 Earnings differentials between selected ethnic groups and British-origin, Canadian-born workers by immigrant status and sex, Canada and MTV, 1991–2006 243
9.1 Demographic characteristics of immigrants in US urban areas 267
9.2 Underemployment rate of immigrants in US urban areas by industry 270
9.3 Regression results: Characteristics associated with underemployment for immigrants in US urban areas 271
10.1 Select characteristics of case study areas 297
11.1 Occupational divisions of Bolivians and Ethiopians in Metropolitan Washington, 2009–2011 310
12.1 Population and banks in the San Francisco Bay area and Vancouver CMA 333

# Preface: Doing Better with Immigrant Integration

AUDREY KOBAYASHI

Immigration presents some of the most pressing social, cultural, and public policy issues in both the United States and Canada. Canada has recently undergone a revamping of its immigration legislation to focus on those newcomers deemed to fit more closely with the Canadian labour market. The United States is in the early stages of a debate on comprehensive immigration reform that will certainly galvanize opinions on the topic, and possibly contribute to a redefinition of who belongs and who does not. Debates in both countries show that tremendous political, economic, and social capital are enmeshed in immigration debates, and the results of those debates are written on the bodies of those who arrive – as well as those who are not given the opportunity to arrive – to establish new homes.

Whatever the topic – changing demographic patterns, education, labour markets, health care, housing – the experiences of immigrants need to be taken into account. In both countries, the recent story is a good news–bad news one. Immigration has helped to shift the dependency ratio, has provided important labour, especially at the lower end of the wage scale, and has helped millions of people for whom conditions in their countries of origin were for whatever reason intolerable. Immigration has enriched the cultural diversity of both countries. On the other hand, there is growing evidence that neither society has come to terms with the needs of immigrants and that their integration is far from optimal. They experience racism. Notwithstanding their high levels of human capital and their contribution to emerging fields such as medicine and technology, they are also disproportionately employed not only in the lowest-paying jobs, but often at levels that do not reflect their educational attainments and international work experience. In housing and economic status, two of the most important indicators of social well-being, immigrants, especially recent immigrants, face significant challenges. Millions of undocumented migrants and those on temporary visas often live in the worst conditions and are exploited horribly.

We should be able to do a better job of providing the social and policy context in which newcomers might find a better welcome, but the challenges are many. Scholars provide more and more evidence that the neoliberal governmentality that regulates contemporary social and political discourse and practices gives us plenty of language to celebrate the role of the state in immigrant integration, but little in the way of substantive change. The result is often offloading responsibility for change on those who face the greatest barriers. Neoliberalism, in other words, contradicts its own ends by withdrawing state support and encouraging disinvestment in social programs. Public opinion polls in both countries also point to a disturbing backlash against difference, a growing distaste for the concept of multiculturalism, and a hardening of intolerance towards the marginalized or disenfranchised. Specific economic circumstances, especially in the wake of the economic downturn and housing crisis in the United States, make it all the more difficult to find the resources for settlement services, and curtail growth in those areas of the economy where immigrants often find themselves, especially in services, agriculture, and construction. Lack of economic opportunity only deepens the difficulties of finding affordable and appropriate housing.

The chapters in this volume propel us to do a better job. The editors have assembled a group of top scholars to provide timely, theoretically and empirically informed analyses at a range of scales from the national level to individual case studies. Overview chapters, some of them comparative, give a big picture of how newcomers are faring in finding housing and in economic integration in both countries, while case studies provide a glimpse of the situation in specific communities. The total picture tells us that there are important structural and historical differences between the two countries, stemming from different immigration histories, especially of non-white migrants, and significant differences in legal systems and public policy. The two countries also have much in common, culturally, economically, and in terms of urban development. Sorting out the strands of difference and convergence is a huge task, and several of the chapters in this book illustrate the importance of doing so.

The experiences of immigrants tell us how a society defines who belongs and who does not, who is representative of the nation and who is other. National immigration policies and practices reflect the normative values of the society as it is; but newcomers also gradually and cumulatively change those values as their experiences comingle with those of the already arrived and the long settled. We do not have to subscribe to the "invasion-succession" models that were favoured by members of the Chicago School during the 1920s to recognize that the process of integration is dynamic, geographically uneven, and expressed differently at different times and for different groups (xv). This set of texts is rich in case studies, which cannot of course cover the entire spectrum of migrant group experiences, but which provide a deeper understanding of a few groups to whom geographers have turned their attention. Place matters, and understanding experiences in place is an important objective of immigration scholarship. Particularly in the area of entrepreneurial activities, these case studies demonstrate the contingent recursivity of the relationship between newcomers and their new environments, and allow several glimpses into the perplexing question of who belongs and how belonging is asserted.

Collectively, these chapters should also enhance our theoretical understanding of the human relationships entangled in the migration process. They speak to the power of place in attracting and confining specific groups, and in sorting people according to the social characteristics that define migrants. They speak of the role of migration in creating urban environments, and of the larger social factors that define, constrain, or facilitate migrant entry into housing and labour markets. And they speak to the very human experiences of those who choose to pick up and move across the globe, as well as the ways in which established residents make room for them. In short, they provide a basis for advancing our ability to do better when it comes to immigrant integration.

# Acknowledgments

The editors wish to thank all contributors for accepting our invitation to participate in this endeavour and for their dedication and hard work on this project. A special thank you goes to Audrey Kobayashi for her friendship and support throughout the years and for her constructive criticism of the first draft of this book. We also thank our editor at the University of Toronto Press, Douglas Hildebrand, for believing in this project and providing the cover photo. Carlos Teixeira expresses his gratitude to Wei Li, a long-time collaborator and friend, for accepting to join forces with him once more to work on another "North-South" project. He also thanks his research assistant Amanda Jones for her help through the different stages of preparing this book. Wei Li thanks Carlos Teixeira for initiating and leading this book project and Wan Yu of Arizona State University for her assistance in drafting the first portion of the Economic Experience section and for data retrieval. She is also indebted to US National Science Foundation grant BCS-0852424, which provides time and resources for her work towards this book and its chapter 12.

# Introduction

# 1 The Housing and Economic Experiences of Immigrants in Canada and the United States

WEI LI AND CARLOS TEIXEIRA

Canada and the United States are known around the world as major destinations for international migrants. The large numbers of immigrants the two countries attract have played an important role in shaping the social, economic, and political landscapes of both countries in the modern "age of migration." These immense inflows have transformed certain metropolitan areas in the United States and Canada into "cities of nations," making them some of the most multicultural places on the planet (Anisef and Lanphier 2003; Kobayashi, Li, and Teixeira 2012).

Cities in North America serve not only as "ports of entry" for immigrants from all over the world, but also as places where the majority of immigrants settle. Many of these cities provide diverse labour markets, but also have high housing costs. As our previous work demonstrates (Teixeira and Li 2009, figure 1.1), economic, housing, and education opportunities are key factors in an immigrant's integration. Moreover, they are the essential elements for a receiving society to be deemed a "welcoming society." Before they can fully settle in a receiving country, most immigrants have two priorities: to search for and establish a dwelling, and a job or opportunity to start a business for income to support themselves and their families. While this appears obvious, no scholarly books focus on the housing and economic experiences of immigrants.

This book is about the housing and economic experiences of immigrants in urban Canada and the United States. Although both countries have been built and defined by immigration, there has been little comparative or comprehensive study of immigrants' housing and economic experiences. Since the mid-1960s, when both countries shifted towards less discriminatory immigration admission policies, new and more heterogeneous immigrant flows have changed the demographic

Figure 1.1 The structure of a welcoming society

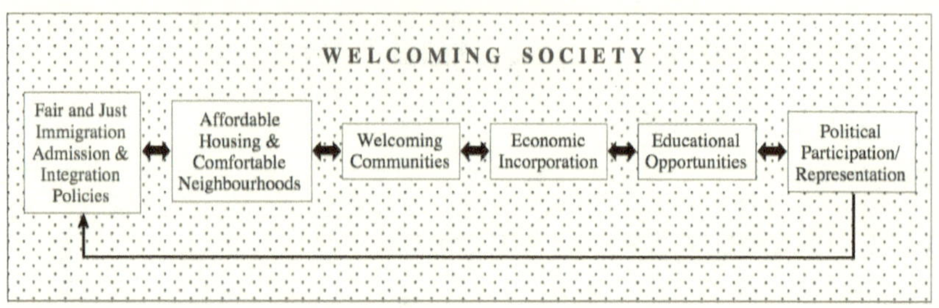


Source: Adapted from Teixeira and Li 2009.

compositions, as well as the business and residential landscapes, of North American cities and their suburbs. Scholars and policy experts have identified access to housing and to jobs as the primary ways in which immigrants achieve social and economic integration in the receiving society, yet the housing and employment or business experiences of many immigrant groups remain under-studied. There has been especially little analysis of these new realities from a bi-national or comparative perspective, although there have been numerous books on housing and economic matters in Canada and the United States (e.g., Bourne 1981; Miron 1988; Clark and Dieleman 1996; Borjas 2001; Hulchanski and Shapcott 2004; Tager 2005; Herman and Smith 2009; Schwartz 2010). To meet this gap, this book offers a collection of mostly original research from leading and emerging researchers from Canada and the United States who study immigration, housing markets, and economic experiences in urban North America. The organizational structure of the book is that we provide introductory remarks in this chapter, followed by an introduction to each part (housing and economic experiences) and the individual chapters. Each part begins with a large-scale data analysis chapter on Canada or the United States, comparatively or separately, followed by case study chapters on specific groups/areas or issues; thus, the contributors use a variety of research methods from macro-level quantitative to small-scale qualitative. A concluding chapter summarizes the major findings and issues addressed in the book and beyond. By drawing together a variety of scholarly perspectives on the main themes and examining either bi-national comparatively or one country comprehensively, this book offers both scholars and students a foundation for classroom study and research reference.

## Housing Experiences

One purpose of this book is to look at the housing experiences of immigrants in urban Canada and the United States. As noted by several of the contributors in both sections of this volume, while there are indeed similarities between Canada and the United States, significant differences also affect immigrants' housing and economic experiences and their integration in the two countries – differences in economic policy, lending, immigration policy, history, racism and discrimination, immigration sources, and so on. For example, Kobayashi (see Preface) rightly reminds us that for immigrants in both countries "the process of integration is dynamic, geographically uneven, and expressed differently at different times and for different groups." Within this context, the numerous "forces" (historical, structural, and racial/cultural) at play in local housing markets are complex and often difficult to untangle, as several of the contributors in this volume demonstrate. The nature of these "forces" as well as their role and impact upon the often volatile, unpredictable, and expensive rental and owner-occupied housing markets in Canada and the United States make it more difficult to discern the dynamics of these markets and make comparisons between the two countries. To complicate matters for those scholars interested in studying and comparing immigrants' experiences in both countries, there is a notable lack of tradition in conducting comparative housing studies among population subgroups, including immigrants, in both countries. Given the geographical and cultural proximity of both countries, the lack of comparative work is remarkable. Not surprisingly, this lack of scholarly work or interest translates into an open field that allows a wider range of analysts, including graduate students and non-scholars (such as housing providers, planners, and civil servants), to venture into conducting such needed comparative studies, research, and planning for better housing outcomes. To partly account for this gap in the housing literature, we should point out the lack of relevant databases and comparable data, including census information, for analysis of the housing situations, trajectories, and integration of immigrants.

In order to fill some of these identified gaps in the housing literature, the contributors to the housing section of this volume (chapters 2 to 7) try to identify differences and similarities between the countries by considering the following interrelated aspects of immigrants housing experiences and integration in urban Canada and the United States: (a) the housing needs and the barriers and challenges immigrants face in their settlement process and in securing affordable housing; (b) the coping strategies used by immigrants to deal with the housing barriers they encounter in local housing markets (rental and ownership); (c) immigrants' housing trajectories or careers and integration; (d) inequality among groups in access to and rates of homeownership in various cities and suburbs, as an indication of immigrants' integration into North American society; (e) the gradual shrinking of the affordable rental housing market and the role and impact of gentrification upon ethnic neighbourhoods; (f) residential segregation, the formation of suburban ethnic clusters, and the suburbanization of immigrant groups; and (g) immigrants' impacts on North American housing markets and the major housing policy implications. Contributors to the housing section of this volume have examined the key housing elements, issues, and themes identified above by using different methods and approaches – quantitative and qualitative – in different contexts in both countries. All the chapters provide timely insights and proceed from an in-depth understanding of immigrants' complex and diverse housing experiences and paths to integration in Canada and the United States.

We have identified three main housing themes reflected in the six chapters on housing – housing inequalities and homeownership; housing careers, social networks, and gentrification; and housing suburban immigrants and policy implications. The first research theme – *housing inequalities and homeownership* – is addressed directly by Darden in chapter 2 and by Haan and Yu in chapter 3. Darden argues that differential rates of homeownership occur among immigrant groups due to historic and ongoing discrimination by place of birth and racism, such that European-born immigrants continue to have the highest homeownership rates in both Canada and the United States. While the role and impact of racial/ethnic discrimination in mortgage lending warrants further research in both countries, this is particularly difficult in Canada where relevant data is unavailable due to the lack of federal government interest in monitoring lending practices, a key example of how housing-related policy is differently politicized in Canada than in the United States.

In another comparative study, Haan and Yu (in chapter 3) compare household formation and homeownership patterns of racialized minority groups in Canada and the United States, comparing nearly identical groups in two different policy contexts, and providing a rare opportunity to look at how context shapes socio-economic outcomes like housing. While in some respects, the process of residential integration appears similar across countries of origin, the stance of these contributors is not the typical economic approach of "how context affects housing behaviour," but examines "whether context affects distinct groups *differently*" (p. 70). These two studies expand our knowledge of differential rates of immigrant homeownership by comparing not just two groups, but several immigrant groups, including racialized minorities in different contexts in Canada and the United States, using the most recent census data available. Given the ongoing recruitment of immigrants and increasing concerns about their socio-economic integration, these two important studies have policy-relevant implications for both Canada and the United States.

Two case studies, both set in Toronto, the main "port of entry" for immigrants in Canada, address the second research theme – *housing careers, social networks, and gentrification*: Ghosh (chapter 4) and Murdie and Teixeira (chapter 5) both focus on particular ethnic groups of immigrants. Ghosh reports on the housing careers of a previously unstudied group, one of the most impoverished "South Asian" subgroups – Sri Lankan Tamils – in Toronto's expensive housing market. Most Tamils enter the country as refugee claimants or family class migrants, bringing with them many forms of capital. Family class migrants are less often studied than refugees, and Ghosh compares their experiences by entry classification. Her results reveal that Tamils rely on strong, pre-existing social networks to locate affordable housing, living in overcrowded housing situations well beyond the initial settlement stage, so that "unlike other 'South Asians,' for this subgroup 'hidden homelessness' is not a temporary issue, but a continuous housing problem" (p. 113). Close familial and social ties define the basis for alliances and mutual assistance, but pragmatism was ultimately paramount for many Tamils; those who failed to contribute to the rent or mortgage were evicted by their own family members. External factors also influenced the geography of settlement; settlement patterns that were typical for Tamils in the 1980s had changed by the 1990s, partly due to gentrification in the urban core, a research topic that is the focus of attention in the work of Murdie and Teixeira (chapter 5).

Despite the rich research literature concerning gentrification in North America, there has been little consideration of the intersection between ethnic groups and gentrification. The available literature shows that "emphasis in gentrification research has been placed much more directly on class and gender than on ethnicity or race" (Murdie and Teixeira, this volume, p. 121). Murdie and Teixeira seek to understand the impact of gentrification on Portuguese residents living in "Little Portugal," the historic site of Toronto's Portuguese community.

Despite Toronto's ethnic and racial diversity, there has been no specific study of the impact of gentrification on the city's ethnic neighbourhoods. This study addresses a major gap in the literature by looking at the negative and positive impacts of gentrification upon one of the most colourful ethnic enclaves of Toronto's inner-city neighbourhoods.

A defining characteristic of gentrification is the loss of affordable housing; this was identified by the majority of low-income Portuguese respondents as the most important negative impact of gentrification in "Little Portugal"; this was also the reason for shifting the dominant settlement pattern of Tamils in Ghosh's study. In contrast, increased property values resulting from gentrification were advantageous to Portuguese homeowners who wished to capitalize on the increased equity in their homes. However, most Portuguese in the neighbourhood (including the homeowners who benefited financially upon the sale of their homes) were well aware of the implications of gentrification for the long-term viability (some would say "survival") of their communities. Murdie and Teixeira note that a subgroup of Portuguese homeowners – the "stayers" (mostly, first-generation seniors) – who remain in Little Portugal tend to have "fixed incomes" and become increasingly vulnerable because they lack the necessary financial resources to maintain their houses and afford the increasingly high property taxes associated with gentrification. Studies of gentrification in North America rarely consider the effects on "land rich but cash poor" immigrant seniors whose ethnic communities are dissolving at a time when their need for support of various kinds is mounting, or on newcomers who may be stranded in former immigrant reception areas.

Two more case studies address the third research theme – *housing suburban immigrants and policy implications*. One study, by Kataure and Walton-Roberts (chapter 6), deals with the suburbs of Toronto and the other, by Yu (chapter 7), focuses on the suburbs of Los Angeles.

Kataure and Walton-Roberts (chapter 6) investigate the advantages and disadvantages of ethnic-enclave housing location choices of a visible minority group – South Asians – in the City of Brampton, a popular destination for immigrants in general and South Asians in particular. Even though the suburbanization of immigrant groups started in Canadian cities a few decades ago and visible minorities are also part of this trend, little is known about the housing behaviour of some of these groups in Canada (e.g., South Asians, Chinese) and their preference for clustering in suburban neighbourhoods and forming suburban ethnic enclaves. The findings from this case study suggest that "suburban living in an ethnic enclave represents a choice rooted in socio-cultural preference" (p. 167), seen by respondents as "beneficial in promoting the successful settlement of immigrants due to the affordability of housing, proximity to family and social networks, the preferences of being in proximity to the co-ethnic community, and access to services in their own language" (p. 162). Thus, the experiences of living in an ethnic enclave for these respondents were generally positive, structured by cultural norms, family traditions, and social and physical environments. However, the authors also noted some concerns among respondents regarding racism and negative stereotyping by members of the majority group, largely associated with the ethnic enclave and housing in the area.

In this vein, the work by Yu (chapter 7) is a relevant counterpoint, focused on the housing experiences and trajectories of ethnoburban Chinese in two neighbourhoods of Los Angeles – the City of Monterey Park ("old" ethnoburb) and Rowland Heights ("new" ethnoburb). In concert with findings from other contributors to this book (see chapters by Ghosh; Murdie and Teixeira; Kataure and Walton-Roberts), Yu notes that Chinese immigrants, like many other immigrant groups in Canada and the United States, manifest a strong desire to move to the suburbs. Particularly since the 1990s, the move to the suburbs by Chinese immigrants has intensified: "when globalization increased the interdependence and interaction between the United States and mainland China, an increasing number of immigrants from China have settled in American suburbs, intent on achieving their 'American dream'" (p. 176). Yet, she argues that these "new Chinese immigrants – ranging from labour migrants to highly skilled professionals – exhibit different housing characteristics and trajectories, which consequently impact the landscapes of the suburban communities they live in and result in housing disparities among Chinese ethnoburbs ['old' versus 'new']" (p. 177).

Regarding the policy implications of the suburbanization of immigrants in both countries and possible avenues for further research, the authors of these two chapters recommend more research focusing particularly on minority groups, including the less disadvantaged ones, and their impact upon suburban housing landscapes and social spaces. More research on the "new" generation of immigrants or second generation, born in Canada and the United States, would increase our understanding of how various immigrant groups integrate over time and the cross-generational transfer of advantage and disadvantage, a subject that aligns with the North American focus on equal opportunity. Kataure and Walton-Roberts note that we need a better understanding of immigrant integration, economic outcomes, living arrangements, and settlement patterns over a span of more than one generation to know whether planning and government policies are well suited to the issues arising and what the future holds.

The current scarcity of comparative research on the housing experiences of immigrants, including minorities, in Canada and the United States and the complex "forces" shaping local housing markets prevents a full understanding of how the housing and economic situation of recent immigrants affects their integration and the role of government policies in promoting (or assuming) their socio-economic inclusion. We hope the housing section of our book will fill some gaps in the literature and open new avenues for further comparative research on immigrants' housing experiences and trajectories and their integration into Canadian and US cities and suburbs.

## Economic Experiences

It is well known that the United States and Canada have different immigrant profiles and economic structures, and that many immigrants take the route of self-employment by starting their own businesses. Table 1.1, based on the two countries' censuses and immigration data, demonstrates some of these differences and profiles. The United States continues to have the largest number of immigrants in the world, but a lower percentage of immigrants in the total population when compared to Canada. The two countries also differ in terms of where their immigrants come from, and what types of immigrants each admit. As a result of different immigration admission policies (Canada emphasizes human capital in the form of a points system, while the United States has continuously emphasized family reunification in addition to seeking wealthy and skilled migrants), the shares of immigrant categories in each country vary significantly. In Canada, immigrants admitted under the economic class (including skilled and business migrants and investors) and their immediate families account for 62.8 per cent, whereas those in the family class represent 22.7 per cent in fiscal year 2011. In the United States, 66.0 per cent were admitted under the family class and 14.0 per cent were admitted as employment-based immigrants in FY2012; this demonstrates a sharp difference in immigrant admission policies and outcomes between the two countries.

Table 1.1 Comparison of Immigration in Canada and the United States, 2011

| | Canada | United States |
|---|---|---|
| Immigrants in total population | 20.6% | 12.8% |
| Total immigrant population | 6,775,800 | 39,286,838 |
| Top three immigrant origin countries and their percentage of total immigrants | India (8.1%); China (8.1%); UK (7.9%) | Mexico (29.6%); China (5.4%); the Philippines (4.5%) |
| Economic vs. family reunification immigrants | 62.8% vs. 22.7% (FY2011) | 14.0% vs. 66.0% (FY2012) |

Sources: Citizenship and Immigration Canada, 2013; Statistics Canada, 2013; US Census Bureau, 2007–2011, American Community Survey Five-year Estimates; US Department of Homeland Security, 2012 Yearbook of Immigration Statistics.

Today's immigrants also come from more diverse source countries. As late as half a century ago, 90.1 per cent of all immigrants to Canada and 75.0 per cent of those to the United States were from Europe. No other major areas of origin exceeded 10 per cent, including Asia (3.4 per cent and 5.1 per cent, respectively), Latin America (1.5 per cent and 9.4 per cent), and North America (4.0 per cent and 9.8 per cent). As of the early 2010s, however, European immigrants decreased to 36.8 per cent and 12.3 per cent, respectively; the share of African immigrants rose to 6.1 per cent and 3.9 per cent, Asians to 40.8 per cent and 28.1 per cent, and Latin Americans to 11.3 per cent and 53.0 per cent. While the North American immigrants' share in all immigrants remained stable in Canada, it declined to 2.1 per cent in the United States. Some non-traditional immigrant groups thus face new challenges as they do not have pre-existing co-ethnic networks and institutions to assist them as new immigrants.

Moreover, it is well documented that it takes time for immigrants to "catch up" with the native-born population in receiving countries in terms of wage levels, regardless of their human capital and credentials (Aydemir and Skuterud 2005; Ferrer and Riddell 2008; Li 2001, 2003). The fact that many immigrants are recent arrivals in both countries adds to the complexity of immigrants' economic experiences. An astounding 51.4 per cent of the foreign-born population in the United States and 41.7 per cent of all landed immigrants in Canada migrated to their respective country in just the past two decades.[1]

Such differences inevitably influence the economic experiences of immigrants. The experiences of investor immigrants (Ley 2010) are different from wage-earning middle-class professionals or working-class immigrants, small-business owners, or day labourers (Waldinger 1994; Wilson 2003). The overwhelming majority of immigrants, however, regardless of where they come from or under what category, need to make a living in their new country by working in the job market or owning their own business. The economic section of this book thus deals primarily with the economic experiences of immigrants who are in the receiving country's job market as wage earners or business owners.

Educational attainment level is one of the cornerstones of Canada's immigrant selection system, which allocates points predominantly on the basis of human capital characteristics; for instance, 25 out of 100 possible points are granted for education, 24 for English- or French-language proficiency, and 21 for work experience. Having relatives in Canada no longer earns an applicant any points. This policy seems to be achieving its goal, as 31.1 per cent of all immigrants living in Canada had at least a bachelor's degree in 2011 (versus 20.3 per cent among the Canadian-born population), compared to 27.4 per cent among the foreign-born population in the United States (which is slightly lower than the 28.3 per cent among all US-born) in the same year. However, a large body of literature attributes underemployment among immigrants in Canada to language challenges, lack of Canadian work experience, devaluation of education and professional credentials obtained in the country of origin, and job market prejudice and discrimination (e.g., Albiom, Finnie, and Meng 2005). The majority of immigrants to the United States do not have a college degree, and so the underemployment problem among US immigrants is not as well explored as it is in Canada. The chapter by Lysenko and Wang (chapter 9) addresses this knowledge gap with analysis of the most recent available data, showing that a large majority of foreign-born people in the US civilian labour force have a good command of spoken English (77.1 per cent) and educational attainment levels that are largely comparable with the US-born population (39.0 per cent having at least a bachelor's degree versus 41.6 per cent among the US-born). These levels of human capital vary significantly by country of origin (table 9.1); however, the underemployment rates (those with more years of education than the industry-specific workforce average) among foreign-born populations closely mirror immigrant groups' human capital levels. Often, the higher the human capital levels are, the higher the underemployment rate (figure 9.1). Similarly, better-spoken English is associated with higher-odds ratios for underemployment. The devaluation of human capital among immigrants indicates an unfortunate situation of "brain waste." The differences between certain minority immigrant groups and the US-born white population (even after controlling for individual characteristics) also points to a racial dimension that needs to be further explored.

The underemployment situation among immigrants inevitably impacts their earning realities, especially among more recent arrivals. Pendakur and Pendakur (in chapter 8) demonstrate that earning disparities between most white and visible minority immigrant groups increased substantially during the past two decades, and the decline of relative earnings was significant, at about 20 percentage points. Specifically, male visible minority immigrants suffered the largest income gaps over the four census periods (1991, 1996, 2001, and 2006), whereas female visible minority immigrants suffered the largest decline over the study period, while both white male and female immigrants fare better than their minority counterparts (figure 8.1). The top three Census Metropolitan Areas, the "MTV" (Montreal, Toronto, and Vancouver), are home to a large majority of immigrants in Canada. Pendakur and Pendakur's analysis finds that the situation in Toronto mirrors the national picture, not surprising given that it is the largest metropolitan area with the highest percentage of immigrants. Immigrants in Montreal fared worse, and visible minority immigrants in Vancouver had a similar downward trajectory, although white immigrants suffered less decline compared to others. When looking at different immigrant groups by national origin, Pendakur and Pendakur note that female visible minority immigrant groups (those from "non-traditional sources" in Africa, Asia, or the Caribbean) fared worse over time when compared to women of British origin, whereas among some of the male visible minority immigrant groups (Arab and West Asian, Black, Caribbean, or African Black origin), the earning gap deteriorated about 10 per cent over the study period. Those of South Asian origin had a minor decline but bigger earning gaps. Pendakur and Pendakur conclude that, overall, visible minority immigrants fared worse than white origin immigrants in Canada, with more deterioration over time.

One way for immigrants to counterbalance their human capital devaluation and lack of work experience and connections in receiving countries, as well as the glass ceiling in the job market, is to start their own business and be their own boss (Kaplan and Li 2006; Lofstrom 2002). Table 1.2, based on the US 2007 Survey of Business Owners, compares the level of business ownership among foreign-born groups and genders. Overall, foreign-born people own 13.6 per cent of all US

Table 1.2 US businesses owned by foreign-born owners by gender and race-ethnicity, 2007

| Race ethnicity | Gender | Owners of respondent firms (number) | Owners of all respondent firms within group (%) | Owners of respondent employer firms (number) | Owners of all respondent employer firms within group (%) |
|---|---|---|---|---|---|
| All owners of respondent firms | All owners | 2,703,014 | 13.6 | 846,367 | 14.3 |
| All owners of respondent firms | Female | 1,018,743 | 13.7 | 264,905 | 14.6 |
| Asian | All owners | 901,658 | 82.3 | 350,668 | 86.1 |
| Asian | Male | 530,858 | 82.4 | 226,746 | 85.9 |
| Black or African American | All owners | 168,401 | 24.1 | 25,217 | 26.0 |
| Black or African American | Male | 99,476 | 26.9 | 16,748 | 27.2 |
| Hispanic | All owners | 644,793 | 55.9 | 151,194 | 54.8 |
| Hispanic | Male | 396,885 | 57.6 | 102,879 | 56.3 |
| White | All owners | 1,637,844 | 9.1 | 471,490 | 8.7 |
| White | Male | 1,057,190 | 9.3 | 338,840 | 8.9 |

Note: Calculations are based on http://factfinder2.census.gov/faces/tableservices/jsf/pages/productview.xhtml?pid=SBO_2007_00CSCBO09&prodType=table (as of 11/12/2014). The "respondent firms" include all firms that responded to the 2007 Survey of Business Owners (SBO) questionnaire, sent to a random sample of businesses selected from a list of all firms operating during 2007 with receipts of $1000 or more, except certain categories classified in the NAICS industries. The "respondent employer firms" refers to all firms with paid employee(s). The denominators for all percentages are total numbers of firms owned by a particular race/ethnic group (foreign-born + US-born). Please refer to http://www.census.gov/econ/sbo/methodology.html?2007 for further details on technical definitions.
Source: US Census Bureau.

businesses, with a slightly higher rate among companies with employees (14.3 per cent). Foreign-born Asians are the only group owning a higher percentage of employer firms (3.5 per cent or higher) among all groups. Foreign-born females as a whole, and Blacks, have a slightly higher percentage in that category, whereas all other groups are lower. Businesses without employees denote self-employment or reliance on unpaid family or other labourers. Whites own the largest number of businesses, owned by foreign-born population followed by Asians, Hispanics, and Africans, each having more than 168,000 businesses.

In general, males have higher business ownership rates than females within each group, reflecting gender differences. Immigrants count as a majority among Asian Americans in the United States at 66.6 per cent according to ACS 2007–2011 (five-year estimates). Therefore, it is not surprising that foreign-born Asians own a majority of the businesses among Asian Americans (82.3 per cent) as a whole. In comparison, 37.5 per cent of the foreign-born Latino/a population owned 55.9 per cent of all US Latino/a-owned businesses. Moreover, this reflects the fact that both foreign-born Asians and Latinos/as have higher business ownership rates than the US-born. Although the percentage of foreign-born blacks is much lower, they nevertheless own about a quarter of all US black-owned businesses.

Such differential business ownership patterns among various racial/ethnic groups and genders reflect different dynamics, and exert differential impacts on the economy and society. Immigrant businesses often have distinct features that reflect their cultural traditions or national characteristics. As such, the morphology of such businesses has changed local commercial landscapes and is reflective of the economic experiences of immigrants. In chapter 10, Oberle combines the traditions of cultural and economic geographies in his observations of the business landscape and community characteristics. Like his prior work in Phoenix (Oberle 2006), but focusing on two Midwest metropolitan areas, Oberle looks into Hispanic business features – including signage, colours, and symbols – for a reflection of their owner's country of origin or for regionally specific characteristics, and examines whether such features can also serve as signifiers of different settlement stages or types. In the emerging Latino gateway of Des Moines, Iowa, he found the Latino business landscape consisting of twenty-six business establishments concentrated along a single thoroughfare that serve Latino/a and non-Latino clientele (including the surrounding area). In the established Latino immigrant gateway of Chicago, however, differences existed between the traditional Latino stronghold of Pilsen, in the city's centre, and the newly emerged outer suburban city of Carpentersville. The seventy-three wide-ranging businesses at the heart of the Pilsen Latino business centre, many of which have signage indicative of Mexican regional identities, span twenty-one city blocks. Carpentersville reflects the new nationwide surge of Latino/a immigration, with 50 per cent of the total 2010 Latino/a total population concentrated in the eastern half of the city. Latino/a-owned businesses there comprise three separate clusters, with a total of 18 business establishments. Interestingly, the anchor stores of both Des Moines and Carpentersville's Latino shopping plazas are supermarkets targeting a largely Latino clientele. The pattern of supermarket-as-anchor for shopping plazas is similar to that at some other immigrant-owned shopping malls (Li 2009), a reflection of contemporary immigrant commercial landscapes. Oberle's comparison of the Midwest and Phoenix, Arizona, illustrates differential regional characteristics of Latino/a businesses and settlements.

While Oberle portrays Latino/a business landscapes in different metropolitan areas, Chacko and Price (chapter 11) compare the businesses owned by two different immigrant groups, one from Africa and one from Latin America, in the Washington, DC, area, a new immigrant gateway in the United States and the primary area of settlement for both immigrant groups. Guided by the literature on immigrant entrepreneurship and social network perspectives, their empirical work argues that the "ethnic sociocommerscapes" of these two groups primarily occur in suburbs in a heterolocal pattern, and that they function not only as business sites but also as important social gathering places for the immigrant population. Chacko and Price argue further that differential local policies have strong impacts on these newly emerged landscapes.

One of the primary gaps in the literature of the immigrant economic experience is the financial dynamics among contemporary immigrant businesses. The chapter by Li and Lo (chapter 12) addresses this issue based on in-depth in-person interviews and focus group discussions with operational and aspiring immigrant entrepreneurs. They analyse immigrant entrepreneurs' access, or lack of access, to financial services offered by formal financial institutions. The authors' work indicates that the financial needs of immigrant entrepreneurs vary depending on the sector, size, and development stage of their business, as well as on the entrepreneur's immigration class. The financial products and services that immigrant entrepreneurs receive depend on the financial institutions they deal with and the metropolitan area in which they are situated. Financial institutions have a long way to go in addressing new immigrants' needs and integration into the receiving society. They ought to offer a larger variety of more tailored products and services to immigrant entrepreneurs in order to fully utilize the latter's human capital and business acumen, allowing them to improve the lives of their families and employees, and ease their way to becoming more economically productive citizens. Since those interviewed in Vancouver were all operating entrepreneurs, whereas most of the focus group participants in the San Francisco Bay Area were aspiring entrepreneurs, it is difficult to draw direct comparisons between the two study areas. But the authors' work nonetheless indicates that the changing nature of immigrants and their businesses, as a result of different financial dynamics, created financial needs that traditional immigrant lending circles can no longer fulfil. The authors outline these needs and, calling for further study of financial dynamics within the new immigration and globalization contexts, explain how the resulting policy interventions could be beneficial.

Collectively, these five chapters provide a set of snapshots that reveal the undervaluation of immigrants' human capital and credentials and point to a racial dimension, given that the problems are more prevalent among visible/racial minorities than white European immigrants in both countries. They also illustrate the various ways immigrant-owned businesses appear, operate, and obtain their financial resources.

This collection of essays constitutes an attempt to examine the housing and economic experiences of immigrants in the United States and Canada. It offers not only a selection of original essays from leading and emerging researchers from both countries, but is also unique in the context of North American studies in that it draws together a variety of scholarly perspectives on the themes of immigration, housing, and economic experiences in urban North America. Aside from the introduction (chapter 1) and the conclusion (chapter 13), this book is comprised of two parts: part 1 is dedicated to the housing experiences of immigrants (chapters 2 to 7) and part 2 focuses mainly on their economic experiences (chapters 8 to 12), with an introduction to each part.

## NOTES

1 Calculations based on the 2011 Canadian Census and 2007–2011 US American Community Survey, hereafter ACS. We are keenly aware that immigrants, international migrants, and the foreign-born population are not the same, and use these terms distinctly when needed, whereas other times we use them interchangeably.

## REFERENCES

Albiom, N., R. Finnie, and R. Meng. 2005. "The Discounting of Immigrants' Skills in Canada: Evidence and Policy Recommendations." *IRPP Choices* 11 (2): 1–26. http://irpp.org.

Anisef, P., and M. Lanphier. 2003. "Introduction: Immigration and the Accommodation of Diversity." In *The World in a City*, ed. P. Anisef and M. Lanphier, 3–18. Toronto: University of Toronto Press.

Aydemir, A., and M. Skuterud. 2005. "Explaining the Deteriorating Entry Earnings of Canada's Immigrant Cohorts, 1966–2000." *Canadian Journal of Economics/Revue canadienne d'économique* 38 (2): 641–72. http://dx.doi.org/10.1111/j.0008-4085.2005.00297.x.

Borjas, G.L. 2001. *Heaven's Door: Immigration Policy and the American Economy*. Princeton, NJ: Princeton University Press.

Bourne, L.S. 1981. *The Geography of Housing*. London: Edward Arnold.

Clark, W.A.V., and F.M. Dieleman. 1996. *Households and Housing: Choice and Outcomes in the Housing Market*. New Brunswick, NJ: Rutgers.

Ferrer, A.M., and W.C. Riddell. 2008. "Education, Credentials and Immigrant Earnings." *Canadian Journal of Economics/Revue canadienne d'économique* 41 (1): 186–216. http://dx.doi.org/10.1111/j.1365-2966.2008.00460.x.

Herman, R.T., and R.L. Smith. 2009. *Immigrant, Inc.: Why Immigrant Entrepreneurs Are Driving the New Economy*. Hoboken, NJ: John Wiley & Sons.

Hulchanski, D., and M. Shapcott. 2004. "Introduction: Finding Room in Canada's Housing System for All Canadians." In *Finding Room: Policy Options for a Canadian Rental Housing Strategy*, ed. D. Hulchanski and M. Shapcott, 3–12. Toronto: CUCS Press, Centre for Urban and Community Studies.

Kaplan, D., and W. Li, eds. 2006. *Landscapes of the Ethnic Economy*. Rowman & Littlefield Publishers, Inc.

Kobayashi, A., W. Li, and C. Teixeira. 2012. "Introduction – Immigrant Geographies: Issues and Debates." In *Immigrant Geographies of North American Cities*, ed. C. Teixeira, W. Li, and A. Kobayashi, xiv–xxxviii. Don Mills, ON: Oxford University Press.

Ley, D. 2010. *Millionaire Migrants: Trans-Pacific Life Lines*. West Sussex, UK: Wiley-Blackwell. http://dx.doi.org/10.1002/9781444319262.

Li, P.S. 2001. "The Market Worth of Immigrants' Educational Credentials." *Canadian Public Policy* 27 (1): 23–38. http://dx.doi.org/10.2307/3552371.

Li, P.S. 2003. "Initial Earnings and Catch-up Capacity of Immigrants." *Canadian Public Policy* 29 (3): 319–37. http://dx.doi.org/10.2307/3552289.

Li, W. 2009. *Ethnoburb: The New Ethnic Community in Urban America*. Honolulu: University of Hawai'i Press.

Lofstrom, M. 2002. "Labour Market Assimilation and the Self-employment Decision of Immigrant Entrepreneurs." *Journal of Population Economics* 15 (1): 83–114. http://dx.doi.org/10.1007/PL00003841.

Miron, J. 1988. *Housing in Postwar Canada: Demographic Changes, Household Formation, and Housing Demand*. Montreal: McGill-Queen's University Press.

Oberle, A. 2006. "Latino Business Landscapes and the Hispanic Ethnic Economy." In *Landscapes of the Ethnic Economy*, ed. David Kaplan and Wei Li, 149–63. New York: Rowman and Littlefield.

Schwartz, A.F. 2010. *Housing Policy in the United States*. 2nd ed. New York: Routledge.

Tager, L., ed. 2005. *Migration and Economy: Global and Local Dynamics*. Walnut Creek, CA: AltaMira Press.

Teixeira, C., and W. Li, eds. 2009. "Immigrant and Refugee Experiences in North American Cities." *Journal of Immigrant & Refugee Studies* 7 (3): 221–7. http://dx.doi.org/10.1080/15562940903150030.

USCIS. 2013. *American Fact Finder*. US Citizenship and Immigration Service. Washington, DC. http://factfinder2.census.gov/faces/nav/jsf/pages/index.xhtml.

Waldinger, R. 1994. "The Making of an Immigrant Niche." *International Migration Review* 28 (1): 3–30. http://dx.doi.org/10.2307/2547023.

Wilson, F.D. 2003. "Ethnic Niching and Metropolitan Labour Markets." *Social Science Research* 32 (3): 429–66. http://dx.doi.org/10.1016/S0049-089X(03)00015-2.

# PART ONE

## The Housing Experiences of Immigrants

# Introduction to Part One: The Housing Experiences of Immigrants

CARLOS TEIXEIRA

This introductory chapter to Part One serves to present a short and updated literature review on immigrants' diverse and complex housing experiences in Canada and the United States, and to situate the work of the six housing chapters that follow.

## Housing and Immigrant Integration

Most immigrants to Canada and the United States continue to be attracted to major urban areas, but those coming since the late 1960s have differed in their country of origin, ethnic background, and economic status (Hou 2006; Hiebert 2009; Brown and Sharma 2010; Schwartz 2010; Murdie and Skop 2012). Immigrants have altered the social and economic diversity of North America's largest urban areas and their suburbs, contributing to greater housing demand and higher prices (Carter 2005; Myers and Liu 2005; Hiebert and Mendez 2008; Moos and Skaburskis 2010; Schwartz 2010; Skop 2012). Their diversity has played a role in affecting dramatic changes in the housing situation in Canada and the United States (Schwartz 2010, 47; Teixeira 2014), altering the range of housing preferences, attitudes, and requirements, which filter through the market (Carter and Vitiello 2012, 92).

As some chapters show (see Ghosh in chapter 4; Murdie and Teixeira in chapter 5; Kataure and Walton-Roberts in chapter 6; and Yu in chapter 7), immigrant settlement patterns vary, with some groups forming ethnic enclaves while others scatter across urban and suburban landscapes; the varied experiences of immigrants present major challenges for social service providers, business leaders, and policymakers concerned about immigrant integration (Alba et al. 1999; Rose 2001;

Sandercock 2003; Bunting et al. 2004; Singer 2008; Lo et al. 2010; Larrivee 2011; Murdie and Skop 2012). The changing geography of immigrant settlement has generated and dispersed housing demand, affecting cities large and small, as well as rural areas.

Some common findings span the Canadian and US literature on housing and immigrants: (a) although a significant percentage of immigrants and refugees exhibit a progressive housing career, with improving income levels, better housing, and rising rates of homeownership over time (Hiebert 2009, 269), many do not, and homeownership rates are declining with successive cohorts of immigrants, due in part to waning income prospects (Borjas 2002; McConnell and Akresh 2008; Carter and Vitiello 2012); (b) refugees and unauthorized immigrants face the most difficult housing circumstances of all newcomers to Canada and the United States, and remain renters for longer periods (Carter and Vitiello 2012); (c) immigrants are more vulnerable than non-immigrants in the housing market, as owners or renters; (d) immigrants in Canada experience higher levels of "core need" (a composite measure of adequacy, suitability, and affordability) than non-immigrants, and some racial minority groups (from African, Caribbean, and Latin American countries) in the United States have greater "acute housing need" in "hot" (i.e., expensive) housing markets, such as in New York and Atlanta (McConnell and Akresh 2008; Schwartz 2010; Carter and Vitiello 2012); and (e) affordability is the single greatest barrier to obtaining suitable, adequate housing, regardless of location (Haan 2007; Murdie 2008a; Hiebert 2009; Leone and Carroll 2010; Schwartz 2010). Moreover, scholars recognize that race and ethnicity remain barriers to the equal treatment of immigrants and minorities in North American housing markets (see Darden in chapter 2 and Haan and Yu in chapter 3 in this volume).

Recent immigrants in both Canada and the United States face a range of challenges in their settlement and housing experiences, including limited financial resources, language barriers, prejudice, and discrimination (see Darden in chapter 2 and Haan and Yu in chapter 3). Despite efforts at amelioration through a range of human rights pursuits, integration policies, and multicultural policies, many immigrants experience discrimination based on their skin colour or ethnoracial or religious backgrounds, exacerbated by tight labour and housing markets (Rosenbaum and Friedman 2007; Ray and Preston 2009; Teixeira and Li 2009; Schwartz 2010; Flippen 2010; Darden and Fong 2012). Immigrant renters are especially at risk of exclusion, marginalization, poverty, even homelessness (Bunting et al. 2004; Leone and Carroll 2010). Yet, as Carter and Vitiello 2012 note, the "diversity of newcomers and of housing markets across North America results in such varied patterns that there is no such thing as a 'typical' immigrant housing experience" (92). In recent years, the role of housing in the integration process has received more attention, prompting examination of the links between access to affordable housing and residential concentrations of newcomers and minorities, as well as successful integration and inclusion. A key element of immigrant integration, especially on arrival, is securing suitable, good-quality, affordable housing. For non-affluent newcomers, this usually means renting in the private sector at prices which are high in relation to their income, or purchasing a home soon after their arrival, even if it strains their financial means, sometimes by relying on support from their ethnic social networks to do so (see Ghosh in chapter 4). Where housing costs are higher, in the urban cores and suburbs of major North American cities, newcomers face a greater challenge (Borjas 2002; Saiz 2007; Schwartz 2010; Teixeira 2014).

In the remainder of this chapter, I discuss three broad themes – some of long-standing note, some newly emerging – that are reflected in the following six chapters: housing inequalities and homeownership; housing careers, social networks, and gentrification; and immigration, suburban settlements, and policy implications.

## Housing Inequalities and Homeownership

For many new immigrants in both countries, becoming homeowners is a "milestone" in their lives, laden with real and symbolic meanings. While high immigration levels have spurred home-buying demand over the past decade, however, the ownership rate for immigrants has been declining since the 1970s in both countries, and it takes immigrants longer to achieve the ownership rates of residents born in the United States and Canada (Carter and Vitiello 2012, 100). Homeownership rates in both countries also vary by ethnicity and class, with immigrants from Europe and Asia exhibiting higher rates of homeownership than newcomers from Africa and Latin America, the latter groups having lower incomes and thus less potential for achieving the "North American dream" (Brown and Webb 2012; Leloup et al. 2011; Flippen 2010; Kochhar et al. 2009; Myers and Liu 2005). Recent immigrants have found themselves arriving at a time of significant change for purchasers in the housing market (Schwartz 2010; Teixeira 2014), with price increases in most major metropolitan centres, low interest rates and greater flexibility in mortgage lending, and lower down payment requirements, especially before 2008, when the housing bubble burst in the United States (Schwartz 2010). These trends have permitted lower-income households to buy homes, although price escalations in some major cities have negated such advantages (Carter and Vitiello 2012; McConnell and Akresh 2010; CMHC 2010; Joint Center for Housing Studies 2005, 2008).

Each of the housing contributors in this volume acknowledges the high priority most immigrants place on owning their home. Darden (chapter 2) and Haan and Yu (chapter 3) present the results of their macro-level comparative studies on inequalities in homeownership attainment by immigrants/minority groups in Canada and the United States. In chapter 4, Ghosh provides a detailed case study of the homeownership experiences of a particular low-income immigrant group, one of the poorest in Canada – Tamils. She examines how their coping strategies, such as accommodating additional family members as renters, help them afford a mortgage and other housing-related expenses. Another immigrant group in Toronto, Portuguese homeowners, particularly seniors on a fixed income, living in the increasingly gentrified "Little Portugal" area, are also pushed to rent part of their houses to cover increasing expenses for maintenance and property taxes (see Murdie and Teixeira in chapter 5).

The rental sector maintains its importance for lower income households, especially in the United States, where recent widespread foreclosures have pushed people back into renting (Mundra 2013; Schwartz 2010; Joint Center for Housing Studies 2008; Saiz 2007). Gateway cities (such as Vancouver, Toronto, San Francisco, Los Angeles, and New York) in both countries have been characterized by low rental vacancy rates, a reflection of high demand and limited stock.

Studies conducted in North America's largest cities suggest that new immigrants and refugees with low incomes experience residential segregation. The housing search process can be stressful for these groups, especially when combined with language barriers, cultural differences, and various forms of discrimination (Yinger 1995; Rose and Ray 2001; Immergluck 2004; Fiedler et al. 2006; Teixeira 2008; Fong and Chan 2010). These constraints contribute to the creation and perpetuation of high levels of involuntary residential concentration of immigrants, and are major barriers to achieving a successful housing career (see Darden in chapter 2 and Haan and Yu in chapter 3). Discriminatory practices by real estate agents, landlords, mortgage lenders, neighbours, and private and public agencies may limit new immigrants' housing options (Myers, Baer, and Choi 1996; Teixeira and Murdie 1997; Novac et al. 2004; Turner and Ross 2005; Smith and Ley 2008; Ray and Preston 2009; Darden and Fong 2012).

The literature suggests that these difficulties have become more acute since the early 1990s in the United States, and since the mid-1990s in Canada, partly due to the reduction in federally subsidized housing stock for low income renters, very low levels of social housing development, relatively high market rents, low vacancy rates, and funding cuts to social assistance and non-governmental organizations that assist new immigrants and refugees (Murdie 2008b; Schwartz 2010; Carter and Vitiello 2012).

In chapter 2, Joe Darden, a long-standing advocate for minority rights in North American cities, notes that homeownership provides immigrants with a "stake in the system," improves the household's "social status," and contributes to "wealth and asset accumulation" (especially in the United States, where the cost of borrowing is tax-deductible). Immigrant groups with low rates of homeownership rates have benefited the least in terms of wealth and social status and are least integrated into receiving societies. Comparative research has shown a higher level of residential segregation in the United States than Canada, and a lower homeowner rate for Black and Hispanic immigrants. As countries that rely heavily on immigration and continue to recruit immigrants, this raises policy questions in both Canada and the United States. Given the similar histories of immigration policies adopted by both countries, notably the active recruitment of immigrants from Europe and the exclusion of immigrants from non-European regions, Darden examines whether "contemporary differences still exist in European and non-European immigrant incorporation as measured by homeownership rates" (p. 45). His findings point to inequalities in terms of homeownership rates, with European-born immigrants continuing to have the highest homeownership rates in both Canada and the United States. He argues that this gap cannot be explained solely by "internal characteristics," such as length of time in the country of settlement or occupation status, but may be due to discrimination against Blacks and Hispanic immigrants. While in the United States there is consensus that mortgage lending discrimination based on race/ethnicity is a major factor impacting homeownership rates, this is not the case in Canada, in large part because of the lack of data (in-depth studies and audit-type surveys administered by the Canadian government) or lending disclosure rules to monitor discriminatory practices. Darden's work fills a major gap in the comparative United States–Canada housing literature by using the most recent census data available to expand our knowledge of differential rates of immigrant homeownership and compare the situation among several immigrant groups in Canada and the United States.

Haan and Yu (in chapter 3) further this analysis by comparing household formation and homeownership patterns of racialized minority groups in Canada and the United States. They question whether the framework typically used to understand the integration experiences of earlier waves (predominantly white immigrants from European countries) is useful for explaining the experiences and situation of more recent, ethnically and racially diverse cohorts of newcomers. Haan and Yu note that members of different groups – whether defined by ethnicity, place of birth, or skin colour – have distinct housing careers and integration experiences in Canadian and US metropolitan housing markets: "Not only do racialized minority groups have very different homeownership rates, but they tend to live in different neighbourhoods ... have access to different amenities ... and even move through the housing market differently as they age" (p. 69).

Taking into account household formation to study residential assimilation, Haan and Yu compare nearly identical groups in two different policy environments, providing a rare opportunity to look at how context shapes socio-economic outcomes like housing. By focusing on large, diverse metropolitan areas, Haan and Yu compare the effects of context across groups, cities, and countries, and assess the housing attainment patterns of several groups: foreign-born black, Asian Indian, mainland Chinese, Filipino, white, and the white native-born reference group, using longitudinal data to measure what percentage of the differences in homeownership attainment can be attributed to household formation, and to compare the extent to which the gaps in housing outcomes changed over a five-year period. Among various immigrant groups, they found evidence of similar levels of household formation, homeownership, and progress over time in Canada and the United States. While newly arrived immigrants were the least likely to form independent households, and the most likely to rent, all immigrant groups showed a gradual and significant increase in demand for separate accommodation that was quite marked after ten years. Haan and Yu argue that immigrant groups in Canada and the United

States have generally kept pace with the progress of the native-born white cohort, with differences shrinking over time, except for Filipino immigrants, who present a unique case, with the lowest levels of housing attainment among all immigrant groups, even after adjusting for covariates.

### Housing Careers, Social Networks, and Gentrification

Immigrants who arrive with limited or no financial resources often live in substandard or poorly maintained housing, or in crowded conditions with other immigrants of the same ethnic background, or with family or ethnic friends (see Ghosh in chapter 4). These are common coping strategies that may in some cases constitute hidden homelessness (Myers, Baer, and Choi 1996; Hiebert et al. 2006; Fiedler et al. 2006; Teixeira and Halliday 2010; Francis and Hiebert 2011). For example, in both Canada and the United States, the incidence of residential crowding among newcomers is higher than among non-immigrants, though it varies by ethnocultural group (see Ghosh in chapter 4 and Kataure and Walton-Roberts in chapter 6). As an economic strategy, many rental households trade space for lower housing costs or share space and pool limited resources to buy property (McConnell and Akresh 2008), practices that may also be part of a "transnational housing strategy" whereby immigrants save money to invest in new housing construction in their home countries. While immigrants' larger households tend to increase demand for larger dwellings (see Ghosh in chapter 4, Kataure and Walton-Roberts in chapter 6, and Yu in chapter 7), the rental sector in both countries "tends to build for the typical US- or Canadian-born rental household requiring one or two bedrooms" and thus largely ignores immigrants' needs, limiting their housing options and further contributing to crowded dwellings (Carter and Vitiello 2012, 101, 102). Quite often, large immigrant families are pushed into homeownership to satisfy their housing needs, despite their limited financial resources, placing some at risk of foreclosure or eviction (see Ghosh in chapter 4). Carter and Vitello (2012, 105) remind us that "social housing provides more opportunities to locate three- and four-bedroom units not available in the private market … However, governments in both Canada and the USA have not funded much new affordable housing stock to meet this need for many years." Senior governments in both countries have been consistent in favouring market housing and homeownership over rental housing.

Housing affordability ranks among the most pervasive and persistent of issues for immigrants and refugees in Canada and the United States (McConnell and Akresh 2008; Schwartz 2010; Teixeira 2014). Since recent immigrants are more likely than native-born residents in Canada and the United States to have poor employment and income prospects (see Pendakur and Pendakur in chapter 8, and Lysenko and Wang in chapter 9) and spend large percentages of their household income on housing (Hiebert 2009; Preston et al. 2009), homelessness is a risk for some of them, particularly visible minority groups and refugees (Murdie 1994; Fiedler et al. 2006; Francis 2009).

New immigrants' integration challenges are amplified when combined with economic disadvantages, limited employment opportunities, high housing costs, language barriers, discriminatory practices, a lack of knowledge about the functioning of the housing market, and limited access to housing services and programs. With immigrant-serving agencies concentrated in the core of major North American cities, their suburban areas are "poorly equipped, in both physical and social terms, to serve their rapidly growing and increasingly diversifying population" (Lo et al. 2010, i). Living in suburban areas with few services prompts immigrants to be self-reliant. Several of the contributors to this volume – Ghosh in chapter 4, Kataure and Walton-Roberts in chapter 6, and Yu in chapter 7 – demonstrate how immigrants in the suburbs of Toronto and Los Angeles "survive" and successfully integrate by relying on their own "community/ethnic social networks" to seek affordable housing in a preferred neighbourhood, obtain a job or financial assistance, and adjust to a new life.

While emigration is inherently a major decision and uprooting, the situation of refugees is typically more pressured and fraught by trauma and scant financial means or preparation for life in North America. Ghosh (in chapter 4) focused her case study on one of the most impoverished "South Asian" refugee subgroups – Sri Lankan Tamils – in Toronto's expensive housing market. Ghosh found that Tamils followed a gendered "culture of migration" in which men tended to emigrate first and, once established in the new country, sponsor female family members. Typically, the first to arrive were young (in their early twenties), suffering familial ruptures as a result of forced emigration from Sri Lanka; many remained separated from their families for decades. Most of her respondents expressed a strong "sense of obligation toward their immediate and extended family members" (p. 106), even to those from the same home village, a cultural norm that survived the migration and settlement process. Most of the first wave who arrived in the 1980s resided in low-income, older highrise apartment buildings located in the inner city. Like many other immigrant groups, once established, these first settlers moved out to the suburbs. The more recent arrivals (in the mid-1990s) followed a different settlement pattern by moving on arrival directly to the inner suburbs of Toronto, bypassing the urban core, primarily because of the high housing costs and gentrification (a topic that is the focus of attention of Murdie and Teixeira in chapter 5).

On arrival in Toronto, most of the Tamils doubled up with their family and friends; regardless of their immigrant admission class or poor financial status, almost none of them used a public shelter. During their entire housing career, this group rarely sought the assistance of a social agency. Most of the Tamil immigrants were later able to purchase homes due to their "composite family structures where the number of earning members is higher than in average Canadian census families" (p. 111). However, almost all the respondents were paying more than 30 per cent of their before-tax income on housing. Ghosh surmises that many Tamils entered homeownership prematurely to take advantage of tax deductions for first-time homebuyers, and household members may have been pressed to remain living in overcrowded conditions to avoid exacerbating financial pressure on those remaining in the home. Yet these social obligations to assist familiars reached their limits in cases where Tamils were evicted by family members for failing to pay their portion of the rent or mortgage.

Gentrification is a factor contributing to higher housing costs and settlement patterns. Whether due to residential conversions, densification, or redevelopment of public housing (notably in the United States), gentrification has become an integral part of urban change in North America. Studies have considered the origins of gentrification, the nature of the process, and its impact (both negative and positive) on local neighbourhoods. Few studies, however, have addressed the intersection between ethnicity and gentrification (Murdie and Teixeira, chapter 5). Murdie and Teixeira explore the impact of gentrification on Portuguese residents, most of them first-generation immigrants, living in "Little Portugal," the historic site of Toronto's Portuguese community. Located immediately west of downtown Toronto, Little Portugal contains most of the city's Portuguese cultural institutions and ethnic businesses. Murdie and Teixeira identified three major negative and positive effects of gentrification: (1) speculative property price increases, loss of affordable housing, and displacement, (2) commercial and industrial displacement, and (3) community resentment and conflict; in contrast, gentrification generates (1) increased property values, (2) stabilization of declining areas and encouragement of further development, and (3) increased social mix.

Loss of affordable housing was identified by the majority of Portuguese respondents as the most serious negative impact of gentrification in Little Portugal. Low-income people, including recent immigrants, were the most severely affected by the escalating house prices, with some long-time renters being forced to leave their apartments due to the cost. Several respondents even questioned the future viability of Little Portugal as an immigrant reception area. These results corroborate Ghosh's findings (chapter 4) on the Tamils' housing situation in Toronto: they too are being pushed to move to suburban Scarborough in search of more affordable housing. In general, the Portuguese regarded the incoming gentrifiers with mixed feelings: they were appreciated for rejuvenating the local housing stock and patronizing local ethnic businesses, but their desire to live in a multicultural neighgourhood like Little Portugal has driven up the cost of housing such that fewer Portuguese (including older residents and newcomers to the area) can afford to continue living there. Portuguese seniors on fixed incomes are the most vulnerable group, lacking the financial resources to maintain their houses and afford the increasing property taxes. With insufficient seniors' housing to accommodate a growing elderly Portuguese population, many of those who decided to stay in Little Portugal rented part of their houses, usually to Portuguese-speaking immigrants, to cope with escalating expenses. Living on fixed incomes (many of them "land rich but cash poor"), these immigrant seniors find themselves living in a former immigrant reception area that is gentrifying and gradually losing its traditional community at a time when they may need it most.

## Immigration, Suburban Settlements, and Policy Implications

In many major metropolitan areas in Canada and the United States (e.g., Toronto, Vancouver, Montreal, Miami, San Francisco, Los Angeles, New York), the "suburbanization of jobs, gentrification of the inner city, and the increasing diversity of the immigrants [have] contributed to the suburbanization of the immigrant social landscape" (Moos and Skaburskis 2010, 734; Brown and Sharma 2010; Leichenko 2001). While these cities remain major immigrant reception sites, their satellite suburbs have become important "social laboratories" for geographers and migration scholars interested in the settlement experiences of contemporary immigrants and their housing circumstances and decision-making (see Kataure and Walton-Roberts in chapter 6 and Yu in chapter 7).

Recent census data in Canada and the United States reveal a new pattern of immigrant dispersal to destinations beyond major urban centres (Broadway 2000; Abu-Laban and Garber 2005; Radford 2007; Teixeira 2007; Brunner and Friesen 2011; Skop 2012), changing the historic geography of immigrant settlement (Qadeer et al. 2010; Murdie and Skop 2012). Rapid population growth and the concentration of immigrants and minorities in the suburbs of major North American cities have led to greater demand for increasingly scarce affordable housing and different housing styles or types to accommodate the needs and preferences of immigrants with different socio-economic backgrounds.

Relatively little is known about recent immigrants' housing experiences in North American suburbs. The few such studies show that recent immigrants and refugees are increasingly settling and concentrating in specific suburban neighbourhoods (see Murdie and Skop 2012; Brunner and Friesen 2011; Preston et al. 2009; Mendez 2008; Fiedler et al. 2006; Leichenko 2001; Alba et al. 1999). While well-to-do immigrants can afford expensive, large houses on generous lots in the suburbs, as South Asians have done in Brampton (see chapter 6) and Chinese in the ethnoburbs of Los Angeles (chapter 7), the scenario for low-income immigrant groups differs, as Ghosh (chapter 4) has shown for the Tamil group and its housing struggles in Toronto suburbs. Distinct pockets of housing-affordability stress and poverty exist in certain suburban areas or neighbourhoods, places often associated with recent immigrant and refugee renters in core housing need or the hidden homeless (see Bratt 2000; Bunting et al. 2004; Smith and Ley 2008; Francis and Hiebert 2011).

In Canada and the United States, "the bifurcation of immigrants' educational and work experiences is reflected in housing demand and market diversity" (Carter and Vitiello 2012, 92). Some immigrants in both countries are underemployed relative to their education and experience (see Pendakur and Pendakur in chapter 8 and Lysenko and Wang in chapter 9). Affluent newcomers, such as business class immigrants with sufficient assets to purchase housing upon arrival, can afford high-priced urban and suburban housing (see Kataure and Walton-Roberts in chapter 6 and Yu in chapter 7), while the non-affluent – including many recent immigrants, refugees, and minorities – are restricted to lower-quality housing in neighbourhoods often characterized by high rates of poverty, a pattern that augments spatial segregation by class and race (see Ghosh in chapter 4) . Thus, diversity translates into "different tenure ratios, different designs and sizes, and a changing metropolitan geography of immigrant housing" in both Canada and the United States (Carter and Vitiello 2012, 92; Ley 2010; Francis and Hiebert 2011; Murdie and Skop 2012).

Research on the new geographies of immigrant settlement in Canadian and American suburbs – particularly the significance of ethnic suburban enclaves, and immigrants' housing choices, trajectories, and integration into a suburban way of life – have moved to the forefront of urban and housing research in the last two decades. The topics of chapters 6 and 7 reflect the increasing interest in ethnic and racial demographic shifts in the suburbs and the impact immigrants are having on the social, cultural, political, and economic landscapes of the suburbs.

In chapter 6, Kataure and Walton-Roberts investigate the advantages and disadvantages of ethnic-enclave housing location choices of a visible minority group – South Asians in Brampton, a popular destination for immigrants in general and South Asians in particular. A fast-growing city, Brampton has the greatest concentration of South Asians in the Toronto Census Metropolitan Area, and the number of South Asians living in ethnic enclaves in Brampton has increased dramatically. Yet very little research deals with their suburbanization, ethnic clustering, and housing preferences. The process of suburbanization of immigrant groups in Canada actually began a few decades earlier, and included visible minority groups, some of which (South Asians, Chinese) were known to cluster in suburban neighourhoods and form suburban ethnic enclaves. Kataure and Walton-Roberts suggest that "ethnic clustering in Canadian suburbs is preference based, which partly explains the strength of continued suburban ethnic concentration" (p. 146). The ethnic enclave of South Asians in Brampton was seen by respondents as "beneficial in promoting the successful settlement of immigrants due to the affordability of housing, proximity to family and social networks, preferences to be in proximity to the co-ethnic community, and access to services in their own language" (p. 162). The experience of living in an ethnic enclave for these respondents was generally positive, structured by cultural norms, family traditions, and social and physical environments, although some signs of racism were noted. The South Asian families preferred to reside in large suburban houses, partly to accommodate extended family members, and Brampton provided them with such "dream [large] houses" at affordable prices. In a similar vein, Yu (chapter 7) focuses on the housing experiences and trajectories among ethnoburban Chinese in two neighourhoods of Los Angeles. Echoing findings by other contributors, Yu notes that Chinese immigrants manifest a strong desire to move to the suburbs in search of the "American dream." Particularly since the 1990s, the move to the suburbs by Chinese immigrants has intensified as "globalization increased the interdependence and interaction between the United States and Mainland China" (p. 176). She argues that "new Chinese immigrants ... exhibit different housing characteristics and trajectories which ... result in housing disparities among Chinese ethnoburbs" (p. 177). The changes brought about by these newcomers to the suburban landscape have policy implications of interest to the development industry (builders, developers, real estate agents, and lending institutions), as well as policymakers and service providers working in Chinese ethnoburbs.

Yu compared different housing types and housing trajectories within the Chinese immigrant group in two Chinese ethnoburbs of Los Angeles that emerged at different times historically. Underlying the emergence of these new Chinese ethnoburbs is a constant inflow of new immigrants from China. Earlier refugees and poorer immigrants from China carved out the old ethnoburbs of moderate-priced housing and established themselves. A relentless search for affordable housing led to residential intensification and single-family homes were converted to flats and rooms added. Old ethnoburbs are spatially dense and compact, usually with multi-unit housing types, now occupied mainly by elderly Chinese and Chinese newcomers, few of whom are homeowners. Over time, a flow of the more prosperous to better housing and neighbourhoods has led to new ethnoburbs that have become more dense, leading in turn to newer ethnoburbs that are geographically dispersed. Superimposed on this process of suburbanization are variations in the wealth of immigrants that permit some to move directly to the newest ethnoburb. A dramatic increase in Chinese immigrants in the last two decades has also affected the city of Richmond, a suburb of Vancouver, where their economic and cultural impact is evident in the form of very large houses. The apparent similarities with their US counterparts in terms of housing experiences and trajectories warrants a comparative "Canada–US" ethnoburban study (see Teixeira 2014).

It is widely understood that housing is a critical indicator of quality of life that significantly influences health, social interaction, community participation, economic activity, and general well-being. The scarcity of comparative research on the housing experiences of immigrants in Canada and the United States prevents a full understanding of how the housing and economic situation of recent immigrants affects their integration and how government policies promote their socio-economic inclusion. We hope this housing section will fill gaps in the literature and open new doors or avenues for further comparative research on immigrants' housing experiences and trajectories and integration into Canadian and US cities and suburbs.

## REFERENCES

Abu-Laban, Y., and J. Garber. 2005. "The Construction of the Geography of Immigration as a Policy Problem: The United States and Canada Compared." *Urban Affairs Review* 40 (4): 520–61. http://dx.doi.org/10.1177/1078087404273443.

Adams, M., and G. Osho. 2008. "Migration, Immigration, and the Politics of Space: Immigration and Local Housing Issues in the United States." *Research Journal of International Studies* 8: 5–12.

Alba, R.D., J.R. Logan, B.J. Stults, G. Marzan, and W. Zhang. 1999. "Immigrant Groups in the Suburbs: A Reexamination of Suburbanization and Spatial Assimilation." *American Sociological Review* 64 (3): 446–60. http://dx.doi.org/10.2307/2657495.

Borjas, G.J. 2002. "Homeownership in the Immigrant Population." *Journal of Urban Economics* 52 (3): 448–76. http://dx.doi.org/10.1016/S0094-1190(02)00529-6.

Bratt, R.G. 2000. "Housing and Family Well-being." *Housing Studies* 17 (1): 12–26.

Broadway, M. 2000. "Planning for Change in Small Towns or Trying to Avoid the Slaughterhouse Blues." *Journal of Rural Studies* 16 (1): 37–46. http://dx.doi.org/10.1016/S0743-0167(99)00038-8.

Brown, L.A., and M. Sharma. 2010. "Metropolitan Context and Racial/Ethnic Intermixing in Residential Space: U.S. Metropolitan Statistical Areas, 1990–2000." *Urban Geography* 31 (1): 1–28. http://dx.doi.org/10.2747/0272-3638.31.1.1.

Brown, L.A., and M. Webb. 2012. "Home Ownership, Minorities, and Urban Areas: The American Dream." *Professional Geographer* 64 (3): 332–57. http://dx.doi.org/10.1080/00330124.2011.601183.

Brunner, L.R., and C. Friesen. 2011. "Changing Faces, Changing Neighbourhoods: Government-Assisted Refugee Settlement Patterns in Metro Vancouver 2005–2009." *Our Diverse Cities* 8: 93–100.

Bunting, T., A. Walks, and P. Filion. 2004. "The Uneven Geography of Housing Affordability Stress in Canadian Metropolitan Areas." *Housing Studies* 19 (3): 361–93. http://dx.doi.org/10.1080/0267303042000204287.

Carter, T. 2005. "The Influence of Immigration on Global City Housing Markets: The Canadian Perspective." *Urban Policy and Research* 23 (3): 265–86. http://dx.doi.org/10.1080/08111470500197797.

Carter, T., and D. Vitiello. 2012. "Immigrants, Refugees, and Housing." In *Immigrant Geographies of North American Cities*, ed. C. Teixeira, W. Li, and A. Kobayashi, 91–111. Don Mills, ON: Oxford University Press.

CMHC. 2010. *CMHC Rental Market Survey*. Ottawa: CMHC.

Darden, J., and E. Fong. 2012. "The Spatial Segregation and Socio-Economic Inequality of Immigrant Groups." In *Immigrant Geographies of North American Cities*, ed. C. Teixeira, W. Li, and A. Kobayashi, 69–90. Don Mills, ON: Oxford University Press.

Fiedler, R., N. Schuurman, and J. Hyndman. 2006. "Hidden Homelessness: An Indicator-Based Approach for Examining the Geographies of Recent Immigrants at-Risk of Homelessness in Greater Vancouver." *Cities* (London, Eng.) 23 (3): 205–16. http://dx.doi.org/10.1016/j.cities.2006.03.004.

Flippen, C.A. 2010. "The Spatial Dynamics of Stratification: Metropolitan Context, Population Distribution, and Black and Hispanic Homeownership." *Demography* 47 (4): 845–68. http://dx.doi.org/10.1007/BF03214588.

Fong, E., and E. Chan. 2010. "The Effect of Economic Standing, Individual Preferences, and Co-Ethnic Resources on Immigrant Residential Clustering." *International Migration Review* 44 (1): 111–41. http://dx.doi.org/10.1111/j.1747-7379.2009.00800.x.

Francis, J. 2009. *"You Cannot Settle Like This": The Housing Situation of African Refugees in Metro Vancouver*. Working paper 09-02. Vancouver: Metropolis British Columbia.

Francis, J., and D. Hiebert. 2011. *Shaky Foundations: Precarious Housing and Hidden Homelessness among Refugees, Asylum Seekers, and Immigrants in Metro Vancouver*. Working paper 11-18.Vancouver: Metropolis British Columbia.

Haan, M. 2007. "The Homeownership Hierarchies of Canada and the United States: The Housing Patterns of White and Non-White Immigrants of the Past Thirty Years." *International Migration Review* 41 (2): 433–65. http://dx.doi.org/10.1111/j.1747-7379.2007.00074.x.

Hiebert, D. 2009. "Newcomers in the Canadian Housing Market: A Longitudinal Study, 2001–2005." *Canadian Geographer/Le Géographe canadien* 53 (3): 268–87. http://dx.doi.org/10.1111/j.1541-0064.2009.00263.x.

Hiebert, D., A. Germain, R. Murdie, V. Preston, J. Renaud, D. Rose, E. Wyly, V. Ferreira, P. Mendez, and M. Murnaghan. 2006. *The Housing Situation and Needs of Recent Immigrants in the Montreal, Toronto, and Vancouver CMAs: An Overview*. Ottawa: Canada Mortgage and Housing Corporation.

Hiebert, D., and P. Mendez. 2008. *Settling In: Newcomers in the Canadian Housing Market, 2001–2005*. Ottawa: Canada Mortgage and Housing Corporation.

Hou, F. 2006. "Spatial Assimilation of Racial Minorities in Canada's Immigrant Gateway Cities." *Urban Studies* (Edinburgh, Scotland) 43 (7): 1191–213. http://dx.doi.org/10.1080/00420980600711993.

Immergluck, D. 2004. *Credit to the Community: Reinvestment and Fair Lending Policy in the United States*. Armonk, NY: M.E. Sharpe.

Joint Center for Housing Studies. 2005. *The State of the Nation's Housing: 2005*. Cambridge, MA: Harvard University Press.

Joint Center for Housing Studies. 2008. *The State of the Nation's Housing: 2008*. Cambridge, MA: Harvard University Press.

Kochhar, R., A. Gonzalez-Barrera, and D. Dockerman. 2009. *Through Boom and Bust: Minorities, Immigrants and Homeownership*. Washington: Pew Hispanic Center.

Larrivee, M. 2011. "Response to the Changing Landscape of Settlement in Greater Vancouver: The Step Ahead Settlement Enhancement Project." *Our Diverse Cities* 8 (Spring): 76–81.

Leichenko, R.M. 2001. "Growth and Change in U.S. Cities and Suburbs." *Growth and Change* 32 (3): 326–54. http://dx.doi.org/10.1111/0017-4815.00162.

Leloup, X., P. Apparicio, and F.D. Esfahani. 2011. "Ethnicity and Homeownership in Montreal, Toronto and Vancouver: Measuring Effects of the Spatial Distribution of Ethnic Groups Using Multilevel Modelling in 1996 and 2001." *Journal of International Migration and Integration* 12 (4): 429–51.

Leone, R., and B.W. Carroll. 2010. "Decentralization and Devolution in Canadian Social Housing Policy." *Environment and Planning C: Government & Policy* 28 (3): 389–404. http://dx.doi.org/10.1068/c09153.

Ley, D. 2010. *Millionaire Migrants: Trans-Pacific Life Lines*. West Sussex, UK: Wiley-Blackwell. http://dx.doi.org/10.1002/9781444319262.

Lo, L., S. Wang, P. Anisef, V. Preston, and R. Basu. 2010. *Recent Immigrants' Awareness of, Access to, Use of, and Satisfaction with Settlement Services in York Region*. Working paper 79. CERIS – The Ontario Metropolis Centre.

McConnell, E.D., and I.R. Akresh. 2008. "Through the Front Door: The Housing Outcomes of New Lawful Immigrants." *International Migration Review* 42 (1): 134–62. http://dx.doi.org/10.1111/j.1747-7379.2007.00116.x.

McConnell, E.D., and I.R. Akresh. 2010. "Housing Cost Burden and New Lawful Immigrants in the United States." *Population Research and Policy Review* 29 (2): 143–71. http://dx.doi.org/10.1007/s11113-009-9134-9.

Mendez, P. 2008. *Immigrant Residential Geographies and the "Spatial Assimilation" Debate in Canada, 1997–2006*. Working paper 08-07. Vancouver: Metropolis British Columbia.

Moos, M., and A. Skaburskis. 2010. "The Globalization of Urban Housing Markets: Immigration and Changing Housing Demand in Vancouver."

*Urban Geography* 31 (6): 724–49. http://dx.doi.org/10.2747/0272-3638.31.6.724.

Mundra, K. 2013. *Minority and Immigrant Homeownership Experience: Evidence from the 2009 American Housing Survey*. Discussion paper no. 7131. Bonn: Institute for the Study of Labor.

Murdie, R.A. 1994. "Blacks in Near-Ghettos? Black Visible Minority Populations in Metropolitan Toronto Housing Authority Public Housing Units." *Housing Studies* 9 (4): 435–57. http://dx.doi.org/10.1080/02673039408720799.

Murdie, R.A. 2008a. "Diversity and Concentration in Canadian Immigration: Trends in Toronto, Montreal and Vancouver, 1971–2006." Toronto: Centre for Urban and Community Studies, University of Toronto. *Research Bulletin* 42.

Murdie, R.A. 2008b. "Pathways to Housing: The Experiences of Sponsored Refugees and Refugee Claimants in Accessing Permanent Housing in Toronto." *Journal of International Migration and Integration* 9 (1): 81–101. http://dx.doi.org/10.1007/s12134-008-0045-0.

Murdie, R.A., and E. Skop. 2012. "Immigration and Urban and Suburban Settlements." In *Immigrant Geographies of North American Cities*, ed. C. Teixeira, W. Li, and A. Kobayashi, 48–68. Don Mills, ON: Oxford University Press.

Myers, D., W.C. Baer, and S.Y. Choi. 1996. "The Changing Problem of Overcrowded Housing." *Journal of the American Planning Association* 62 (1): 66–84. http://dx.doi.org/10.1080/01944369608975671.

Myers, D., and C. Liu. 2005. "The Emerging Dominance of Immigrants in the U.S. Housing Market, 1970–2000." *Urban Policy and Research* 23 (3): 347–66. http://dx.doi.org/10.1080/08111470500197920.

Novac, S., J. Darden, D. Hulchanski, and A. Seguin. 2004. "Housing Discrimination in Canada: Stakeholder Views and Research Gaps." In *Finding Room: Policy Options for a Canadian Rental Housing Strategy*, ed. D. Hulchanski and M. Shapcott, 135–46. Toronto: CUCS Press, Centre for Urban and Community Studies.

Preston, V., R.A. Murdie, J. Wedlock, S. Agrawal, U. Anucha, S. D'Addario, M.J. Kwak, J. Logan, and A.M. Murnaghan. 2009. "Immigrants and Homelessness: At Risk in Canada's Outer Suburbs." *Canadian Geographer/Le Géographe canadien* 53 (3): 288–304. http://dx.doi.org/10.1111/j.1541-0064.2009.00264.x.

Qadeer, M., S. Agrawal, and A. Lowell. 2010. "Evolution of Ethnic Enclaves in the Toronto Metropolitan Area, 2001–2006." *Journal of International Migration and Integration* 11 (3): 315–39. http://dx.doi.org/10.1007/s12134-010-0142-8.

Radford, P. 2007. "A Call for Greater Research on Immigration Outside of Canada's Three Largest Cities." *Our Diverse Cities* 3 (Summer): 47–51.

Ray, B., and V. Preston. 2009. "Geographies of Discrimination: Inter-Urban Variations in Canada." *Journal of Immigrant & Refugee Studies* 7: 228–49. http://dx.doi.org/10.1080/15562940903150055.

Rose, D., and B. Ray. 2001. "The Housing Situation of Refugees in Montreal Three Years after Arrival: The Case of Asylum Seekers Who Obtained Permanent Residence." *Journal of International Migration and Integration* 2 (4): 493–529. http://dx.doi.org/10.1007/s12134-001-1010-3.

Rose, J. 2001. "Contexts of Interpretation: Assessing Immigrant Reception in Richmond, Canada." *Canadian Geographer/Le Géographe canadien* 45 (4): 474–93. http://dx.doi.org/10.1111/j.1541-0064.2001.tb01497.x.

Rosenbaum, E., and S. Friedman. 2007. *The Housing Divide: How Generations of Immigrants Fare in New York's Housing Market*. New York: New York University Press.

Saiz, A. 2007. "Immigration and Housing Rents in American Cities." *Journal of Urban Economics* 61 (2): 345–71. http://dx.doi.org/10.1016/j.jue.2006.07.004.

Sandercock, L. 2003. *Integrating Immigrants: The Challenge for Cities, City Governments, and the City-Building Professions.* Working paper 03-20. Vancouver: Metropolis British Columbia.

Schwartz, A.F. 2010. *Housing Policy in the United States*. 2nd ed. New York: Routledge.

Singer, A. 2008. "Twenty-First-Century Gateways: An Introduction." In *Twenty-First Century Immigrant Gateways: Immigrant Incorporation in Suburban America*, ed. A. Singer, C. Bretell, and S. Hardwick, 3–30. Washington, DC: The Brookings Institution.

Skop, E. 2012. *The Immigration and Settlement of Asian Indians in Phoenix, Arizona 1965–2011: Ethnic Pride vs. Racial Discrimination in the Suburbs.* Lewiston, NY: Edwin Mellen Press.

Smith, H., and D. Ley. 2008. "Even in Canada? The Multiscalar Construction and Experience of Concentrated Immigrant Poverty in Gateway Cities." *Annals of the Association of American Geographers* 98 (3): 686–713. http://dx.doi.org/10.1080/00045600802104509.

Teixeira, C. 2014. "Living on the 'Edge of the Suburbs' of Vancouver: A Case Study of the Housing Experiences and Coping Strategies of Recent Immigrants in Surrey and Richmond." *Canadian Geographer/Le Géographe canadienne* 58 (2): 168–87.

Teixeira, C. 2008. "Barriers and Outcomes in the Housing Searches of New Immigrants and Refugees: A Case Study of 'Black' Africans in Toronto's Rental Market." *Journal of Housing and the Built Environment* 23 (4): 253–76. http://dx.doi.org/10.1007/s10901-008-9118-9.

Teixeira, C. 2007. "Residential Experiences and the Culture of Suburbanization: A Case Study of Portuguese Homebuyers in Mississauga." *Housing Studies* 22 (4): 495–521. http://dx.doi.org/10.1080/02673030701387622.

Teixeira, C., and B. Halliday. 2010. "Introduction: Immigration, Housing and Homelessness." In *Newcomer's Experiences of Housing and Homelessness in Canada*, ed. C. Teixeira and B. Halliday, 3–7. Montreal: Association for Canadian Studies.

Teixeira, C., and W. Li, eds. 2009. "Immigrant and Refugee Experiences in North American Cities." *Journal of Immigrant & Refugee Studies* 7 (3): 221–7. http://dx.doi.org/10.1080/15562940903150030.

Teixeira, C., and R.A. Murdie. 1997. "The Role of Ethnic Real Estate Agents in the Residential Relocation Process: A Case Study of Portuguese Homebuyers in Suburban Toronto." *Urban Geography* 18 (6): 497–520. http://dx.doi.org/10.2747/0272-3638.18.6.497.

Turner, M.A., and S.L. Ross. 2005. "How Racial Discrimination Affects the Search for Housing." In *The Geography of Opportunity: Race and Housing Choice in Metropolitan America*, ed. X. de S. Briggs, 81–100. Washington, DC: Brookings Institution Press.

Yinger, J. 1995. *Close Doors, Opportunities Lost: The Continuing Cost of Housing Discrimination*. New York: Russell Sage Foundations.

# 2 Homeownership among Immigrants in Canada and the United States: Similarities and Differences

JOE T. DARDEN

The housing search process is the first activity immigrants engage in after arriving in their new country. All immigrants seek access to housing, whether rented or purchased. Most immigrants find that access to housing tends to be a function of the *differences* between their characteristics and those of the larger society (such as income, educational attainment, etc.). Because housing access is unequal, based on both the internal characteristics of immigrants and external forces such as discrimination in housing and mortgage lending, differential rates of homeownership occur among *immigrant*[1] groups by place of birth.

## *Objectives*

The objectives of this chapter are fourfold:

1. To determine the extent of *inequality* in homeownership rates by place of birth in both countries.
2. To assess the similarities in the patterns of inequality of homeownership rates among European and non-European immigrants in both countries.
3. To assess whether the internal characteristics of immigrants from Europe (the reference group) are sufficiently different from the characteristics of non-European immigrants to explain their traditional homeownership advantage (higher rates) in both Canada and the United States.
4. To assess whether external forces (e.g., discrimination in mortgage lending) may be more of a factor in the United States than in Canada in explaining any differences in homeownership rates between

immigrants born in Europe or of European origin and those born in other countries and of non-European origin.

### *Countries of Birth of Immigrants in the United States and Canada*

In 2010, the foreign-born or immigrant population in the United States reached about 40 million and represented 13 per cent of the total population. The highest percentage (37%) came from Central America (with 29.3% born in Mexico). By comparison, 28% of the foreign-born population were born in Asia, 12% in Europe, 4% in Africa, 2% in Northern America (consisting of Canada, Bermuda, Greenland, and St Pierre and Miquelon), and less than 1% in Oceania, which consists of Australia, New Zealand, Melanesia, Micronesia, and Polynesia (US Bureau of the Census 2012).

In 2006, immigration in Canada had a far-reaching impact on the country's population growth. It was responsible for two-thirds of Canada's population growth in the inter-census period of 2001–6 (Chui, Tran, and Maheux 2009). Most immigrants settled in Canada's largest Census Metropolitan Areas. The 2006 portrait of the foreign-born population was a diverse one, reflecting the waves of immigrants from different regions around the world.

In 2006, Canada had 6,186,950 immigrants, constituting 19.8 per cent of the total Canadian population. Most immigrants were born in Asia (40.8%). Immigrants born in Europe constituted 36.8% of all immigrants. The rest were composed of those born in Central and South America (6.1%), Africa (6.0%), the Caribbean and Bermuda (5.1%), the United States (4%), and Oceania (0.9%).

### *Changes in Immigration Policy and the Diversity of Immigrants*

Immigrants from non-European regions entered the United States and Canada after both countries changed their racially restrictive immigration policies. A wave of immigrants from non-European countries began to arrive in the United States after amendments to the 1965 Immigration Act abolished national origins quotas favouring Europeans. In 1960, 75 per cent of all immigrants were born in Europe due to the original, restrictive immigration policies. Eventually, the change in immigration policy resulted in greater racial and ethnic diversity in the characteristics of immigrants. By 2010, the racial composition of the immigrants were as follows: Hispanics constituted 47%, Asians consisted of 23.5%, non-Hispanic Whites made up 20.9%, Blacks comprised 7.6%, and immigrants consisting of two or more races composed the remaining 1.0% (US Bureau of the Census 2010).

In Canada, after 1967, the immigrant pool became more ethnically and racially diverse. By 2006, Canada's total non-White, non-European visible minority immigrant population was 3.3 million, or 54.3 per cent of the total immigrant population.

Given the similarities in the historical policies of each country towards the recruitment of immigrants from Europe and the exclusion of immigrants from Non-European regions, the goal here is to examine whether contemporary differences still exist in European and non-European immigrant incorporation as measured by homeownership rates.

## Conceptual or Theoretical Framework

To achieve the objectives stated above, I have adopted the theoretical framework of *differential incorporation*. Differential incorporation describes a process in which the majority native-born population *differentially* incorporates immigrant groups based on place of birth and characteristics into the mainstream society (Henry 1994). Incorporation is conceptualized on the basis of equal access to the rewards generated and distributed by the economic and political systems in Canada and the United States (Breton, Isajiw, Kalbach, and Reitz 1990).

Differential incorporation is also conceptualized as a two-way process relating to the *internal characteristics* of the immigrant group and the external *forces* the group encounters. Although many other characteristics may also be related to homeownership rates, the *internal characteristics* examined in this chapter include length of time in the country, educational attainment level, occupational status level, income level, and family type or marital status. These internal characteristics define each immigrant group demographically and by socio-economic status. It is expected that these characteristics will be highly related to immigrant homeownership rates such that any differences that exist between the rates for European immigrants and other immigrants can be attributed largely to differences in immigrant internal characteristics.

The remaining differences can be explained by examining *external forces*, the other process of differential incorporation. These forces are imposed on some immigrant groups by the majority native-born population via the housing market institutions, policies, and practices which they control. These forces may involve discrimination in mortgage lending and home purchase based on place of birth despite the *internal characteristics* of the immigrant group (Darden 2004; Lieberson 1980). Some forces may restrict some immigrant groups from becoming homeowners to a greater extent than other immigrants even if the groups have similar internal characteristics. Thus, discrimination in mortgage lending is also an important factor in explaining differences in homeownership rates (Haan 2005).

Consistent with differential incorporation, in theory, the more similar the immigrant group's characteristics are to those of the native-born population, the higher the probability of the group attaining homeownership rates at a higher level than more dissimilar groups. Also, the more similar an immigrant group's characteristics are to immigrants born in Europe, the smaller the gap in homeownership rates. A test of these hypotheses will provide an answer to the question as to whether the traditional European immigrant advantage in homeownership is similar or greater in Canada than in the United States.

This study controls the place of birth of immigrant groups by examining the homeownership rates of immigrants from the same places of birth who have settled in Canada and the United States. This comparative approach will reveal whether "place of birth," which has historically provided an inherent advantage to immigrants born in Europe, continues to exist in both Canada and the United States based on the most recent data. If so, to what extent does it exist? Is the "place of birth" in Europe advantage greater in Canada compared to the United States? Finally, does settlement in Canada offer a greater opportunity for immigrant homeownership regardless of place of birth compared to settlement in the United States?

Differential incorporation has important policy implications, since both Canada and the United States continue to use immigration policy to recruit immigrants. The ability of immigrants to become easily incorporated based on their education, occupation, income, and marital status and other internal characteristics is a positive factor in immigration policy (Darden 2009). Homeownership is one of the most important criteria of a successful incorporation process.

### *Implications for Immigrant Integration by Place of Birth in Canada and the United States*

Homeownership is a very important indicator of immigrant integration into the social and spatial structure of the country of destination, that is, the society at large. Homeownership provides immigrants with a stake in the system. It has been related to wealth and asset accumulation. Most immigrants view it as an investment, providing a path to social status, security, permanency, and stability in the new country of settlement (Carter and Vitiello 2012). Homeownership also creates a sense of strong attachment to one's home and neighbourhood (Manturuk, Lindblad, and Quercia 2010). When immigrants buy a home, it normally means they are able to fit into the social fabric of the country and that the immigrant family is in the new country to stay (Darden and Kamel 2000). Homeownership means that the immigrants are committed to the neighbourhood, the community, and the new country they now call home.

In turn, homeownership offers the immigrant benefits that renting does not. One such benefit is wealth accumulation via equity in the home and the ability and right to transfer the home to the children in the long run (i.e., intergenerational transfer) (Haurin, Hendershott, and Wachter 1996). Economically speaking, federal, provincial, and municipal policies in Canada and federal policy in the United States have favoured homeowners over renters (Carter and Vitiello 2012). In fact, in the United States, a homeowner can deduct mortgage interest and taxes paid to the government. Presently, a homeowner in the United States can deduct the interest on mortgages worth up to a total of one million dollars on the family's principal residence or on the family's second home (Lowenstein 2006). Also, the homeowner can deduct up to $100,000 on a home-equity loan.

Canadian federal income tax policy does not allow a deduction from the taxable income for interest on loans incurred by the taxpayer's primary residence. However, there is an indirect method in Canada for making interest on mortgage loans for a primary residence tax deductible via an asset swap. This action involves the homebuyer selling the existing property, purchasing a house in full or in part by the sale, getting a mortgage on the house, and finally buying back the original house with the money from the mortgage (Singleton v. Canada 2001). The Supreme Court ruled in 2001 that transactions in asset swaps are legal, but must be regarded as distinct ways of making home mortgages tax deductible.

Although providing tax deductions for mortgage interest was intended to increase homeownership in the United States, the evidence suggests that it has not had that effect. Homeownership rates in Australia and England are not significantly different than rates in the United States (Lowenstein 2006). Moreover, the analysis of the data in this chapter may show that the rates in Canada may also be as high as rates in the United States.

Nevertheless, overall homeownership does provide benefits to homeowners above those provided to renters. Green and White (1997) found that in the United States, homeownership has had a significant impact on the success of the children of homeowners. The decision to stay in school by teenagers and not become a high school dropout is more prevalent among students who reside in owner-occupied housing. Furthermore, the daughters of homeowners have a much lower incidence of teenage pregnancy (Harkness and Newman 2003). Homeowners are more likely than renters to participate in the political process. Finally, research on crime and homeownership shows that homeowners are less likely than renters to be victims of crime (Glaeser and Sacerdote 1999).

In sum, homeownership has enabled the immigrant groups with the highest homeownership rates to accumulate more wealth, transfer the wealth to the next generation, improve the household's social status, boost the educational performance of the household's children, induce greater political participation, and reduce the probability of exposure to crime. Above all, homeownership provides the sense of satisfaction and belonging to the new community and denotes a sense of social status and permanency. By contrast, immigrant groups with the lowest homeownership rates have benefited the least in terms of wealth and social status and are least integrated into the mainstream society. Thus, one can conclude that immigrants with the highest homeownership rates (e.g., Europeans or immigrants of European origin) have received more social and political benefits in Canada and the United States. This pattern is consistent with the theory of differential incorporation.

### *Factors Related to Homeownership Rates within the Framework of Differential Incorporation*

Regardless of place of birth and regardless of country of settlement, immigrant homeownership rates are expected to *increase* with (1) length of time in the country of settlement, (2) educational attainment level, (3) occupational status level, (4) income level, and (5) marital status as measured by the percentage of married couples. However, even when some immigrant groups have similar characteristics as described by the five variables, homeownership rates may still differ due to *external factors* such as discrimination based on place of birth.

Prior research has indicated that place of birth matters after controlling for socio-economic and demographic characteristics of homeowners. For example, White immigrants from Europe, especially those born in Northern and Western Europe, have higher homeownership rates in part because historically policies and practices in Canada and the United States gave a homeownership advantage to them (Darden 2004; Haan 2005). They are also least likely to face discrimination in mortgage lending and housing sales. By contrast, immigrants born in non-European countries, especially Africa, the Caribbean, and Central America, have lower homeownership rates in part because they have more dissimilar internal characteristics and are more likely than European immigrants to face discrimination (Darden and Teixeira 2009).

Finally, studies in the United States show that discrimination in mortgage lending may be a contributing factor, even after holding constant the level of credit worthiness (Munnell, Browne, McEneaney, and Tootell 1992). Most previous studies of immigrant homeownership rates have been based on a limited number of immigrants groups, conducted in a single country, or not based on the most recent data available. This chapter will expand our knowledge of differential rates of immigrant homeownership by comparing several groups in Canada and the United States using the most recent data available.

### *The Difficulty of Conducting Comparative Studies*

Ideally, a comparative study of homeownership rates would use data for the same census years and the time of arrival of immigrants into Canada and the United States to enable an analysis of each selected immigrant group from the same country of origin at the time of their first entry into Canada and the United States. The next step would be to follow them over time to determine when they first purchased their home. However, such *comparable* immigrant and homeownership data for both Canada and the United States are not available. This study uses the best and most recent data available to achieve the objectives stated in this chapter.

The Canadian analysis is based on Statistics Canada's 2006 Public Use Microdata Files (Statistics Canada 2008). These files provide comprehensive demographic and socio-economic data on the native- and foreign-born population, including place of birth by housing tenure (owned and rented) for the nation as a whole and for Census Metropolitan Areas (CMAs).

The Canadian analysis includes the six largest CMAs with the highest percentage of immigrants, namely, Montreal, Toronto, Vancouver, Ottawa, Edmonton, and Calgary. These CMAs contain significant numbers of non-European immigrant groups and have served as *gateways* for immigrants, especially Toronto, Montreal, and Vancouver.

Data for the United States were obtained from the US Bureau of the Census American Community Survey, 2006–2010. The American Community Survey provides data on selected characteristics of the native- and foreign-born populations of the nation as a whole and of metropolitan areas. The percentage of owner-occupied housing units and other characteristics related to homeownership are included.

For the United States, this chapter examines ten Primary Metropolitan Statistical Areas (PMSAs). The areas selected are the top ten areas with the highest percentage of foreign-born (i.e., immigrants). They also have the largest concentrations of non-European immigrant groups. The areas are New York City, Los Angeles, San Francisco, Miami, Chicago, Washington, DC, Houston, Dallas, Boston, and San Diego. Many of these areas are well established *gateways* for the continuous entry of immigrants; however, Dallas and Washington, DC, both with one million or more immigrants, emerged as immigrant gateways more recently (Singer 2009). All have populations of more than 600,000 foreign-born (US Bureau of the Census, 2012).

### *Immigrant Homeownership by Place of Birth in Canada*

Computation of the data from the 2006 Census Public Use Microdata files reveals that immigrants born in Europe had the highest mean homeownership rate, 76.1 per cent (table 2.1).The mean rate for immigrants born in the United States was 73.8% and for immigrants born in Oceania the rate was 71.7%. All these immigrants have higher homeownership rates than immigrants born in non-European countries that are not predominantly white. It is worth noting that the characteristics of these immigrants are most similar ethnically and racially (including skin colour) to the characteristics of the majority population among the native-born in Canada.

Homeownership rates among immigrants born in non-European countries were highest among immigrants born in Asia (65.4%)

Table 2.1 Mean homeownership rates by place of birth and ratio of equality with homeownership rate with immigrants born in Europe, Canada, 2006

| Place of birth | Mean homeownership rate | Ratio of equality with immigrants born in Europe |
|---|---|---|
| Europe | 76.1 | 1.00 |
| United States | 73.8 | 0.96 |
| Oceania | 71.7 | 0.94 |
| Asia | 65.4 | 0.86 |
| Caribbean | 60.8 | 0.80 |
| South America | 58.2 | 0.76 |
| Central America | 48.8 | 0.64 |
| Africa | 47.7 | 0.63 |

Note: The mean homeownership rates are based on the six CMAs in table 2.2. The ratio of equality is determined by dividing the rate for each immigrant group by the rate for immigrants born in Europe. A ratio of 1.00 = equality of homeownership.

Source: Computed by the author from data obtained from Statistics Canada (2008).

(table 2.1). Ranked next to Asia-born immigrants were immigrants born in the Caribbean (60.8%), South America (58.2%), Central America (48.8%), and Africa (47.7%). Immigrants born in Central America and Africa are the only immigrants (among the study groups) in Canada that had fewer homeowners than renter households.

### *The Inequality Gap in Homeownership Rates: Assessing the European Advantage*

To assess the inequality gap in homeownership rates by place of birth, we compare the immigrants born in Europe to immigrants born in non-European countries. With a ratio of 1.00 = equality to European-born immigrants, the homeownership rate gap reveals a four-tier structure of inequality. The least inequality (tier 1) is found among immigrants born in the United States and Oceania, with ratios of 0.96 and 0.94 respectively. In the second tier are immigrants born in Asia and the Caribbean, with ratios of 0.86 and 0.80 respectively. Immigrants born in South America are in tier 3 with a ratio of 0.76. In tier 4 are immigrants born in Central America and Africa, with ratios of 0.64 and 0.63 respectively (table 2.1). The immigrants with the smallest gaps in inequality have similar characteristics to the immigrants born in Europe. These characteristics include language, income, education, occupation, and skin colour. The gap in ratios (with few exceptions) increases as differences in characteristics increase.

### *Variation in Homeownership Rates by Census Metropolitan Area in Canada*

The homeownership rates for immigrants born in Europe ranged from 66.9% in Montreal to 80.5% in Calgary (table 2.2). The fact that immigrants born in Europe, Oceania, and the United States had the highest homeownership rates reflected a clear *divide* related to the European advantage. All of these households were ethnically, racially, and culturally related to the majority native-born population in Canada. The results are consistent with the differential incorporation process – that is, the greater the similarity between the immigrant group and the native-born population, the higher the probability of homeownership and the greater the incorporation into the larger society in each of the Census Metropolitan Areas.

Among immigrants born in non-European countries, those born in Asia had the highest homeownership rates. The rates ranged from 47.1% in Montreal to 77.9% in Calgary. Immigrants born in the Caribbean, a group with a long history in Canada (Darden 2004; Darden and Teixeira 2009), had rates that ranged from 40.5% in Montreal to 76.1% in Calgary. The rates for immigrants born in South America ranged from 39.8% in Montreal to 67.6% in Calgary. The homeownership rate for immigrants born in Africa ranged from 33.5% in Montreal to 57.5% in Calgary. In addition to Calgary, African immigrant homeowners exceeded renters only in Toronto (57.1%) and Vancouver (53.2%) (table 2.2).

Place of settlement in Canada "matters" in immigrant homeownership. There is a lower probability of becoming a homeowner if immigrants settle in Montreal and a higher probability of homeownership if immigrants settle in Calgary. This may be related to the fact that many immigrants decide to rent in Montreal rather than buy since Montreal is one of the most affordable cities in Canada for rental housing. For example, in 2010 the median rent for a two-bedroom apartment in Montreal was only $650. That compares to $1400 in Vancouver and

Table 2.2 Homeownership rates and place of birth by metro area in Canada, 2006

| | Metropolitan area | | | | | | | |
|---|---|---|---|---|---|---|---|---|
| Place of birth | Montreal | Toronto | Vancouver | Ottawa | Edmonton | Calgary | Mean | Ratio |
| Europe | 66.9 | 79.8 | 72.5 | 76.8 | 80.1 | 80.5 | 76.1 | 1.03 |
| Oceania | 59.5 | 68.1 | 74.6 | 66.7 | 80.5 | 80.7 | 71.6 | 0.97 |
| United States | 58.2 | 72.3 | 66.9 | 75.6 | 73.6 | 80.1 | 71.1 | 0.96 |
| Asia | 47.1 | 69.3 | 64.5 | 67.5 | 66.2 | 77.9 | 65.4 | 0.88 |
| Caribbean | 40.5 | 62.4 | 66.7 | 50.7 | 68.4 | 76.1 | 60.7 | 0.82 |
| South America | 39.8 | 63.8 | 59.1 | 54.9 | 63.9 | 67.6 | 58.2 | 0.78 |
| Central America | 29.5 | 46.0 | 40.2 | 48.3 | 64.1 | 64.8 | 48.8 | 0.66 |
| Africa | 33.5 | 57.1 | 53.2 | 36.9 | 48.2 | 57.5 | 47.7 | 0.64 |

Source: Computed by the author from data obtained from Statistics Canada (2008).

Table 2.3 Mean homeownership rates by place of birth and ratio of equality with homeownership rate with immigrants born in Europe, United States, 2010

| Place of birth | Mean homeownership rate | Ratio of equality with immigrants born in Europe |
|---|---|---|
| Europe | 65.8 | 1.00 |
| Canada | 61.5 | 0.93 |
| Asia | 59.7 | 0.90 |
| Oceania | 52.6 | 0.79 |
| South America | 50.2 | 0.76 |
| Caribbean | 49.8 | 0.75 |
| Africa | 40.7 | 0.61 |
| Central America | 39.7 | 0.60 |

Note: The mean homeownership rates are based on the 10 primary metropolitan areas with the highest percentage of foreign-born population. A total of 51% of all foreign-born in the United States reside in these 10 primary metropolitan areas. The ratio of equality is determined by dividing the rate for each immigrant group by the rate for immigrants born in Europe. A ratio of 1.00 = equality of homeownership.
Source: Computed by the author from data obtained from the US Census Bureau (2010).

$1050 in Toronto and Calgary (Federation of Canadian Municipalities 2010). Conversely, Calgary has the distinction of being the only very large CMA where all the immigrant groups examined have more homeowners than renters.

### *Immigrant Homeownership by Place of Birth in the United States*

Computation of the data from the US Bureau of the Census (2010) revealed that immigrants born in Europe had the highest homeownership rate, 65.8 per cent (table 2.3). The group ranked second was born in Canada, with a homeownership rate of 61.5%. Asian-born immigrants ranked third, with a mean homeownership rate of 59.7%, compared to 52.6% for immigrants born in Oceania. Immigrants with the lowest homeownership rates were born in South America (50.2%), the Caribbean (49.8%), Africa (40.7%), or Central America (39.7%). Thus, the stratification of homeownership rates, with immigrants born in Europe at the top, followed by immigrants born in Asia and Oceania in the middle and immigrants born in Africa and Central America at the bottom, is similar to that in Canada. Also similar was the fact that, as in Canada, most immigrants born in Central America and Africa were still characterized as primarily renters, that is, as not yet achieving the status of homeowners.

### *The Inequality Gap in Homeownership Rates: Assessing the European Advantage*

To assess the inequality gap in homeownership rates by place of birth, we compare immigrants born in Europe to immigrants born in non-European countries. With a ratio of 1.00 = equality with those immigrants born in Europe, the analysis revealed a homeownership rate gap reflecting a three-tier hierarchy. Following immigrants born in Europe were immigrants born in Canada, with a ratio of 0.93, and Asia, with a ratio of 0.90. In the second tier were immigrants born in Oceania, South America, and the Caribbean. They had ratios of 0.79, 0.76, and 0.75 respectively (table 2.3). In the third and last tier were immigrants born in Africa (0.61) and Central America (0.60).

Some similarities existed in the homeownership hierarchy in Canada and the United States. In both countries, the inequality gap in homeownership rates was greatest for immigrants born in Central America and Africa. More important was the fact that immigrants in the United States and Canada who were born in Europe had the highest homeownership rate compared to all other immigrant groups. That is the European advantage, and it existed in both countries.

### *Variation in Homeownership Rates by Selected Primary Metropolitan Statistical Areas (PMSAs)*

The homeownership rate for immigrants born in Europe ranged from 56.2% in Los Angeles to 68.8% in Washington, DC. Only immigrants born in Europe and Canada had homeownership rates that exceeded 50% in all ten PMAs. The rates for immigrants born in Canada ranged from 51.2% in New York to 68.4% in Boston. Homeownership rates for immigrants born in Asia (the non-European group with the highest homeownership rates) ranged from 49.6% in New York to 70.1% in Washington, DC. Homeownership rates for immigrants born in Central America (the group with the lowest rates) ranged from 25.6% in Boston to 54.8% in Chicago (table 2.4). Similarly to Canada, *place matters* in homeownership. There is a higher probability for all immigrants to become homeowners if they settle in Miami and a lower probability if they settle in New York.

Table 2.4 Homeownership rates and place of birth by primary metropolitan area in the United States, 2010

| | Metropolitan area | | | | | | | | | | | |
|---|---|---|---|---|---|---|---|---|---|---|---|---|
| Place of birth | Boston | Chicago | Dallas | Houston | Los Angeles | Miami | New York | San Diego | San Francisco | Washington, DC | Mean | Ratio |
| Europe | 65.4 | 74.6 | 67.4 | 69.9 | 56.2 | 76.9 | 56.3 | 63.6 | 58.8 | 68.8 | 65.8 | 1.00 |
| Canada | 68.4 | 65.8 | – | – | 62.3 | – | 51.2 | – | 59.7 | – | 61.4 | 0.93 |
| Asia | 50.0 | 65.4 | 63.3 | 64.8 | 50.8 | 67.2 | 49.6 | 58.1 | 57.7 | 70.1 | 59.7 | 0.90 |
| Oceania | 61.0 | – | – | – | 46.3 | – | 38.9 | – | 55.2 | 51.7 | 50.6 | 0.76 |
| South America | 31.9 | 60.2 | 54.4 | 59.6 | 47.0 | 59.0 | 39.1 | 45.3 | 43.6 | 62.3 | 50.2 | 0.76 |
| Caribbean | 34.4 | 56.1 | 59.4 | 62.4 | 46.5 | 60.6 | 32.8 | 35.6 | 53.0 | 58.0 | 49.8 | 0.75 |
| Africa | 41.6 | 41.3 | – | – | 43.6 | – | 32.8 | – | 44.4 | – | 40.7 | 0.62 |
| Central America | 25.6 | 54.8 | 47.9 | 44.9 | 32.3 | 43.4 | 19.4 | 40.3 | 41.9 | 47.1 | 39.7 | 0.60 |

Source: Computed by the author from data obtained from US Bureau of the Census (2010).

## Comparison of Homeownership Rates in Canada and the United States

In this section I assess the significance of "place" of settlement on homeownership rates by comparing the rates for the same immigrant group (those who settled in Canada and those who settled in the United States). For immigrants born in Europe, those who settled in Canada had a higher homeownership rate (76.1% vs. 65.8%). The homeownership difference with the advantage for immigrants who settled in Canada was even higher for immigrants born in Oceania. The homeownership rate for those who settled in Canada was 71.6%, compared to only 52.6% for those who settled in the United States, a gap of 19 percentage points. Settlement in Canada also offered an advantage to those immigrants born in Asia. The rate was 65.4% in Canada compared to 59.7% in the United States, a gap of 5.7 percentage points. Other immigrants also experienced higher homeownership rates in Canada. The rate for immigrants born in South America was 58.2%, compared to 48.9% in the United States, a gap of 9.3 percentage points. The rate for immigrants born in the Caribbean was 60.8% in Canada and 49.8% in the United States, a difference of 11 percentage points. Although immigrants born in Central America and Africa had the lowest homeownership rates in Canada and the United States, those immigrants who settled in Canada had higher rates. For those immigrants born in Central America, the rate was 48.8% in Canada, compared to 39.7% in the United States, a difference of 9.1 percentage points. For immigrants born in Africa, the rate in Canada was 47.7%, compared to 39.7% in the United States, a gap of 8 percentage points. These differences in Canada and the United States based on metropolitan areas are also reflected in the two countries as a whole. Immigrant homeownership rates are 14 percentage points higher on average in Canada than in the United States (table 2.5). The rates are higher in Canada for every immigrant group in the study.

In sum, immigrants from Europe compared to immigrants from non-European countries have the highest homeownership rates in both Canada and the United States. The question based on the theory of differential incorporation is whether the higher rates can be totally explained by differences in the *internal characteristics* of immigrants born in Europe compared to immigrants born in non-European regions, or whether external forces (i.e., discrimination in mortgage lending) are necessary to explain the differences. The next section will examine possible reasons for the differences in homeownership rates of European immigrants compared to those born in non-European countries.

Table 2.5 National homeownership rates by place of birth and ratio of equality with homeownership rates with immigrants born in Europe – Comparison of United States in 2010 and Canada in 2006

| | Rate | Ratio | Rate | Ratio |
|---|---|---|---|---|
| Place of birth | United States | United States | Canada | Canada |
| Europe | 68.1 | 1.00 | 78.3 | 1.00 |
| Canada | 72.4 | 1.06 | – | – |
| Asia | 58.2 | 0.85 | 60.1 | 0.76 |
| Oceania | 55.8 | 0.81 | 74.5 | 0.95 |
| South America | 49.9 | 0.73 | 59.6 | 0.76 |
| Caribbean | 48.3 | 0.70 | 57.1 | 0.72 |
| Africa | 41.4 | 0.60 | 47.1 | 0.60 |
| Central America | 40.0 | 0.58 | 50.8 | 0.64 |
| United States | – | – | 73.8 | 0.94 |
| Mean | 48.6 | | 62.6 | |

Note: The ratio of equality is determined by dividing the rate for each immigrant group by the rate for immigrants born in Europe. A ratio of 1.00 = equality of homeownership.
Sources: Computed by the author from data obtained from the US Bureau of the Census (2010) and Statistics Canada (2008).

### *Expected and Observed Ranking of Immigrant Groups Based on Differential Incorporation Theory*

Based on differential incorporation theory, immigrants born in Europe compared to immigrants born in non-European countries should have the following characteristics:

1. the highest number years of residency;
2. the highest educational attainment;
3. the highest occupational status;
4. the highest median income; and
5. the highest percentage of married households.

The observed ranking of Canadian immigrant groups was based on Statistics Canada census data. It shows that European immigrants exceed other immigrants only in the category of the number of years of residency in Canada. They do not have the highest level of educational attainment. Only 18.6 per cent of their immigrants have bachelor degrees or a higher level of education. This level was much lower than for immigrants born in the United States (36.4%), Africa (33.6%), and Asia (32.1%).

Likewise, European immigrants do not have the highest occupational status compared to other immigrants. Instead, immigrants born in Europe had 28.4% of its group in the highest occupations, including management, business, finance, and administrative occupations. This was lower than the percentage of immigrants born in the United States and Africa. For both immigrant groups, the percentage was 30.1. European immigrants also did not have the highest median income. Instead, their median income was $30,048, which was slightly lower than the $30,068 for immigrants born in the United States. Finally, European immigrants did not have the highest percentage of married couples. Instead, immigrants with the highest percentage of married couples were born in Asia.

Similarly to the pattern in Canada, in the United States (2010) European immigrants had the longest number of years of residency. However, this was the only characteristic that exceeded the level of other immigrant groups. European immigrants did not have the highest level of educational attainment. Of Asian immigrants, 48.5% had bachelor's degrees or a higher level of education, compared to 36.4% of European immigrants. European immigrants did not have the highest occupational status. Instead, immigrants born in Europe were ranked third, with only 41.8% of its group in management, business, science, and art occupations. They were exceeded by immigrants born in Canada (56.8%) and Asia (46.6%). European immigrants also did not earn the highest income. The median income for full-time year-round male workers was $51,377. They were exceeded by male immigrants from Canada ($73,539) and Oceania ($51,473). Finally, European immigrants did not have the highest percentage of married households. With only 52%, they were exceeded by Asian immigrants (63%).

In sum, the data show that the highest homeownership rate found among European immigrants cannot be explained by their internal characteristics. The only characteristic among the five in which European immigrants exceeded other immigrants is in the length of time in their country of settlement. On socio-economic and demographic characteristics the highest homeownership rate of European immigrants cannot be explained. Thus, it is important to examine the other component of differential incorporation, that is, external forces, which include discrimination in mortgage lending.

Mortgage-lending discrimination occurs via a process of differential acceptance and rejection of loan applications based on the characteristics of the applicants. Racial/ethnic discrimination is a factor when other characteristics (e.g., income, education, occupation, or marital status) are controlled (Haurin, Herbert, and Rosenthal 2007). Researchers have found that after controlling for loan applicant characteristics, African American loan applicants were more likely to have loan applications rejected by at least 8 percentage points than comparable White applicants (Carr and Megbolugbe 1993; Ladd 1998). The broad consensus among researchers in the United States is that mortgage-lending discrimination based on race/ethnicity is a factor impacting homeownership rates. Such discrimination occurs on the front end via denial of the mortgage loan and the differences in the type of loan received (prime versus sub-prime). Discrimination also occurs on the rear end via default rates and foreclosures. Berkovec, Canner, Gabriel, and Hannan (1998) found that African American mortgage default rates were higher than White default rates after controlling for a variety of characteristics (Haurin et al. 2007).

Comparative research between the United States and Canada has revealed a higher level of residential segregation in the United States and a lower homeowner rate for Black and Hispanic immigrants (Darden and Fong 2012). Research continues to show that discrimination in the housing market restricts non-White groups disproportionately to housing in the central city, where more limited supplies of single-family housing exist. Such restrictions lower the supply of available housing for ownership, thus reducing homeownership rates (Haurin et al. 2007; Turner, Ross, Galster, and Yinger 2002). Herbert (1997) has argued that the reduced minority access to single-family detached housing lowers minority homeownership rates because homeowners and single-family housing are complements (Haurin et al. 2007).

When one compares homeownership rates of White European immigrants and non-White, non-European immigrants in the United States, race is as much or more of a factor than income. For example, Herbert, Haurin, Rosenthal, and Duda (2005) found that the overall homeownership rates of high-income White households is nearly 10 percentage points higher than for similar African American and other minority (Asian and Hispanic) high-income households. The difference was not due to the internal characteristic – income – but to differential access to neighbourhoods where single-family housing is more prevalent (Haurin et al. 2007).

It appears that even when authors include several characteristics beyond the number presented here, the gap in homeownership by race cannot be explained by the array of internal characteristics. For example, Flippen (2001) examined such characteristics as inheritances, age, number of children, health, cognitive ability, self-employment, retirement status, expected years of life remaining, region, and urban location. This study of the most comprehensive list of characteristics ever controlled found that African Americans and Hispanics were still significantly less likely than Whites to be homeowners. There was still a 6 percentage point residual that could not be explained by such a comprehensive list of variables (Flippen 2001; Haurin et al. 2007).

Also the Great Recession, plus continued discrimination in mortgage lending, impacted immigrants of colour differentially. The results were a greater loss of wealth to foreclosure. The Center for Responsible Lending analysed the demographics of the foreclosure crisis in 2010 to find that nearly 8 per cent of both African Americans and Latinos have lost their homes to foreclosures, compared to 4.5 per cent of Whites (Weller, Fields, and Agbede 2011). Furthermore, a home owned by a Black family was found to be 76 per cent more likely to go through foreclosure than a home owned by a White family. Easing constraints to homeownership in a way that protects against the risk of default through malicious non-bank lending practices is particularly critical to addressing the homeownership gaps among immigrant groups (ibid.).

In sum, discrimination in mortgage lending begins with differential treatment experienced by non-White immigrants and the non-White native-born in the housing search process. Numerous studies conducted in the United States suggest that race/ethnicity has a significant effect on whether a household is rejected for a mortgage loan even after controlling for a variety of demographic and economic variables. The evidence presented by individual researchers has been strengthened by a United States government-funded study administered by the US Department of Housing and Urban Development. The most recent Housing and Urban Development–funded study based on paired tests demonstrated that in several large metropolitan areas, real estate agents in general engage less than they did in 1977 in the most blatant forms of housing discrimination (refusing to meet with a Black, Hispanic, or Asian home-seeker or provide information about available units), but the form of housing discrimination that persists today is providing information about *fewer* housing units to non-White home-seekers compared to White home-seekers (Turner et al. 2013). This discriminatory practice raises the costs of the housing search and restricts the homeownership options for Blacks, Hispanics, and Asian home-seekers. Earlier studies have indicated that Whites also receive more favourable treatment than Hispanics and Blacks in the form of information about financial resources available (Coulson and Kang 2001). Lenders also advised Whites on how to improve their credit score. Moreover, the explanatory variables that are most important in explaining the Asian immigrant/White European immigrant homeownership rates depend on the particular Asian immigrant group – for instance, Chinese versus South Asians. In sum, internal characteristics alone do not explain the gap in homeownership rates between European immigrants and those born in non-European countries.

## Conclusions

This chapter has examined differential homeownership rates between European and non-European immigrants in two predominantly White countries with a history of providing recruitment preferences and homeownership advantages to immigrants born in Europe while excluding immigrants born in non-European countries. The objective was to examine the contemporary homeownership rate gap within the theoretical context of differential incorporation. The results, using census data, revealed that European-born immigrants continue to have the highest homeownership rates in both Canada and the United States. However, such high rates cannot be totally explained by internal characteristics alone (e.g., length of time in the country, educational attainment, occupational status, and marital status). Immigrants born in non-European countries exceeded immigrants born in Europe on all socio-economic and demographic variables except length of residency in the country.

Based on the paired test or audit method to determine racial/ethnic discrimination in housing the United States (Darden 2004; Turner et al. 2002; Turner et al. 2013), individual studies by researchers and analyses of home mortgage disclosure data revealed (Apgar and Calder 2005) that discrimination against Black, Hispanic, and Asian immigrants was also important in explaining the gap in European versus non-European homeownership rates despite differences in immigrant internal characteristics. Schwartz (2010) provides the most up-to-date evidence that discrimination in mortgage lending via predatory lending continued even during the Obama administration and Turner et al. (2013) provides the most recent data on racial discrimination in housing.

*Limitations of the Study*

As with any comparative study, there are data limitations. This study is limited by the fact that the data for the United States are more recent than the data for Canada. Since the data for homeownership in Canada is limited to 2006 (i.e., a snapshot census for one point in time) and the data for the United States is rolling sample data extending from 2006 to 2010, the two databases are not comparable. Whereas the data for the United States extends through the Great Recession of 2007–9, the data for Canada does not. Thus, it is not possible to determine the exact extent of the differences in homeownership rates by comparing pre-recession 2006 data for Canada with post-recession 2010 data for the United States.

It is known that the recession had a negative impact on homeownership rates in the United States. During the period, homeownership rates dropped to 67.1 per cent for all households. The recession was very harmful to all households and all immigrant groups, but the impact was more severe for non-White households, especially immigrants from Africa and Central America (Weller, Fields, and Agbede 2011).

Also, comparable databases do not exist in both countries to measure the extent of discrimination in housing and mortgage lending based on the *paired test method.* In the United States, as a result of the Fair Housing Act of 1968 and supplementary civil rights legislation, the US Department of Housing and Urban Development has funded several nationwide studies since 1977 to assess the extent of racial discrimination using the *paired test* or audit method. The most recent study was published in 2013 (Turner et al. 2013). This is the gold standard in measuring racial discrimination. Such studies have documented the existence of racial discrimination in housing and mortgage lending after controlling for other factors (Turner et al. 2002).

In 1975, the US Congress enacted the Home Mortgage Disclosure Act. This act requires major financial institutions, including banks and saving and loan associations, to provide data to the public on where they have made mortgage loans and to which racial/ethnic groups by census tracts (neighbourhood). Such data allows for measurement of the acceptance and rejection of mortgage loans by neighbourhood racial composition and race and class of applicant. With these two databases, one can assess the extent of racial/ethnic discrimination against immigrant groups by race/ethnicity. Unfortunately, databases that are similar to the ones described above are not available in Canada.

The most data that exist on housing discrimination in Canada were collected via the complaint method by Human Rights Commissions. This is the weakest method among those available and is less effective in detecting discrimination compared to the most effective audit method (Darden 2004). Given the limitations of the audit method to compare data on discrimination in Canada and the United States, it is possible that Haan and Yu (chapter 3) emphasized differences in homeownership rates within the context of culture and household formation. The authors emphasized the external barriers of discrimination less in explaining the homeownership gap between Blacks, Asian Indians, mainland Chinese, Filipinos, and Whites.

The reader should notice the link between the chapter by Haan and Yu, with its emphasis on culture and household formation, and this chapter. Unlike Haan and Yu, this chapter compares similarities and differences in homeownership rates in Canada and the United States among immigrants born in Europe with those born in Asia, Oceania, South America, Caribbean, Africa, and Central America. It concludes that the differences in rates between immigrants born in Europe and those born in other countries cannot totally be explained by internal characteristics (i.e., culture and household formation). Instead, discrimination in housing and mortgage lending must also be considered.

This chapter is also limited in scope. It was beyond its scope to determine which explanatory dependent variable is the most and least significant predictor of homeownership (e.g., income, education, marital status, etc.). For example, a logit regression analysis would have provided more certainty about *which* variable among the array of internal characteristics contributed most to the differential advantage in European immigrant homeownership rates compared to the rates for non-European immigrants. The purpose of this chapter was to focus instead on whether internal characteristics *alone* could explain the gap in homeownership rates. The argument presented is that a *total* explanation of the homeownership rate gap requires an examination of both internal and external characteristics (e.g., discrimination in housing).

## Policy Implications

This study is important nevertheless to policymakers in both countries as they assess how to integrate non-European immigrants into their societies. Such immigrants constitute the wave of the future. This chapter has identified the extent of inequality in homeownership by immigrant place of birth. The results are not equal. To bring about equality, policymakers must introduce policies to assist immigrants, especially those who remain primarily renters, and to remove the barriers to homeownership experienced by such immigrants based on race/ethnicity. It is also important for the Canadian government to provide the type of databases as exist in the United States that allow for the audit method to be used to measure racial/ethnic discrimination. Only then can a truly comparative study be done to assess the role of external forces in contributing to immigrant differences in homeownership rates in both countries.

## NOTES

1 Immigrant and foreign-born will be used interchangeably in this chapter. The US Census Bureau uses the term *foreign-born* to refer to anyone who is not a US citizen at birth. This includes naturalized citizens, lawful permanent residents, temporary migrants (such as foreign students), humanitarian migrants (such as refugees), and undocumented migrants. The term *native-born* refers to anyone born in the United States, Puerto Rico, or a US Island Area, as well as those born abroad of at least one US citizen parent.

Statistics Canada defines immigrants as persons residing in Canada who were born outside of Canada, excluding temporary foreign workers, Canadian citizens born outside Canada, and those with student or working visas.

## REFERENCES

Apgar, W., and A. Calder. 2005. "The Dual Mortgage Market: The Persistence of Discrimination." In *Mortgage Lending: In the Geography of Opportunity*, ed. X. Briggs, 101–23. Washington, DC: The Brookings Institution.

Berkovec, J.A., G.B. Canner, S.A. Gabriel, and T.H. Hannan. 1998. "Discrimination, Competition, and Loan Performance in FHA Mortgage Lending." *Review of*

*Economics and Statistics* 80 (2): 241–50. http://dx.doi.org/10.1162/003465398557483.

Breton, R., W. Isajiw, W. Kalbach, and J. Reitz. 1990. *Ethnic Identity and Equality: Varieties of Experiences in a Canadian City*. Toronto: University of Toronto Press.

Carr, J., and I. Megbolugbe. 1993. *A Research Note on the Federal Reserve Bank of Boston Study on Mortgage Lending*. Washington, DC: Federal National Mortgage Association, Office of Housing Research.

Carter, T., and D. Vitiello. 2012. "Immigrants, Refugees, and Housing." In *Immigrant Geographies of North American Cities*, ed. C. Teixeira, W. Li, and A. Kobayashi, 91–111. Don Mills, ON: Oxford University Press.

Chui, T., K. Tran, and H. Maheux. 2009. *2006 Census. Immigration in Canada: A Portrait of the Foreign-Born Population. 2006 Census: Findings*. Ottawa: Statistics Canada.

Coulson, N.E., and M. Kang. 2001. *The Heterogeneity of Asian-American Homeownership*. University Park: Pennsylvania State University Press.

Darden, J.T. 2004. *The Significance of White Supremacy in the Canadian Metropolis of Toronto*. New York: Edwin Mellen Press.

Darden, J.T. 2009. "The African Diaspora: Historical and Contemporary Immigration and Employment Practices in Toronto." In *The African Diaspora in the United States and Canada at the Dawn of the 21st Century*, ed. J. Frazier, J.T. Darden, and N. Henry, 49–60. Binghamton: Global Academic Publishing.

Darden, J.T., and E. Fong. 2012. "The Spatial Segregation and Socioeconomic Inequality of Immigrant Groups." In *Immigrant Geographies of North American Cities*, ed. C. Teixeira, W. Li, and A. Kobayashi, 69–90. Don Mills, ON: Oxford University Press.

Darden, J.T., and S. Kamel. 2000. "Black and White Differences in Homeownership Rates in the Toronto Census Metropolitan Area: Does Race Matter?" *Review of Black Political Economy* 28 (2): 53–76. http://dx.doi.org/10.1007/s12114-000-1017-6.

Darden, J.T., and C. Teixeira. 2009. "The African Diaspora in Canada." In *The African Diaspora in the United States and Canada at the Dawn of the 21st Century*, ed. J. Frazier, J.T. Darden, and N. Henry, 13–34. Binghamton: Global Academic Publishing.

Federation of Canadian Municipalities. 2010. *No Vacancy: Trends in Rental Housing in Canada*. Ottawa: Federation of Canadian Municipalities.

Flippen, C.A. 2001. "Residential Segregation and Minority Home Ownership." *Social Science Research* 30 (3): 337–62. http://dx.doi.org/10.1006/ssre.2001.0701.

Glaeser, E.L., and B. Sacerdote. 1999. "Why Is There More Crime in Cities?" *Journal of Political Economy* 107 (S6): S225–58. http://dx.doi.org/10.1086/250109.

Green, R., and M. White. 1997. "Measuring the Benefits of Homeowning: Effects on Children." *Journal of Urban Economics* 41 (3): 441–61. http://dx.doi.org/10.1006/juec.1996.2010.

Haan, M. 2005. "The Decline of Immigrant Home-Ownership Advantage: Life-Cycle, Declining Fortunes, and Changing Housing Careers in Montreal, Toronto, and Vancouver, 1981–2001." *Urban Studies* (Edinburgh, Scotland) 42 (12): 2191–212. http://dx.doi.org/10.1080/00420980500331983.

Harkness, J., and S. Newman. 2003. "Differential Effects of Homeownership on Children from Higher and Lower Income Families." *Journal of Housing Research* 14 (1): 1–19.

Haurin, D.R., P. Hendershott, and S. Wachter. 1996. "Wealth Accumulation and Housing Choices of Young Households: An Exploratory Investigation." *Journal of Housing Research* 7: 33–57.

Haurin, D.R., C.E. Herbert, and S.S. Rosenthal. 2007. "Homeownership Gaps among Low-Income and Minority Households." *Cityscape: A Journal of Policy Development and Research* 9 (2): 5–51.

Henry, F. 1994. *The Caribbean Diaspora in Toronto: Learning to Live with Racism.* Toronto: University of Toronto Press.

Herbert, C.E. 1997. *Limited Choices: The Effect of Residential Segregation on Homeownership among Blacks.* Cambridge, MA: Harvard University Press.

Herbert, C.E., D.R. Haurin, S.S. Rosenthal, and M. Duda. 2005. *Homeownership Gaps among Low-Income and Minority Borrowers and Neighborhoods.* Washington, DC: U.S. Department of Housing and Urban Development, Office of Policy and Research.

Ladd, H. 1998. "Evidence on Discrimination in Mortgage Lending." *Journal of Economic Perspectives* 12 (2): 41–62. http://dx.doi.org/10.1257/jep.12.2.41.

Lieberson, S. 1980. *A Piece of the Pie: Black and White Immigrants since 1880.* Berkeley: University of California Press.

Lowenstein, R. 2006. "Who Needs the Mortgage Interest Deduction?" *New York Times*, 5 March.

Manturuk, K., M. Lindblad, and R. Quercia. 2010. "Friends and Neighbors: Homeownership and Social Capital among Low- to Moderate-Income Families." *Journal of Urban Affairs* 32 (4): 471–88. http://dx.doi.org/10.1111/j.1467-9906.2010.00494.x.

Munnell, A., L. Browne, J. McEneaney, and G. Tootell. 1992. *Mortgage Lending in Boston: Interpreting HMDA Data.* Boston: Federal Reserve Board of Boston.

Schwartz, A. 2010. *Housing Policy in the United States.* 2nd ed. New York: Routledge.

Singer, A. 2009. *The New Geography of United States Immigration*. Washington, DC: The Brookings Institution.

Singleton v. Canada, 2 S.C.R. 1046, 2001 SCC 61 (Supreme Court of Canada 2001).

Statistics Canada. 2008. *2006 Public Use Microdata Files (POMF)*. Ottawa: Minister of Industry.

Turner, M., S. Ross, G. Galster, and J. Yinger. 2002. *Discrimination in Metropolitan Housing Markets: National Results from Phase 1 HDS 2000*. Washington, DC: The Urban Institute.

Turner, M., R. Santos, D. Levy, D. Wissoker, C. Aranda, and R. Pitingolo, R. 2013. *Housing Discrimination against Racial and Ethnic Minorities 2012*. Washington, DC: The Urban Institute.

US Bureau of the Census. 2010. *American Community Survey, 2006–2010*.

US Bureau of the Census. 2012. *Profile of the Foreign-Born Population in the United States: 2010*. Washington, DC: US Department of Commerce.

Weller, C.E., J. Fields, and F. Agbede. 2011. "The State of Communities of Color in the U.S. Economy." http://www.americanprogress.org/issues/economy/report/2011/01/21/8881/the-state-of-communities-of-color-in-the-u-s-economy/.

# 3 Household Formation and Homeownership: A Comparison of Immigrant Racialized Minority Cohorts in Canada and the United States

MICHAEL HAAN AND ZHOU YU

Since the late 1960s, both Canada and the United States have been attracting large numbers of immigrants from Asia, Latin America, Africa, and the Middle East. By 1971, more newcomers to Canada cited areas other than Europe as their previous residence (Troper 2003), and in 2006 nearly half of all residents in some Canadian cities identified themselves as a visible minority.[1] In the United States, four states have recently become majority-minority states where less than 50 per cent of the population is White of non-Hispanic origin.

Shortly after these new immigrant groups began to arrive in Canada and the United States, researchers started questioning whether the frameworks used to understand the integration experiences of earlier waves of (predominantly white) immigrants could be used to illuminate those of more recent and diverse cohorts of newcomers. The emerging research established that unexplained differences seem to surface quickly between groups on nearly all outcomes of interest. Whether it be earnings (Lewin-Epstein and Semyonov 1992; Picot and Sweetman 2005), segregation (Fong and Wilkes 2003; Funkhouser 2000), interaction with the host society (Fong 1992; Lieberson and Waters 1987), or intermarriage (Alba and Golden 1986; Kalbach 2002), it seems that racialized groups – whether defined by ethnicity, place of birth, or skin colour – have different integration experiences.

There are few instances where these differences are as pronounced as in Canadian and US metropolitan housing markets. Not only do racialized minority groups have very different homeownership rates, but they tend to live in different neighbourhoods (Fong 1994; Fong and Wilkes 2003), have access to different amenities (Myles and Hou 2004), and even move through the housing market differently as they age (Haan

2007; Myers and Lee 1998). Taken together, this research suggests that members of different groups have distinct housing careers (Haan 2010). Controlling for housing relevant characteristics typically brings groups closer together, but unexplained differences almost always remain (Myers 1999; Painter, Yang, and Yu 2003) and pose a major hurdle for immigrants and racialized minority groups in both countries (Farley 1996; Fong and Gulia 1999).

The purpose of this chapter is to examine the extent to which disparities in housing attainment stem from differences in the effect of context on different groups. Considerable research exists on how context affects housing behaviour on the whole, but much less research looks at whether context affects distinct groups *differently*. We rely on microdata to control for other confounding factors and examine the effect of context across groups, metropolitan areas, and countries.

**The Context of Reception: Canadian and US Housing Markets**

With the exception of Latino immigrants, Canada and the United States have been drawing newcomers from the same regions and experiencing similar immigration trends. With a variety of housing types, both countries have also witnessed a substantial increase in rates of household formation after the Second World War (although these rates have recently plateaued (Skaburskis 1994; Yu and Myers 2010).

That said, there are important economic differences between the two countries. First, US mortgage interest is often tax-deductible, which lowers the cost of owning relative to renting, especially for high-income earners. Second, until recently the United States had more aggressive lending policies and purchasing incentives. Third, housing prices appreciated more rapidly in the United States than in Canada for the study period, thereby providing an additional incentive for US residents to see housing as an investment. Fourth, immigration policies differ significantly between the two countries: while Canada favours highly skilled immigrants, most recent immigrants to the United States come through family ties. These factors all point to the prospect of large differences between immigrant groups in Canada and the United States, in terms of both household formation and homeownership attainment.

This study builds on our previous work, where we compared immigrant housing behaviour in Toronto and Los Angeles (Yu and Haan 2012). We concluded our study by suggesting that immigrant incorporation and settlement processes were not only culturally or contextually specific, but that group behaviour was *itself* context-specific, which points to the presence of an interaction effect between groups and their host society. Here we test this prospect more explicitly, by including two metro-level controls, and by including more birth and arrival cohorts. This, when considered alongside the comparative aspect of our study, allows us to look at the effect of different contexts while beginning to "control for culture."

## Research Questions

We investigate three specific questions arising from the discussion above:

1. What are the overall housing attainment patterns of foreign-born Blacks, Asian Indians, mainland Chinese, Filipinos, Whites,[2] and the White native-born reference group between 2001 and 2006 in Canada and between 2000 and 2005 in the US?[3]
2. After controlling for human capital and other factors (particularly income and metropolitan contextual variables), to what extent do these gaps change?
3. What proportion of the differences in homeownership attainment between groups can be attributed to household formation? To what extent have the gaps in housing outcomes changed over the five-year period relatively to the native-born White reference cohort?

Our broader interest is in comparing the United States and Canada. We hypothesize that (1) given the similarities between the chosen groups across countries, there should be similar patterns of housing outcomes by group (e.g., Filipinos in the United States should have similar housing behaviour to Filipinos in Canada, etc.); and (2) any observed differences that do exist are largely the result of group reactions to specific context.

## Moving In to Move Up the Homeownership Hierarchy?

Previous research shows that racialized minority groups encounter different levels of access to owner-occupied housing in Canada and the United States. While the Chinese move quickly into homeownership in both countries (Haan 2007; Painter, Yang, and Yu 2003), some other groups do not, especially in the United States (e.g., Coulson 1999; Krivo 1995). Although access to owner-occupied housing hinges heavily on labour market success, some groups have ownership rates that are higher or lower than expected based on what typical socio-economic indicators can predict. This suggests that there is room in current analytical models for additional characteristics to refine explanations of housing decisions.

One such addition, proposed recently by Yu and Myers (2010), involves jointly modelling homeownership attainment and household formation. Most studies either ignore the role of household formation or treat household formation decisions as separate from homeownership attainment. This practice would be acceptable if all immigrant groups had a similar propensity for forming independent households. In reality, there are large between-group variations, even after adjusting for all relevant variables. Therefore, ignoring household formation distorts our understanding of the residential assimilation process.

Analytically, adjusting for household formation entails jointly modelling homeownership attainment alongside a person's position within the household (Harvard University Joint Center for Housing Studies 2010; Yu and Myers 2010). This, when coupled with a double-cohort design (described more fully below), results in a more accurate representation of individual and group-specific housing careers.

## Data and Methods

### *The Double Cohort Design*

Typically, housing models are estimated on a single cross-section of data, which suffers from the limitation of confounded duration, period, and immigration cohort effects. With a "double-cohort" design (Myers and Lee 1998), longitudinal birth cohorts can be created by placing people from the same age groups in times A and B by adding five years to an individual's age at time B. Changes in the effect of other characteristics on homeownership over time can be measured by interacting variables of interest with a year-of-observation indicator.

This study focuses on the immigrant racialized minority groups[4] with the largest presence in both countries: Blacks, Asian Indians, Chinese, and Filipinos. We also include non-Hispanic White immigrants in the analysis. Three specific birth (born 1946–55, 1956–65, and 1966–75)[5] and arrival cohorts (native-born and arrived in 1976–85, 1986–95 in Canada)[6] are tracked over time from 2000 to 2005 in the United States and from 2001 to 2006 in Canada, using native-born, non-Hispanic[7] Whites as a common reference group. Data from both countries have been carefully examined and matched to ensure comparability. For Canada we use data from the 2001 and 2006 censuses, and for the United States data from the 2000 census and the 2005 American Community Survey. Although the data sets are not perfectly matched, this arrangement ensures a five-year observation period for each country.

The unit of analysis for this study is individuals, and the sample is limited to persons who are working thirty hours or more per week. In this way, we can examine the variations in household formation between immigrant groups, and between immigrants and the native-born reference group. In separate robustness tests we adjusted the sample criteria to twenty-five hours and to thirty-five hours, and the results were largely consistent.

### *Analytical Models*

Given our premise that household formation is a major factor behind homeownership propensities, it is necessary to change the outcome variable from a dichotomous owner-renter variable estimated on households to a trichotomous variable estimated on individuals.

Expressed more formally, the model is as follows:

HS = Age + Sex + Region + Educ + Marital Status + Migstat + Income + Vismin + Price at q25 + Median Rent

Where:

*HS* = Householder status (2 = head, owned dwelling, 1 = head, rented dwelling, 0 = non-head).

*Age* = Age/birth cohort, coded as 25–34, 35–44 (ref.), and 45–54 at time 1. At time 2, five years are added to each group.

*Sex* = Sex of respondent (1 = Male, 0 = Female).

*Region* = Dummy variables to control for region-specific homeownership propensities (Ontario = Reference Group).

*Educ* = Indicators to control for attainment (< High school = Ref).

*Migstat* = Immigrant status of respondent (native-born, arrived 1975–1984, 1985–1994, arrived 1985–1994 = Reference group).

*Income* = Personal income adjusted for inflation using Consumer Price Index (2000 basket, Canada, All Items, (http://cansim2.statcan.ca) and US Consumer Price Index, All Items), logged.

*Marital Status* = Marital status of respondent.

*Vismin* = Racialized minority status.

*Price at Q25* = Value of owner-occupied housing at 25th percentile in Census Metropolitan Area in Canada or Metropolitan Statistical Areas in the United States, adjusted for inflation, logged.
*Rent* = Median rent in Census Metropolitan Area in Canada or Metropolitan Statistical Areas in the United States, logged.
*Unemployment* = unemployment rates in Census Metropolitan Area in Canada or Metropolitan Statistical Areas in the US.

Further details about the variables used in this chapter, and their sample means, are provided in tables 3.1a (Canada) and 3.1b (United States).

### *Modelling Homeownership and Household Formation*

As demonstrated by Haurin and Rosenthal (2007) and Yu and Myers (2010), traditional homeownership models may overlook variable rates of household formation as an important factor in homeownership attainment due to a sample selection bias whereby a household's tenure choice cannot be observed if that household has not yet been formed. Although there are several sample-selection correction models, this study uses a three-outcome multinomial logit model. This allows for the comparison of non-head, non-owners to renter-heads (household heads that live in a rented dwelling) and owner-heads (heads that own their dwelling).

Of central interest in the regressions are the *vismin* coefficients. Convergence with the native-born in the models above can be defined as the difference between the *vismin* coefficient main effects, which denote disparities with the native-born at a point in time. Comparing *vismin* coefficients over time indicates the degree to which a group gains on the native-born over time. Across countries, these coefficients denote the relative gap that a group has with their respective native-born populations. It can therefore be thought of as a comparative indicator of residential success in the host society.

### *The Sample*

Our five immigrant racialized minority groups have either a large or growing presence in each country. Native-born whites of non-Hispanic origin are also included as a reference group. For immigrants, the sample includes two arrival cohorts, namely, those who came to Canada in 1977–86 and 1987–96 or to the United States in 1976–85 and

Table 3.1a Descriptive statistics (2001, 2006) and coding information, Canada

| | | 2001 | 2006 |
|---|---|---|---|
| *Socio-demographic characteristics* | | | |
| Non-head (reference group [RG]) | Dichotomous ↓ | 41.8% | 39.3% |
| Renter-head | | 18.5% | 14.7% |
| Owner-head | | 39.7% | 46.0% |
| Male (RG) | | 57.3% | 57.0% |
| Female | | 42.7% | 43.0% |
| Age 25–34 or 30–39* | | 30.9% | 31.9% |
| Age 35–44 or 40–49* (RG) | | 39.0% | 40.7% |
| Age 45–54 or 50–59* | | 30.1% | 27.4% |
| Single | | 19.8% | 15.3% |
| Married/common-law (RG) | | 70.0% | 72.7% |
| Once married | | 10.2% | 12.0% |
| *Immigration characteristics* | | | |
| Born in Canada (RG) | Dichotomous↓ | 89.3% | 89.0% |
| Arrived 1976–85 | | 4.1% | 4.2% |
| Arrived 1986–95 | | 6.6% | 6.9% |
| Speak English/French well (RG) | | 99.3% | 99.3% |
| Speak English/French not well | | 0.7% | 0.7% |
| *Education characteristics* | | | |
| Less than high school | Dichotomous↓ | 13.9% | 9.9% |
| High school and trades (RG) | | 62.0% | 65.2% |
| University degree | | 24.2% | 25.0% |
| Logged individual income | Continuous | 10.59 | 10.62 |
| *Visible minority characteristics* | | | |
| White native-born (RG) | Dichotomous↓ | 88.4% | 88.2% |
| White immigrant | | 5.0% | 5.0% |
| Blacks | | 1.3% | 1.4% |
| Asian Indian | | 2.7% | 2.8% |
| Mainland Chinese | | 1.4% | 1.4% |
| Filipino | | 1.2% | 1.2% |
| *Geographic characteristics* | | | |
| Logged housing value at Q1 | Continuous↓ | 11.88 | 12.17 |
| Logged median rent | | 6.76 | 6.76 |
| Metropolitan unemp. rate | | 5.6% | 4.3% |
| Ontario (RG) | Dichotomous↓ | 41.3% | 40.3% |
| Western Canada | | 28.7% | 29.5% |
| Quebec | | 24.1% | 24.2% |
| Atlantic Canada | | 5.9% | 6.0% |
| Total observations (unweighted) | | 1,165,392 | 1,170,226 |
| Total observations (weighted) | | 5,478,205 | 5,382,329 |

* In the latter year

Source: 2001 and 2006 Census of Canada Confidential Master Files.

Table 3.1b Descriptive statistics (2000, 2005) and coding information, US top 20 metros

| | | 2000 | 2005 |
|---|---|---|---|
| *Socio-demographic characteristics* | | | |
| Nonhead (reference group [RG]) | Dichotomous↓ | 39.4% | 37.9% |
| Renter-head | | 18.6% | 14.7% |
| Owner-head | | 41.9% | 47.4% |
| Male (RG) | | 57.8% | 58.2% |
| Female | | 42.2% | 41.8% |
| Age 25–34 or 30–39* | | 30.9% | 30.9% |
| Age 35–44 or 40–49* (RG) | | 37.7% | 38.7% |
| Age 45–54 or 50–59* | | 31.4% | 30.5% |
| Single | | 22.2% | 16.4% |
| Currently married (RG) | | 63.3% | 67.0% |
| Once married | | 14.6% | 16.6% |
| *Immigration characteristics* | | | |
| Born in the US (RG) | Dichotomous↓ | 86.7% | 86.0% |
| Arrived 1976–85 | | 7.4% | 7.8% |
| Arrived 1986–95 | | 6.0% | 6.2% |
| Speak English well (RG) | | 99.8% | 99.8% |
| Speak English not well | | 0.2% | 0.2% |
| *Education characteristics* | | | |
| Less than high school | Dichotomous↓ | 1.7% | 1.4% |
| High school and trades (RG) | | 55.1% | 52.8% |
| University degree | | 43.2% | 45.7% |
| Logged individual income | Continuous | 10.72 | 10.82 |
| *Minority characteristics* | | | |
| White native-born (RG) | Dichotomous↓ | 90.0% | 89.0% |
| White immigrant | | 3.5% | 3.8% |
| Blacks | | 1.9% | 2.2% |
| Asian Indian | | 1.4% | 1.5% |
| Mainland Chinese | | 1.9% | 2.1% |
| Filipino | | 1.3% | 1.4% |
| *Geographic characteristics* | | | |
| Logged housing value at Q1 | Continuous↓ | 11.76 | 12.21 |
| Logged median rent | | 6.69 | 6.81 |
| Metropolitan unemp. rate | | 4.5% | 5.3% |
| Pacific division (RG) | Dichotomous↓ | 21.6% | 21.2% |
| Mountain division | | 3.3% | 3.5% |
| West North Central | | 4.4% | 4.5% |
| West South Central | | 9.4% | 9.6% |
| East North Central | | 16.7% | 16.5% |
| Middle Atlantic division | | 24.2% | 24.2% |
| South Atlantic division | | 15.2% | 15.5% |
| New England division | | 5.3% | 4.9% |
| Total observations (unweighted) | | 965,270 | 181,191 |
| Total observations (weighted) | | 21,798,254 | 20,295,050 |

* In the latter year

Source: 2000 US Decennial Census 5% and 2005 American Community Survey Public Use Microdata files.

1986–95, and remained in Canada or the United States over the early 2000s. These two arrival cohorts are large and have a relatively stable membership of respondents who have lived in the destination country for an average of 17.5 years. Since after an initial period of adjustment immigrants begin to accelerate their housing careers (Myers and Lee 1998), this interval allows us to observe them over an important stage of residential assimilation.

Our analysis centres on the cohorts aged 25–34, 35–44, and 45–54 in 2000/2001 and 30–39, 40–9, and 50–9 in 2005/2006. We use this selection criterion for both native-born residents and immigrants because they are the mainstay in the workforce and in the housing market (Kendig 1990; Miron 1988). We also limit our sample to those who worked at least thirty hours per week so that all the observations in our study are not dependents and are theoretically able to form independent households if they so choose.

Our data sources include US Decennial Census Public Use Microdata for 2000, American Community Survey Public Use Microdata for 2005 (Ruggles, Sobek, Alexander, Fitch, Goeken, Hall, King, and Ronnander 2003), and the confidential Canadian census microdata for both 2001 and 2006.

### *The Study Areas*

This analysis is conducted in the twenty most populated Canadian Census Metropolitan Areas and the twenty most populated Metropolitan Statistical Areas in the United States. The twenty US metropolitan areas are grouped into eight different regions,[8] to capture possible regional variations in housing behaviour. To capture regional disparities, Canadian data are divided into four regions.[9]

Although on average housing costs slightly more in the United States than in Canada, financing has also (until recently) been easier to obtain, presumably levelling the differences in opportunity structures. As a result, immigrants face some of the same frontiers in each country as they make their decisions to form independent households and buy or rent a home.

### *The Five Immigrant Groups*

The first group selected for analysis is non-Hispanic White immigrants. This group (drawn mostly from Eastern Europe or Great Britain in Canada, and mostly from Canada and Europe in the United States) resembles the native-born White reference group most closely, so we hypothesize that it should have little difficulty with residential assimilation and reach similar levels of household formation soon after arrival.

Black immigrants to the United States and Canada come from a variety of regions, predominantly from Africa and the Caribbean. Early evidence has shown that they face numerous challenges in both countries (Darden and Kamel 2000; Freeman 2002).

In contrast to White and Black immigrants, Asian immigrants came from just a handful of destinations. They tended to have higher educational levels and exhibit more rapid economic advancement than average immigrants in both destination countries. This was most prominent among those who came from India, mainland China, and the Philippines.

The differences in residential trajectories across countries have not yet been examined. Established theories of residential assimilation typically hypothesize that in the long run differences in residential attainment between immigrants and native-born White majorities will narrow (Massey 1985). Accordingly, for our residential behaviour comparison we have selected native-born, non-Hispanic Whites as the native-born reference group in each country.

## Descriptive Findings

Age is an important determinant of housing attainment for several reasons. As individuals age they become more fully engaged in the labour market, accumulate more wealth, become more likely to have a family of their own, and aspire to have their own dwelling. As a result, we should expect to see an increase in both homeownership and independent household formation, both across birth cohorts and census years.

Tables 3.2a and 3.2b shows that this expectation is met in Canada and the United States (with some minimal differences between the two countries). Both per capita homeownership rates (defined as the percentage of persons in a group who own) and headship rates (defined as the percentage of persons in a group who are the householders/heads of households) rise across birth cohorts at both observation points. Furthermore, the same cohorts experience increases in both measures over time, with young cohorts making the largest improvement.

Table 3.2a Per capita homeownership and headship rates by birth cohort in Canadian Census Metropolitan Areas, 2001 and 2006

| | 2001 | #obs. (weighted) | 2006 | #obs. (weighted) |
|---|---|---|---|---|
| *Per capita homeownership rates* | | | | |
| Birth cohorts | | | | |
| Age 25–34 or 30–39* | 0.25 | 1,867,764 | 0.39 | 1,875,510 |
| Age 35–44 or 40–9* | 0.44 | 2,357,278 | 0.48 | 2,384,786 |
| Age 45–54 or 50–9* | 0.50 | 1,822,785 | 0.52 | 1,610,070 |
| *Headship rates* | | | | |
| Birth cohorts | | | | |
| Age 25–34 or 30–9* | 0.50 | 1,842,758 | 0.57 | 1,875,510 |
| Age 35–44 or 40–9* | 0.60 | 2,321,545 | 0.61 | 2,384,786 |
| Age 45–54 or 50–9* | 0.64 | 1,799,404 | 0.65 | 1,610,070 |

* In the latter year
Note: The headship rate is the number of householders/heads in a given group divided by the number of all persons in that group. The per capita ownership rate is the number of households with a homeowner in a given group divided by the number of all persons in that group.
Source: 2001 and 2006 Census of Canada Confidential Master Files.

Table 3.2b Per capita homeownership and headship rates by birth cohort in top 20 US metropolitan areas, 2000 and 2005

| | 2000 | #obs. (weighted) | 2005 | #obs. (weighted) |
|---|---|---|---|---|
| *Per capita homeownership rates* | | | | |
| Birth cohorts | | | | |
| Age 25–34 or 30–9* | 0.26 | 6,738,947 | 0.40 | 6,263,452 |
| Age 35–44 or 40–9* | 0.46 | 8,215,436 | 0.49 | 7,849,024 |
| Age 45–54 or 50–9* | 0.52 | 6,843,871 | 0.53 | 6,182,574 |
| *Headship rates* | | | | |
| Birth cohorts | | | | |
| Age 25–34 or 30–9* | 0.53 | 7,830,583 | 0.59 | 7,349,194 |
| Age 35–44 or 40–9* | 0.63 | 9,163,770 | 0.63 | 8,763,156 |
| Age 45–54 or 50–9* | 0.65 | 7,192,514 | 0.64 | 6,520,769 |

* In the latter year
Note: Per capita homeownership and headship rates are defined as in table 3.2a above.
Source: 2000 US Decennial Census 5% and 2005 American Community Survey Public Use Microdata Files.

Tables 3.2a and 3.2b display data for everyone, regardless of their immigration status. However, to reflect the additional challenges faced by immigrants specifically, it is necessary to examine an immigrant cohort's progress separately.

Table 3.3a lists the homeownership and headship rates for the native-born and two cohorts of immigrants to Canada. As expected, there is a gradual increase in per capita homeownership rates and headship with time spent in Canada. That said, even though those who have been in Canada for more than two decades have similar homeownership rates with the native-born, differences in household formation are evident from the headship rates. For immigrants these rates remain below those of the native-born, suggesting that there may be differences in how households form in the two groups.

Table 3.3b shows a similar pattern of increases in the United States, with the most recent cohort placed behind more established immigrants and the native-born.

As with birth cohorts, the similarities between Canada and the United States are striking. In every instance the difference in the homeownership or headship rates between countries is less than four percentage points.

Table 3.3a Per capita homeownership and headship rates in Canadian Census Metropolitan Areas, 2001 and 2006

| | 2001 | #obs. (weighted) | 2006 | #obs. (weighted) |
|---|---|---|---|---|
| *Homeownership rates* | | | | |
| Arrival cohorts | | | | |
| Native-born | 0.40 | 5,398,438 | 0.47 | 5,301,450 |
| Arrived 1976–85 | 0.40 | 342,196 | 0.46 | 347,957 |
| Arrived 1986–95 | 0.30 | 223,073 | 0.41 | 220,958 |
| *Headship rates* | | | | |
| Arrival cohorts | | | | |
| Native-born | 0.59 | 5,398,438 | 0.61 | 5,301,450 |
| Arrived 1976–85 | 0.55 | 342,196 | 0.58 | 347,957 |
| Arrived 1986–95 | 0.53 | 223,073 | 0.56 | 220,958 |

Note: Per capita homeownership and headship rates are defined as in table 3.2a above.
Source: 2001 and 2006 Census of Canada Confidential Master Files.

Table 3.3b Per capita homeownership and headship rates in top 20 US metropolitan areas, 2000 and 2005

| | 2000 | #obs. (weighted) | 2005 | #obs. (weighted) |
|---|---|---|---|---|
| *Per capita homeownership rates* | | | | |
| Arrival cohorts | | | | |
| Native-born | 0.44 | 18,896,612 | 0.49 | 17,463,879 |
| Arrived 1976–85 | 0.37 | 1,298,605 | 0.43 | 1,251,352 |
| Arrived 1986–95 | 0.26 | 1,603,037 | 0.37 | 1,579,819 |
| *Headship rates* | | | | |
| Arrival cohorts | | | | |
| Native-born | 0.61 | 18,896,612 | 0.63 | 17,463,879 |
| Arrived 1976–85 | 0.57 | 2,328,620 | 0.60 | 2,213,487 |
| Arrived 1986–95 | 0.54 | 2,961,635 | 0.58 | 2,955,753 |

Note: Per capita homeownership and headship rates are defined as in table 3.2a above.
Source: 2000 US Decennial Census 5% and 2005 American Community Survey Public Use Microdata Files.

Perhaps it is unsurprising that age and immigrant arrival cohorts fare similarly in each country. Still, there is considerable research that gives reason to expect differences by racialized minority group. The United States has a long history of differential treatment of racialized minorities, with Blacks being particularly disadvantaged (Massey 1990). Canada has a different history of race relations, but research that focuses on attainment across cohorts shows striking similarities both within and across countries (Haan 2007). This is particularly true when racialized minority groups are composed entirely of immigrants, as is the case for our study.

Tables 3.4a (Canada) and 3.4b (United States) show homeownership rates by racialized minority groupings in Canada and the United States over the study period. Here we present per capita homeownership, and only include individuals who work more than thirty hours per week.

First, consistently with previous studies (e.g., Haan 2005; Painter, Yang, and Yu 2003), we find that in Canada the native-born hold the highest level of homeownership. The rate is also relatively high for White and Chinese immigrants, followed by Asian Indians. Blacks and Filipinos have much lower rates. Five years later, we detect substantial

Table 3.4a Per capita homeownership and headship rates in Canadian Census Metropolitan Areas, 2001 and 2006

| | 2001 | | #obs. (weighted) | 2006 | | #obs. (weighted) |
|---|---|---|---|---|---|---|
| | Per capita ownership rates | Headship rates | | Per capita ownership rates | Headship rates | |
| White NB | 0.41 | 0.59 | 5,398,438 | 0.47 | 0.61 | 5,301,450 |
| Immigrants | | | | | | |
| White | 0.38 | 0.57 | 302,515 | 0.46 | 0.59 | 298,049 |
| Black | 0.23 | 0.61 | 79,675 | 0.33 | 0.64 | 82,644 |
| Asian Indian | 0.32 | 0.48 | 69,968 | 0.43 | 0.52 | 72,437 |
| Mainland Chinese | 0.36 | 0.47 | 48,088 | 0.45 | 0.53 | 49,866 |
| Filipino | 0.25 | 0.44 | 65,023 | 0.33 | 0.47 | 65,919 |
| Total | 0.40 | 0.58 | 5,963,706 | 0.46 | 0.61 | 5,870,365 |

Note: This table reports those who were 25–54 in 2001 and 30–59 in 2006. Sample only includes those who worked more than 30 hours per week. Immigrants only include those who arrived between 1977 and 1996. Per capita homeownership and headship rates are defined as in table 3.2a above.
Source: 2001 and 2006 Census of Canada Confidential Master Files.

Table 3.4b Per capita homeownership and headship rates in top 20 US metropolitan areas, 2000 and 2005

| | 2000 | | #obs. (weighted) | 2005 | | #obs. (weighted) |
|---|---|---|---|---|---|---|
| | Per capita ownership rates | Headship rates | | Per capita ownership rates | Headship rates | |
| White NB | 0.44 | 0.61 | 18,876,439 | 0.49 | 0.63 | 17,438,240 |
| Immigrants | | | | | | |
| White | 0.36 | 0.61 | 768,071 | 0.43 | 0.62 | 766,900 |
| Black | 0.24 | 0.59 | 422,247 | 0.33 | 0.65 | 451,784 |
| Asian Indian | 0.34 | 0.59 | 306,525 | 0.44 | 0.59 | 310,466 |
| Mainland Chinese | 0.35 | 0.51 | 422,669 | 0.44 | 0.57 | 421,087 |
| Filipino | 0.28 | 0.44 | 292,602 | 0.36 | 0.50 | 291,095 |
| Total | 0.42 | 0.61 | 21,798,254 | 0.47 | 0.62 | 20,295,050 |

Note: This table reports those who were 25–54 in 2000 and 30–59 in 2005. Sample only includes those who worked more than 30 hours per week. Immigrants only include those who arrived between 1976 and 1995.Per capita homeownership and headship rates are defined as in table 3.2a above.
Source: 2000 US Decennial Census and 2005 American Community Survey Public Use Microdata Files.

increases for all groups. In many cases per capita homeownership rates rise by more than ten percentage points. White, mainland Chinese, and Asian Indian immigrants in particular have almost converged with the native-born White reference group.

Second, Canadian data reveal no discernible pattern between headship and homeownership. In 2001, native-born Whites, White immigrants and Black immigrants have the three highest levels of independent household formation, yet blacks have the lowest ownership rate and White native-born respondents have the highest. The same continues to be true for 2006, even while all groups increased their independent household formation rates.

Correspondingly, in the United States it is White native-born, White immigrant, Chinese, and Asian Indians with the highest per capita ownership rates, and Blacks and Filipinos with the lowest. Five years later, there is once again an across-the-board increase, but little reduction in the gaps between groups, except for the convergence between Whites, Chinese, and Asian Indians with the native-born. Also similar to Canada is the lack of a discernible relationship between ownership and headship in the United States.

To further illustrate these parallels, consider figure 3.1, which plots the proportion of individuals by homeownership and headship status. The darkest bar on the bottom of each figure represents per capita homeownership rates of individual groups in the two years. Per capita homeownership here can also be understood as owner householder share of the population. The grey colour bars in the middle represent per capita rentership rates, or renter householder share of the population. The grey bars and the dark bars combined together represent headship rates. Notice how in each instance the differences between racialized minority groups is evident within countries, but not across countries for the same group. In other words, these results show similar housing experiences by group in each country, at least as measured by homeownership and household formation.

While headship rates are slightly higher in the United States for some groups, per capita rentership rates for the same groups tend to also be larger. Per capita homeownership rates and household formation rates have increased over the five-year period for all groups, while per capita rentership rates decreased. For native-born Whites, headship rates have been steady over the five-year period. The increase in homeownership was the result of the decline in rentership, suggesting that it is largely White renters (native-born and immigrant) who bought homes over time. In contrast, many racialized minority immigrants have larger

Figure 3.1 Headship and homeownership rates by race/ethnicity and immigrant status over the early 2000s

Canada

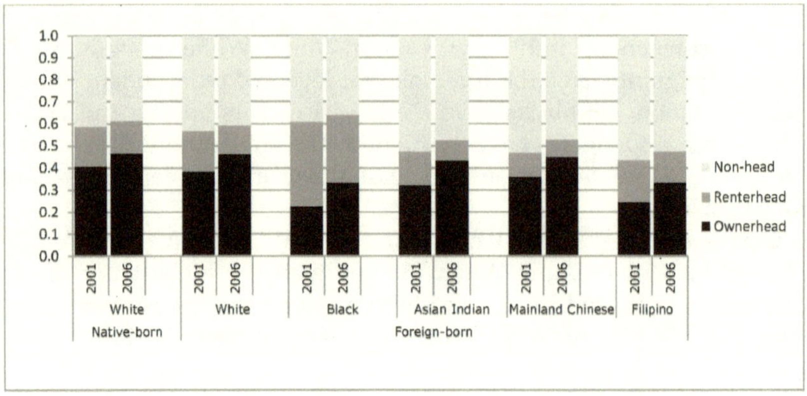


The U.S.

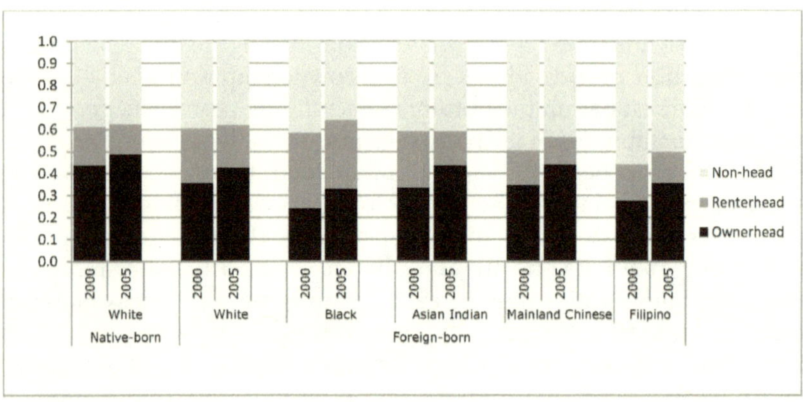


increases in headship rates than native-born Whites. In other words, they are not only buying homes, but also forming more independent households.

There are two main findings thus far in this chapter. First, there are large variations between racialized minority groups in terms of both ownership and headship status, without any clear link between the two characteristics. In fact, the group with the highest rate of household formation (Blacks) has the lowest ownership rate, and the group with the second-highest rate of household formation (White native-born) has the highest ownership rate. On the other end of the spectrum, several groups with low rates of household formation have high ownership rates (such as the Chinese), whereas others (like Filipinos) do not. Second, although there is an increase in both ownership and household formation for all groups in both countries over time, there is very little change in the gaps between immigrant groups. However, the differences in per capita homeownership have shrunk between immigrants and the native-born White reference group, suggesting convergence and a gradual residential assimilation among immigrants.

### *Multivariate Results*

One problem with the findings thus far is that they may be driven by factors that have not yet been accounted for, like differences in income, education, employment status, and other influences on residential behaviour. For this reason we present a series of multinomial logit models to control for the influence of numerous other housing-relevant individual characteristics. We first show overall models for each country and year, and then we run the same models for each racialized minority group.

Estimation results in relative risk ratios (RRR) of the determinants are presented as relative risk ratios in tables 3.5a and 3.5b. RRRs are commonly used with multinomial logit models to make results more intuitive than raw regression coefficients. Each coefficient reflects the effect of a particular characteristic on one of the three types of household status, relative to the probability of being a non-householder. There are two columns for each model, showing the probabilities of being a renter or owner householder. Separate models are estimated each year, and in all cases the baseline group is the probability of being a non-householder, which is omitted from the table.

First, in table 3.5a we see that males are more likely to be householders in general, and owner householders in particular. Income, educational attainment, and English (English/French in Canada) proficiency are all positively associated with household formation. Married-couple households are the least likely to form renter households and have a stronger propensity of becoming homeowners.As we might expect, higher housing price encourages renter household formation and deters owner household formation, whereas the

Table 3.5a Relative risk ratios (RRRs) of the determinants of household formation in Canadian Census Metropolitan Areas, 2001 and 2006

| | Canada | | | |
|---|---|---|---|---|
| | 2001 | | 2006 | |
| Obs (weighted): | 5,478,205 | | 5,382,329 | |
| Log likelihood: | −1,020,924 | | −996,974 | |
| Pseudo R2: | 0.158 | | 0.146 | |
| Variables | Renter householder | Owner householder | Renter householder | Owner householder |
| Male gender (omitted: female) | 2.485*** | 5.003*** | 2.730*** | 4.384*** |
| Personal income (log) | 1.012*** | 1.592*** | 0.966*** | 1.274*** |
| *Birth cohorts* (omitted: Age 35–44 or 40–49†) | | | | |
| Age 25–34 or 30–39† | 1.078*** | 0.515*** | 1.128*** | 0.707*** |
| Age 45–54 or 50–59† | 0.932*** | 1.197*** | 1.078*** | 1.147*** |
| *Visible minority groups* (omitted: native-born non-Hispanic White) | | | | |
| Immigrants: non-Hispanic White | 0.855*** | 0.959*** | 0.836*** | 0.952*** |
| Black | 1.584*** | 0.827*** | 1.572*** | 0.844*** |
| South Asian Indian | 0.639*** | 0.760*** | 0.565*** | 0.889*** |
| Mainland Chinese | 0.419*** | 0.956* | 0.381*** | 0.844*** |
| Filipino | 0.675*** | 0.619*** | 0.694*** | 0.646*** |
| *Education* (omitted: high school dip. w/ college) | | | | |
| No high school diploma | 1.044*** | 0.755*** | 1.167*** | 0.726*** |
| College degree or better | 1.160*** | 1.182*** | 0.977** | 1.300*** |

| | | | | |
|---|---|---|---|---|
| *Marital status* (omitted: married) | | | | |
| Never married | 3.482*** | 0.639*** | 7.736*** | 1.241*** |
| Formally married | 14.037*** | 3.860*** | 20.924*** | 5.035*** |
| *Language proficiency* (omitted: does not speak English/French) | | | | |
| Speak English/ French | 1.091* | 0.928* | 1.112** | 0.939* |
| *Immigrant status* (omitted: immigrants arrived between 1987 and 1996) | | | | |
| Immigrants arrived between 1977 and 1986 | 0.624*** | 1.244*** | 0.659*** | 1.108*** |
| *Metropolitan housing price and rent* | | | | |
| The 25th percentile housing price (log) | 1.277*** | 0.616*** | 1.503*** | 0.761*** |
| Median rent (log) | 0.260*** | 1.327*** | 0.261*** | 1.328*** |
| Metropolitan unemployment rates | 0.002*** | 6.976*** | 0.003*** | 0.910 |
| *Regions in Canada* (omitted: Ontario) | | | | |
| Western Canada | 1.115*** | 1.095*** | 1.010 | 1.101*** |
| Quebec | 1.126*** | 0.928*** | 1.149*** | 1.024*** |
| Atlantic Canada | 0.897*** | 0.961*** | 0.882*** | 0.986 |

* $p < 0.05$ **$p < 0.01$ ***$p < 0.001$ Two-tailed tests

† In the latter year

Note: Non-head is the baseline group.

The reference group for gender is “female”; for nativity/visible minority, “native-born non-Hispanic White”; for educational attainment, “high school dip. w/ college”; for marital status, “currently married”; for language proficiency, “does speak English/French”; for immigrant status, “the native-born”; for regions in Canada, “Ontario.”

Table 3.5b Relative risk ratios (RRRs) of the determinants of household formation in top 20 US metropolitan areas, 2000 and 2005

| | United States | | | |
|---|---|---|---|---|
| | 2000 | | 2005 | |
| Obs (weighted): | 24,151,215 | | 22,616,348 | |
| Log likelihood: | –20,474,638 | | –20,404,801 | |
| Pseudo R2: | 0.197 | | 0.116 | |
| Variables | Renter householder | Owner householder | Renter householder | Owner householder |
| Male gender (omitted: female) | 4.052*** | 7.791*** | 2.203*** | 2.396*** |
| Personal income (log) | 1.201*** | 2.298*** | 0.943*** | 1.713*** |
| *Birth cohorts* (omitted: age 35–44 or 40–9†) | | | | |
| Age 25–34 or 30–9† | 1.080*** | 0.528*** | 1.228*** | 0.760*** |
| Age 45–54 or 50–9† | 0.838*** | 1.209*** | 0.828*** | 1.141*** |
| *Visible minority groups* (omitted: native-born non-Hispanic White) | | | | |
| Immigrants: non-Hispanic White | 1.534*** | 0.754*** | 1.794*** | 0.864*** |
| Black | 2.288*** | 0.788*** | 2.256*** | 0.843*** |
| Asian Indian | 1.787*** | 0.669*** | 2.711*** | 0.893*** |
| Mainland Chinese | 0.881*** | 0.947*** | 1.375*** | 1.047*** |
| Filipino | 0.808*** | 0.651*** | 0.968*** | 0.662*** |
| *Education* (omitted: high school dip. w/ college) | | | | |
| No high school diploma | 1.134*** | 0.765*** | 1.180*** | 0.762*** |
| College degree or better | 1.253*** | 1.167*** | 1.148*** | 1.166*** |

| | | | | |
|---|---|---|---|---|
| *Marital status* (omitted: married) | | | | |
| Never married | 3.107*** | 0.736*** | 5.565*** | 1.391*** |
| Formally married | 7.832*** | 2.612*** | 10.652*** | 3.137*** |
| *Language proficiency* (omitted: does not speak English) | | | | |
| Speak English | 0.687*** | 1.801*** | 1.209*** | 2.769*** |
| *Immigrant status* (omitted: immigrants arrived between 1986 and 1995) | | | | |
| Immigrants arrived between 1976 and 1985 | 0.787*** | 1.191*** | 0.648*** | 1.212*** |
| *Metropolitan housing price and rent* | | | | |
| The 25th percentile housing price (log) | 1.848*** | 0.505*** | 2.105*** | 0.664*** |
| Median rent (log) | 0.320*** | 1.487*** | 0.253*** | 1.356*** |
| Metropolitan unemployment rates | 1.008*** | 0.960*** | 0.684* | 4.876*** |
| *Regions in the US* (omitted: Pacific Division) | | | | |
| Mountain Division | 0.831*** | 1.239*** | 0.860*** | 1.133*** |
| West North Central | 1.139*** | 1.052*** | 1.087*** | 1.074*** |
| West South Central | 0.936*** | 1.013*** | 0.850*** | 1.133*** |
| East North Central | 1.191*** | 1.081*** | 1.159*** | 1.086*** |
| Middle Atlantic Division | 1.062*** | 1.001 | 1.031*** | 1.071*** |
| South Atlantic Division | 0.743*** | 1.421*** | 0.635*** | 1.320*** |
| New England Division | 1.718*** | 0.785*** | 1.732*** | 0.903*** |

* $p < 0.05$ ** $p < 0.01$ *** $p < 0.001$ Two-tailed tests

† In the latter year

Note: Non-head is the baseline group. The reference groups are the same as in table 3.5a, except: for language proficiency, “does Speak English”; for regions in the US, the Pacific Division.

opposite is true in places of high rent cost. The results on the covariates in Canada largely mirror the findings in the United States, further increasing our confidence about the comparability of the results in the two countries.

In the United States, males are once again more likely to be householders, especially owner householders. Income positively predicts both headship and ownership. The propensity to be a household head increases with age, as does ownership. In most instances, married individuals are less likely to be both renter-heads and owner-heads (three of the four RRRs for never and formerly married are positive). Language fluency also has a sizable impact on ownership status.

The trend towards homeownership shown in table 3.3b persists here, with a negative propensity for independent rental household formation relative to the most recent arrival cohort. Increases in cheap housing (defined as those at the first quartile) push ownership propensities downward and rental propensities upward. As the first quartile becomes more expensive, the risk of renting rises and the risk of owning falls, with the median rent eliciting the expected opposite effect (rental propensities drop while ownership rises).

### *Overall Differences across Racialized Minority Groups*

After controlling for the covariates, how have immigrant groups fared relative to native-born Whites in both countries? To what extent did immigrants improve their housing outcomes over the five-year period? (Keep in mind that as the native-born White cohort grows older, they also improve their housing outcomes, suggesting that assimilation for immigrants is a "moving target.")

First, racialized immigrant groups have largely similar performances in the two countries and over time. This is true even after adjusting for other covariates. At least, immigrants have kept up with the progress of native-born Whites. In some cases, immigrants have shrunk the differences in housing attainment from native-born Whites. However, there are large variations between groups.

Second, White immigrants have a similar propensity for household formation to native-born Whites. The finding is particularly evident in Canada. In the United States, White immigrants fare slightly worse than native-born Whites in terms of per capita homeownership.

However, they are more prone to forming renter households than the native-born.

Third, Blacks have very high rates of household formation in both study areas. The reason behind their low homeownership seems to be their high rate of renter household formation. This is not necessarily a sign of distress, if we take household formation into consideration.

Fourth, Asian Indian immigrants appear to be the exception to the rules of similarity, with substantial differences in the two countries. Asian Indian immigrants are more likely to form renter households than Whites in the United States than in Canada. They are more likely to form independent households and have higher propensity for forming owner households over time, even though they are still lower than native-born Whites in per capita homeownership. The differences may reflect the diverse origins of Asian Indian immigrants in the two countries.

Fifth, Chinese in both countries have low rates of household formation, forming far fewer renter households than, for example, Black immigrants. While it is unclear whether these rates are the result of cultural preference or market pressure, we can say that these housing "high achievers" stand out much less when the unit of analysis switches to individuals.

Finally, Filipinos have relatively low propensity for household formation in both countries.

To illustrate how the earlier unadjusted differences between groups have changed with the addition of covariates, and to compare racialized minority groups across countries, figure 3.2 plots the RRRs from the regressions in tables 3.5a and 3.5b.

In terms of rentership, Black immigrants have *much* higher renter-head propensities than all other groups in every instance but 2005 in the United States. In Canada, it is the Chinese with the lowest propensities to be renter-heads, followed closely by Asian Indians. In the United States the results are more dispersed.

Figure 3.2 also illustrates the attenuation of homeownership disparities between nearly all groups and the native-born over time spent in the host society. It is interesting to note that for many in the United States, but not Canada, this occurred alongside an increase in the propensity to be a renter-head, suggesting that the increase in owners came from the stock of non-heads. In Canada, increases in ownership largely occurred alongside relative decreases in rentership.

Figure 3.2 Relative risk ratios by race/ethnicity: Assessing variable household formation

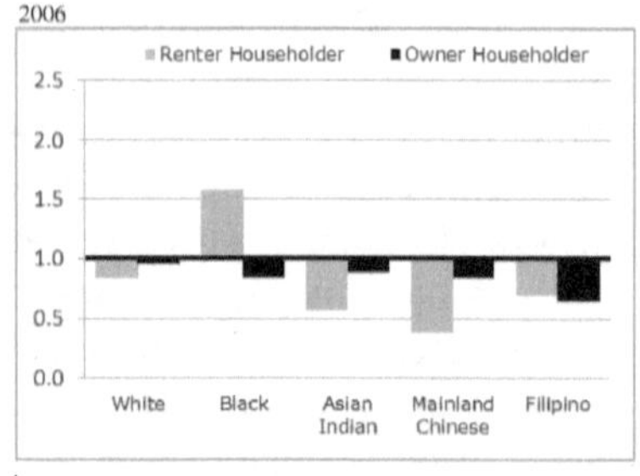


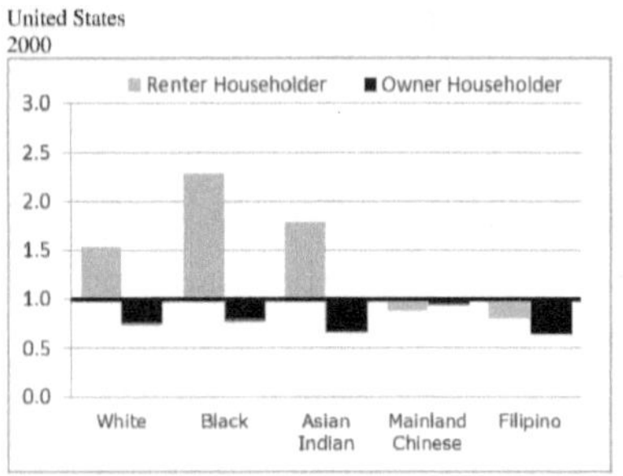


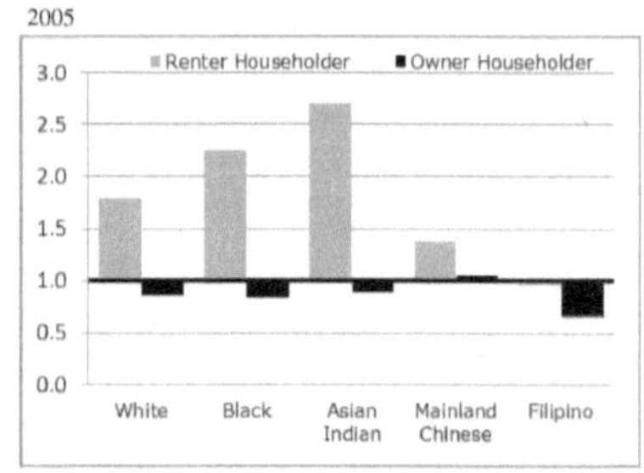


## Conclusions

We find evidence of similar levels of household formation, homeownership, and progress over time in Canada and the United States. This suggests that, despite the many differences between the countries, the process of residential integration is similar for many. At the same time, there is reason to believe that context shapes housing behaviour. The trends for Asian Indians and mainland Chinese provide some support for the existence of differences across countries.

This study is innovative for several reasons. First, it overcomes the limitation of the conventional measure of homeownership, which is measured at the household level and ignores household formation (Yu and Haan 2012; Yu and Myers 2010). Newly arrived immigrants are least likely to form independent households, and most likely to reside in rental units. Moreover, immigrant groups also have variable rates of household formation over time, reflective of their differences in socio-economic status and possibly cultural norms.

Second, most existing studies rely on cross-sectional analysis to study residential assimilation. This is problematic because assimilation is longitudinal in nature and there are substantial variations between immigrant arrival cohorts. Evidently, more recent immigrant arrivals have lower socio-economic status and worse housing outcomes than earlier arrivals in both the United States and Canada (Borjas 2002; Haan 2005). The size of each arrival cohort also changes significantly over time. Therefore, it is necessary to treat residential assimilation as a process instead of an outcome at one point in time (e.g., Myers and Lee 1996; Yu and Myers 2007).

Third, all immigrant groups show a gradual but significant increase in demand for both renter- and owner-occupied housing over time. By and large, immigrants in both countries have kept up with the progress of the native-born White cohort over the five-year period. In some immigrant groups, the differences even shrunk.

Fourth, Filipino immigrants present a unique case. The Filipinos have the lowest levels of housing attainment among all immigrant groups, even after adjusting for other covariates. To the extent that this group clusters in cities where rental housing is more abundant (Yu and Myers 2007), further analysis is necessary.

Finally, there is very little research that compares racialized immigrant minorities across countries. There are nearly no comparative studies on housing, so we feel that our contribution here is especially noteworthy. Most research looks at the behaviour of groups *or* the effect of environmental factors, but few consider these effects simultaneously.

An object for future study is to examine the inter-cohort variations within each immigrant group. This would help quantify the differences between arrival cohorts, and further identify how culture and context intersects. As Latino immigrants increase their presence in Canada, future study should also examine the extent to which they have fared differently from their US counterparts.

## NOTES

1 Although normally the term "racialized minority" is used to describe all non-White groups, here we use it in shorthand to describe *immigrant* ethno-racial groups only. Native-born racialized minority individuals are excluded from the samples used in this study. Similarly, the US term "household head" is used in place of the term "householder" or the Canadian term "primary household maintainer."

2 Whites in this analysis refer to non-Hispanic whites.

3 Traditional homeownership rates are measured at the household level and refer to the percentage of households that live in owner-occupied housing units.
4 We do not include Hispanics or Latinos in the analysis because of their relatively small population size in Canada.
5 1945–54, 1955–64, and 1965–74 in the United States.
6 Native-born and arrived in 1975–84, 1985–94 in the United States.
7 Hispanics or Latino immigrants are excluded from the study. The cohort size of Hispanic immigrants is too small in Canada, which makes it difficult to compare them with those in the United States.
8 The sample includes the top 20 metropolitan areas based on the population size. We base it on Census Bureau information and group them into 8 regions. Pacific Division includes four metropolitan areas: San Francisco–Oakland–Vallejo, CA, Riverside–San Bernardino, CA, Los Angeles–Long Beach, CA, and Seattle–Everett, WA. Mountain Division includes Phoenix, AZ. West North Central Division includes Minneapolis–St Paul, MN. West South Central Division includes Dallas–Fort Worth, TX, and Houston–Brazoria, TX. East North Central includes Chicago, IL, Detroit, MI, and St Louis, MO/IL. Middle Atlantic Division includes New York–northeastern NJ, Philadelphia, PA/NJ, and Pittsburgh, PA. South Atlantic Division includes Washington, DC/MD/VA, Atlanta, GA, Baltimore, MD, and Tampa–St Petersburg–Clearwater, FL.
9 Canadian Census Metropolitan Areas are coded according to the following criteria: West of Ontario: Vancouver, Victoria, Calgary, Edmonton, Winnipeg, Saskatoon, Regina, and Kelowna; Ontario: Toronto, Ottawa-Gatineau, Hamilton, London, Kitchener, St Catharines–Niagara, Oshawa, Windsor, and Barrie; Quebec: Montreal, Quebec, Sherbrooke; Atlantic: Halifax, St John's.

This research was made possible by Ryerson University's New Faculty Developemt Fund. My sincere thanks to them and all respondents and key informants who have sacrificed their valuable family time for this research. I gratefully acknowledge the contributions of Raymond Mark Garrison and Carlos Teixeira for their editorial comments and valuable insights.

## REFERENCES

Alba, R.D., and R.M. Golden. 1986. "Patterns of Ethnic Marriage in the United States." *Social Forces* 65 (1): 202–23. http://dx.doi.org/10.1093/sf/65.1.202.

Borjas, G.J. 2002. "Homeownership in the Immigrant Population." *Journal of Urban Economics* 52 (3): 448–76. http://dx.doi.org/10.1016/S0094-1190(02)00529-6.

Coulson, N.E. 1999. "Why Are Hispanic- and Asian-American Home-ownership Rates So Low? Immigration and Other Factors." *Journal of Urban Economics* 45 (2): 209–27. http://dx.doi.org/10.1006/juec.1998.2094.

Darden, J.T., and S.M. Kamel. 2000. "Black and White Differences in Homeownership Rates in the Toronto Census Metropolitan Area: Does Race Matter?" *Review of Black Political Economy* 28 (2): 53–76. http://dx.doi.org/10.1007/s12114-000-1017-6.

Farley, R. 1996. "Racial Differences in the Search for Housing: Do Whites and Blacks Use the Same Techniques to Find Housing?" *Housing Policy Debate* 7 (2): 367–85. http://dx.doi.org/10.1080/10511482.1996.9521225.

Fong, E. 1992. *Racial Interaction Patterns of American and Canadian Neighborhoods. A Paper Presented in American Sociological Association Annual Meeting*. ASA.

Fong, E. 1994. "Residential Proximity among Racial Groups in U.S. and Canadian Neighborhoods." *Urban Affairs Quarterly* 30 (2): 285–97. http://dx.doi.org/10.1177/004208169403000206.

Fong, E., and M. Gulia. 1999. "Differences in Neighborhood Qualities among Racial and Ethnic Groups in Canada." *Sociological Inquiry* 69 (4): 575–98. http://dx.doi.org/10.1111/j.1475-682X.1999.tb00887.x.

Fong, E., and R. Wilkes. 2003. "Racial and Ethnic Residential Patterns in Canada." *Sociological Forum* 18 (4): 577–602. http://dx.doi.org/10.1023/B:SOFO.0000003004.78713.2e.

Freeman, L. 2002. "Does Spatial Assimilation Work for Black Immigrants in the US?" *Urban Studies* (Edinburgh, Scotland) 39 (11): 1983–2003. http://dx.doi.org/10.1080/0042098022000011326.

Funkhouser, E. 2000. "Changes in the Geographic Concentration and Location of Residence of Immigrants." *International Migration Review* 34 (2): 489–510. http://dx.doi.org/10.2307/2675911.

Haan, M. 2005. "The Decline of the Immigrant Home-Ownership Advantage: Life-Cycle, Declining Fortunes and Changing Housing Careers in Montreal, Toronto and Vancouver, 1981–2001." *Urban Studies* (Edinburgh, Scotland) 42 (12): 2191–212. http://dx.doi.org/10.1080/00420980500331983.

Haan, M. 2007. "The Homeownership Hierarchies of Canada and the United States: The Housing Patterns of White and Non-White Immigrants of the Past Thirty Years." *International Migration Review* 41 (2): 433–65. http://dx.doi.org/10.1111/j.1747-7379.2007.00074.x.

Haan, M. 2010. "The Sources of Differentiation in the Immigrant Housing Market: Insights from the Longitudinal Survey of Immigrants to Canada." In *Canadian Immigration: Economic Evidence for a Dynamic Policy Environment*, ed. T. McDonald, E. Ruddick, A. Sweetman, and C. Worswick, 235–56. Montreal: McGill-Queen's University Press.

Harvard University Joint Center for Housing Studies. 2010. *The State of the Nation's Housing 2010*. http://www.jchs.harvard.edu/research/publications/state-nations-housing-2010.

Haurin, D.R., and S.S. Rosenthal. 2007. "The Influence of Household Formation on Homeownership Rates across Time and Race." *Real Estate Economics* 35 (4): 411–50. http://dx.doi.org/10.1111/j.1540-6229.2007.00196.x.

Kalbach, M.A. 2002. "Ethnic Intermarriage in Canada." *Canadian Ethnic Studies / Études ethniques au Canada* 34: 25–39.

Kendig, H.L. 1990. "A Life Course Perspective on Housing Attainment." In *Housing Demography: Linking Demographic Structure and Housing Markets*, ed. D. Myers, 133–56. Madison: University of Wisconsin Press.

Krivo, L.J. 1995. "Immigrant Characteristics and Hispanic-Anglo Housing Inequality." *Demography* 32 (4): 599–615. http://dx.doi.org/10.2307/2061677.

Lewin-Epstein, N., and M. Semyonov. 1992. "Local Labor Markets, Ethnic Segregation, and Income Inequality." *Social Forces* 70 (4): 1101–19. http://dx.doi.org/10.1093/sf/70.4.1101.

Lieberson, S., and M.C. Waters. 1987. "The Location of Ethnic and Racial Groups in the United States." *Sociological Forum* 2 (4): 780–810. http://dx.doi.org/10.1007/BF01124384.

Massey, D.S. 1985. "Ethnic Residential Segregation: A Theoretical Synthesis and Empirical Review." *Sociology and Social Research* 69: 315–50.

Massey, D.S. 1990. "American Apartheid: Segregation and the Making of the Underclass." *American Journal of Sociology* 96 (2): 329–57. http://dx.doi.org/10.1086/229532.

Miron, J.R. 1988. *Housing in Postwar Canada: Demographic Change, Household Formation, and Housing Demand*. Kingston: McGill-Queen's University Press.

Myers, D. 1999. "Cohort Longitudinal Estimation of Housing Careers." *Housing Studies* 14 (4): 473–90. http://dx.doi.org/10.1080/02673039982731.

Myers, D., and S.W. Lee. 1996. "Immigration Cohorts and Residential Overcrowding in Southern California." *Demography* 33 (1): 51–65. http://dx.doi.org/10.2307/2061713.

Myers, D., and S.W. Lee. 1998. "Immigrant Trajectories into Homeownership: A Temporal Analysis of Residential Assimilation." *International Migration Review* 32 (3): 593–625. http://dx.doi.org/10.2307/2547765.

Myles, J., and F. Hou. 2004. "Changing Colours: Spatial Assimilation and New Racial Minority Immigrants." *Canadian Journal of Sociology* 29 (1): 29–58. http://dx.doi.org/10.2307/3341944.

Painter, G., L. Yang, and Z. Yu. 2003. "Heterogeneity in Asian American Home-ownership: The Impact of Household Endowments and Immigrant

Status." *Urban Studies* (Edinburgh, Scotland) 40 (3): 505–30. http://dx.doi.org/10.1080/0042098032000053897.

Picot, G., and A. Sweetman. 2005. *The Deteriorating Economic Welfare of Immigrants and Possible Causes: Update 2005*. Ottawa: Statistics Canada.

Ruggles, S., J.T. Alexander, K. Genadek, R. Goeken, M.B. Schroeder, and M. Sobek. 2010. *Integrated Public Use Microdata Series: Version 5.0*. [Machine-readable database.] Minneapolis: University of Minnesota.

Ruggles, S., M. Sobek, T. Alexander, C.A. Fitch, R. Goeken, P.K. Hall, M. King, and C. Ronnander. 2003. *Integrated Public Use Microdata Series: Version 3.0*. Minneapolis: Historical Census Projects, University of Minnesota.

Skaburskis, A. 1994. "Determinants of Canadian Headship Rates." *Urban Studies* (Edinburgh, Scotland) 31 (8): 1377–89. http://dx.doi.org/10.1080/00420989420081211.

Troper, H. 2003. "To Farms or Cities: A Historical Tension between Canada and Its Immigrants." In *Host Societies and the Reception of Immigrants*, ed. Jeffrey Reitz, 509–31. La Jolla, CA: Center for Comparative Immigration Studies.

Yu, Z., and M. Haan. 2012. "Cohort Progress toward Household Formation and Homeownership: Young Immigrant Cohorts in Los Angeles and Toronto Compared." *Ethnic and Racial Studies* 35 (7): 1311–37. http://dx.doi.org/10.1080/01419870.2011.602089.

Yu, Z., and D. Myers. 2007. "Convergence or Divergence in Los Angeles: Three Distinctive Ethnic Patterns of Immigrant Residential Assimilation." *Social Science Research* 36 (1): 254–85. http://dx.doi.org/10.1016/j.ssresearch.2006.01.001.

Yu, Z., and D. Myers. 2010. "Misleading Comparisons of Homeownership Rates When the Variable Effect of Household Formation Is Ignored: Explaining Rising Homeownership and the Homeownership Gap between Blacks and Asians." *Urban Studies* 47 (12): 2615–40. http://dx.doi.org/10.1177/0042098009359956.

# 4 How Are Sri Lankan Tamils Doing in Toronto's Housing Markets? A Comparative Study of Refugee Claimants and Family Class Migrants

SUTAMA GHOSH

Each year, thousands of immigrants arrive in Canada from diverse source countries and associated political economies of migration. Entering Canada under three main admissions classes – economic, refugee, and family class – newcomers bring with them various forms of capital (socio-economic-cultural) and networks (local and transnational) which profoundly affect their settlement experiences. Also, the recently arrived immigrants are almost an exclusively urban population, settling predominantly in three gateway cities: Toronto, Montreal, and Vancouver (TMV). The urban geography of these big cities is often unique, creating distinct and diverse inter- and intra-urban contexts. Therefore, despite important nation-wide similarities of housing and economic experiences, settlements of new Canadians are also city-specific. More importantly, as outlined by Li and Teixeira (in chapter 1 of this book), several other interconnected factors – both internal (e.g., immigrant's time of arrival, admissions class, educational and financial capital, social networks, and cultural identities and associated practices) and external (e.g., economic and political conditions in the sending and receiving countries, at the global and local scales) to the group – create distinct immigrant experience (see also chapters 2 and 3). Researchers therefore agree that the complex interplay of these factors not only creates inter-group settlement needs and aspirations (e.g., choice of residential locations), but also determines the kinds of barriers (e.g., competition from other groups and discrimination) they may face in order to access housing and employment (see chapters 1 and 2).

Despite a general recognition regarding the diversity of immigrant experiences, there are many unknowns. One reason for this is that early research initiatives often grouped non-European immigrants to Canada based on their phenotypical similarities (e.g., South Asian, Black, and Chinese). Consequently, policy decisions were also taken at a highly aggregated level. It has been argued earlier that such policies often ignore the subgroup-level needs which may be more acute (e.g., Bangladeshis) than the broader needs of the larger population category under which they are subsumed (i.e., "South Asian") (Ghosh 2007, 2012).

The main purpose of this chapter is to report on the housing experiences of Sri Lankan Tamils in Toronto, one of the most impoverished "South Asian" subgroups (Ornstein 2001). Most Sri Lankan Tamils entered Canada as refugee claimants (RC) and family class migrants (FCM). Although the refugee class has attracted much research attention, FCMs are a relatively less examined group. Because most FCMs receive assistance from their friends and family already living in Canada, in their migration and resettlement processes, it is assumed that they have a relatively easier transition into the new society. By contrast, previous research has concluded that RCs are the most disadvantaged immigrant class, as most of them arrive in Canada with little human, financial, and social capital. Consequently, they are more likely (than FCMs) to face numerous barriers in the urban milieu. The study presented in this chapter, however, shows that in the case of Tamils, the housing experiences of the two cohorts (RCs and FCMs) are very similar. Both groups face difficulties with respect to not only accessing but also sustaining quality housing in Toronto. This chapter begins with a review of the current literature on the housing situations of new Canadians. This is followed by a brief overview of the Toronto context and the research design and methods. Next, the research findings are presented and discussed, and finally conclusions are drawn.

## Literature Review

The current state of knowledge on the settlement experiences of new Canadians concludes that in comparison to the earlier arrivals, those who entered Canada since the mid-1990s have been particularly vulnerable in the housing and labour markets of large Census Metropolitan Areas (CMA) (see chapter 1). They are earning less and living longer in housing that is unaffordable, inadequate, and overcrowded (see Murdie and Logan 2011). Although for many new Canadians affordability remains the most important housing challenge, limited access to adequate and suitable accommodation and neighbourhoods are also high on the list of housing barriers. Within this broader landscape, however, the housing experiences of new Canadians remain extremely diverse, often attributed to their heterogeneity with respect to place of birth, time of arrival, admissions class, "race," forms of capital, and associated urbanisms (see chapter 1).

As has been outlined by Li and Teixeira (chapter 1), among the admissions classes, recently arrived refugees have been identified as the most disadvantaged population (see, e.g., Carter and Vitiello 2012; Hiebert 2009; Murdie 2010; Preston et al. 2006; Rose 2010; Teixeira 2008). Living predominantly in unaffordable housing, they are more likely to live in crowded spaces. Although most refugees qualify for social housing in Canada, the waiting lists are extremely long, particularly in cities like Toronto where there is an increasing demand for affordable housing. Significant differences have also been found within refugees to Canada. Murdie (2010), for instance, found that upon their arrival in the Toronto CMA, sponsored refugees[1] were more successful than RCs, primarily due to their pre-existing social networks. By the fourth year of their stay in Canada, however, the gap between these groups had narrowed (Hiebert 2009, 278; Murdie 2010). Although FCMs are the second largest group of immigrants to Canada, they are a less studied cohort. FCMs are those who entered Canada to join their sponsoring family. Hiebert (2009, 275) found that among all admissions classes, "the family class group is the most distinct, with by far the highest share of households comprising families with unrelated persons and, especially multiple families." More important, unlike in other admissions classes, among FCMs the percentage of composite family households did not drop over time. Hiebert (2009) speculates that one positive impact of this may be that joint-income pools enable this cohort to purchase homes faster than others.

Studies have found that housing experiences of new Canadians differ racially, and differences also exist within visible minorities as well. Preston et al. (2006), for example, demonstrate that "South Asians"[2] are the most precariously housed visible minority group. Besides logging the highest rates of residential overcrowding (32%), almost all are renters (80%), close to a quarter pay more than 30 per cent and over one-half were pay more than 50 per cent of their before-tax income on rent. Hiebert's (2009) longitudinal study further reports that although, upon their arrival, overcrowding is common among visible minorities, unlike in other groups (e.g., East Asians) "South Asians" tend to live longer in overcrowded conditions. Hiebert (2009) speculates that perhaps overcrowding is a strategy adopted by this "group" to achieve homeownership (also see Haan 2006, 2011).[3]

The findings of these large-scale quantitative studies suggest a need and scope for developing an in-depth understanding of immigrant and refugee housing experiences. Towards that goal, the first step is to nuance the highly aggregated findings about these "groups" (e.g., "South Asians") which are often arbitrarily defined by the Canadian census.[4] Disaggregating results on "South Asians" by country of birth (COB) is particularly important for several reasons. First, most members of this group self-identify nationally and not phenotypically (Ghosh 2012). Second, since the end of the colonial period, "South Asian" nation states (India, Pakistan, Sri Lanka, and Bangladesh) have undergone differential socio-economic, cultural, and political changes, as a result of which the contexts of emigration in these countries has become increasingly diverse. Third, as several recent studies have shown, "South Asian" subgroups often experience differential migration and settlements, as a result of which they also integrate differentially into Canadian society (Ferdinands 2001; Oliveira 2004; Ghosh 2007; Murdie and Ghosh 2010). This study addresses these issues by examining the experiences of Sri Lankan Tamils as a subgroup of "South Asians," as well as by comparing the experiences of the RCs and the FCMs – the foremost admissions classes under which they arrive.

Another important drawback in the literature is that in some studies "models" of immigrant residential behaviours have been created. This approach is conceptually problematic because it presupposes that there is a "rational logic" in housing "choice." I argue that in reality housing decisions may be more relational than logical in nature. As a corollary to that, in many housing career studies immigrants' economic conditions are considered as "constraining" factors circumscribing their housing desires, while cultural practices are regarded as aspects of "choice," defining their housing preferences. It is important to recognize, however, that finances together with social and cultural obligations may "constrain" a household in accessing its desired dwelling and neighbourhood. Also, women and men may be differentially affected by these obligations, resulting in gender differences in housing experiences.

## The Toronto Context

In 2006, Toronto attracted 40 per cent of all immigrants to Canada (Statistics Canada 2006). According to Canada Mortgage and Housing Corporation (CMHC 2010), the housing situations of Canadians in the province of Ontario, and especially in the Toronto CMA, are particularly acute. In Toronto, almost a fifth of all households live in unacceptable[5] housing, and among those, recently arrived immigrants have higher-than-average incidences of housing need (Hulchanski 2007; Murdie and Ghosh 2010). Arguably, the removal of rent control led to increased competition among immigrants over low-quality housing, such that even forty-year-old high-rise rental buildings that were difficult to rent out became "source[s] of productive wealth for the capitalist class" (Merrifield 2006, 139). Since the mid-1990s, Toronto's social housing sector has been deteriorating as well (Hackworth 2009, 257). Under neoliberal governance, newcomers to Toronto, many of whom are visible minorities, became a part of the "hidden homeless" (Fiedler et al. 2006). Preston et al. (2009) asserted that with little knowledge of Toronto's tight rental housing markets, newcomers often pay substantial security deposits, and "double-up" with friends and family in order to secure an apartment. Many pay more than 50 per cent of their before-tax income on rent, having little left to meet the other primary necessities of life. The issue of affordability, however, has been further complicated by the existing labour market conditions. Non-recognition of foreign credentials, gatekeeping within professional services, lack of full-time work, and lower earnings have severely circumscribed recent immigrants' economic successes (Fuller and Vosko 2008; Chen et al. 2012). As a result, immigrants who came to Toronto after the mid-1990s are predominantly renters and living in unacceptable housing.

## Research Design and Methods

In order to develop a wholistic understanding of the housing situations of Sri Lankan Tamils in Toronto, three main research themes, *migration*, *settlement patterns*, and *housing situations*, were explored at three levels: subgroup, household, and individual. Housing situation is considered to be both a process and an outcome of migration and settlement patterns. Given the nature of enquiry of this empirical study, a triangulation method (i.e., combining quantitative and qualitative approaches) was deemed to be the most appropriate research methodology. Two secondary and two primary sources were used. The secondary data sources were publicly available census data (1996 to 2006) and PRDS (Permanent Residents Data System)[6] data. Quantitative analysis of these data generated information at the group level on *migration and settlement patterns*. The primary data sources were five key informant interviews with settlement workers and community leaders selected by purposive sampling, and 134 semi-structured interviews with Tamil households – 74 were RCs and 60 were FCMs.[7] The respondents for the semi-structured interviews were recruited through surname-sampling and screening interviews. By contacting households with common Tamil surnames living all across the Toronto CMA, a significant variety in responses was captured. All respondents had lived in Canada for at least ten years, by which time, ideally, most should experience progress in employment and housing (Murdie 2008). The adult who was most knowledgeable about the housing career of the family was selected for the interview. All open-ended answers were thematically coded and analysed. The qualitative data (key informant and semi-structured interviews) were used to explore all three research themes and their interconnections at the household level.

### *The Respondents*

In table 4.1 the demographic and socio-economic conditions of the households are summarised. The table includes information on their length of stay in Toronto, current marital status, household size and structure, educational achievements, occupations, and total household income before taxes. As shown in table 4.1, at the time of the interview, most respondents had lived in Toronto for more than ten years on average – enough time to have developed a "progressive" housing career. Although very few respondents had lived in another Canadian city before moving to Toronto, over three-quarters who did first went to Montreal,[8] entering Canada under Quebec's immigration policy. The range of their stay in Montreal varied widely (less than a month to over a decade). Still, few differences were noted between the RCs and the FCMs in this regard. Since the overall cost for rental housing in Montreal is relatively lower than in Toronto, it was expected that those who lived in Montreal for some time would be more dissatisfied with their housing situation upon arrival in Toronto, particularly in terms of cost and the availability of space.

Table 4.1 Demographic and socio-economic conditions of respondents at the time of interview

| Variables | | | RC<br>*N* = 74 | FCMs<br>*N* = 60 |
|---|---|---|---|---|
| Length of stay in Toronto | 20 years or more | | 20% | 50% |
| | 15 to 20 years | | 42% | 27% |
| | 10 to 15 years | | 38% | 15% |
| | 5 to 10 years | | 10% | 7% |
| Average age | 25–54 (87%) | | | |
| Marital status | Heterosexual married couples (91%) | | | |
| Average number of adults per household | | | 4.1 | 4.3 |
| Household structure | Nuclear | | 34% | 18% |
| | Composite | | 66% | 82% |
| Education level of the respondent | No schooling and some schooling | | – | 4% |
| | High school | | 28% | 40% |
| | Community college, technical and trade school | | 24% | 21% |
| | Some university and bachelor's degree | | 42% | 34% |
| | MA, doctoral, and professional | | 5% | 2% |
| Occupation | Professional managerial | | 7% | 16% |
| | Manufacturing and technical | | 12% | 16% |
| | Services | | 61% | 33% |
| | Student | | 1% | 15% |
| | Homemaker | | 5% | 1% |
| | Unemployed/searching for employment | | 10% | 13% |
| | Retired | | 3% | – |
| | More than one job | | 10% | |
| Total household income before taxes 2010 | Low < $30,000 | | 25% | 26% |
| | Middle: | Lower $30,000–$49,999 | 34% | 23% |
| | | Middle $50,000–$79,999 | 27% | 27% |
| | | Upper $80,000–$99,999 | 4% | 9% |
| | High > $100,000 | | 10% | 15% |

Source: Semi-structured interviews with respondent households

As shown in table 4.1, the semi-structured interviews suggested that most respondents were in the core working age (25 to 54), and almost all were in heterosexual married relationships. The size of a typical Tamil household was larger (4.3) than the Toronto average (3.1). Most Tamils surveyed for this study had at least a high school diploma, about 20 per cent had attended a community college, and more than 40 per cent had an undergraduate degree. Almost all respondents were employed, and quite a few held multiple jobs. The service sector (e.g., retail, janitorial, restaurant, and hospitality) was the largest employer, although some were also engaged in manufacturing, professional, and managerial occupations. Whereas close to one half of all households were middle class, about 40 per cent had very low income (less than $30,000), and close to 20 per cent were in the high-income category. Although few differences were found between the RCs and FCMs, there were comparatively more RCs in the low–middle income bracket than the FCMs. Factors such as length of stay in Toronto, a relatively high participation of women in the workforce, and composite household structures with adult working offspring living at home may account for these differences.

## Research Findings

In this section of the chapter the major findings of this research are presented. As has been mentioned earlier, housing situation is considered to be both a process and an outcome of several factors, including migration and settlement patterns. The section begins by outlining the reasons for and processes of migration of Tamil respondents to Canada and Toronto.

### *Migration*

According to the landing records (PRDS data), at least 60,000 Sri Lankan Tamils had entered Canada between 1980 and 2005. Of these, more than one half entered as refugee claimants. It is difficult to estimate exactly how many Sri Lankan Tamils are currently residing in Canada. This is because publicly available census data on Sri Lankan Tamils – that is, Tamil-speaking persons of Sri Lankan origin[9] – are not available. Based on language data (mother-tongue single response and home language) there were more than 100,000 Tamil landed immigrants in Canada in 2006, and almost all of them lived in Toronto (Statistics Canada 2006).

Although Tamils have been emigrating from Sri Lanka since the 1950s,[10] the Tamil diaspora in Canada is largely made up of refugees and asylum seekers leaving Sri Lanka from the early 1980s onwards in the wake of intense ethnic violence. Corroborating these facts, a majority of the respondents interviewed for this study came to Canada as refugees. Most of them had arrived before 1995 and almost all claimed refugee status upon reaching. Among those who had entered Canada under the family class,[11] most had arrived after 1995. This difference in their time of arrival may be explained by the fact that it was only after living and working in Canada for a few years that the earlier migrants were able to sponsor their relatives to Canada (about mid-1990s onwards).

Further corroborating the PRDS data, many interviewees came to Canada directly from Sri Lanka, and almost all had first landed in Toronto. While many Tamil men came alone to Canada (i.e., as a one-person household) as RCs, many FCMs arrived as a family – either nuclear (i.e., couple with or without children) or composite (e.g., with grandparents, parents, married siblings, spouses, and children).

Research findings under the migration theme suggest *six* important issues:

(1) As noted with respect to some other "South Asian" subgroups (e.g., Punjabi Sikhs), Tamils seem to follow a gendered "culture of migration." Men emigrate at first and after establishing themselves in the new country, they sponsor female family members.

(2) Unlike some other "South Asian" subgroups (e.g., Indian Bengalis, Bangladeshis), however, among Tamils the head of the household (i.e., typically the father) was rarely the first one to immigrate to Canada. Most often the first one to arrive was a very young (in his early twenties) single male, who was "expected to work hard and eventually sponsor the rest of the family to Canada" (#Ambika).[12]

(3) Many Tamil households experienced ruptures in their family lives as a result of forced emigration from Sri Lanka and phased immigration to Canada. Since the whole family could not arrive at once due to the limited financial circumstances of the sponsor, various family members remained separated for decades. As #Shanthi stated: "We [Shanthi and her husband] just sponsored my sister to Canada ... Before that we had sponsored my brother in law and his family ... Now we are settled." Although #Shanthi's "family" – comprising her own siblings as well as her in-laws – is now reunited, the settlement process took close to twelve years, during which the family members had never seen each other.

(4) Most respondents expressed a sense of obligation towards their immediate and extended family members (familial ties), fellow

villagers (social ties), as well as those with whom they had no prior contact (symbolic ties). For instance, #Mohan [single male] came to Canada when he was twenty years old. As he explains, "I am still struggling in Canada since I came as a twenty years old ... When I came [to Canada], I was alone and responsible for the betterment of everyone in my family, people from my village, friends – everyone."

(5) The role of institutional ties in the migration process seems to be of little significance among these respondents. This may be because many of them had fled from rural areas of Sri Lanka, with limited educational and professional backgrounds.

(6) Sri Lankan Tamils, like other recently arrived immigrants from "South Asia," seem to foster and maintain strong transnational ties, and these ties seemingly play a significant role in their migration to Canada. For instance, as in other South Asian subgroups, information about Canadian immigration laws and potential struggles in the initial settlement phase was conveyed by the former to the latter migrants over transnational social spaces (e.g., mainly telephone calls). Also, various social and symbolic ties and associated obligations are maintained long after the migration process was completed, with important effects on their housing careers – discussed below (Ghosh 2007).

### *Settlement Patterns*

Households who arrived in the 1980s predominantly settled in St James Town, an inner-city Toronto neighbourhood composed of eighteen low-income high-rise apartment buildings. Some of these buildings are owned privately, while others are publicly owned (Toronto Community Housing). Built in the postwar decades (Caulfield 1994, 58), most of these buildings are now in a state of rapid physical deterioration, mainly due to disinvestment and neglectful maintenance. Many Tamil interview respondents who had first settled in this area eventually moved out to the suburbs, resettling mainly in Scarborough and Markham.

Unlike the earlier migrants, and following the general "South Asian" trend, Tamils arriving in the mid-1990s settled directly in the inner suburbs of Toronto (Doucet 1999),[13] primarily in Scarborough. This "choice" of residential location is a result of several interconnected factors. Gentrification and a lack of affordable housing in downtown Toronto pushed the newcomers out of the city core (see also chapter 5 of this book). Compared to the available private rental accommodations in

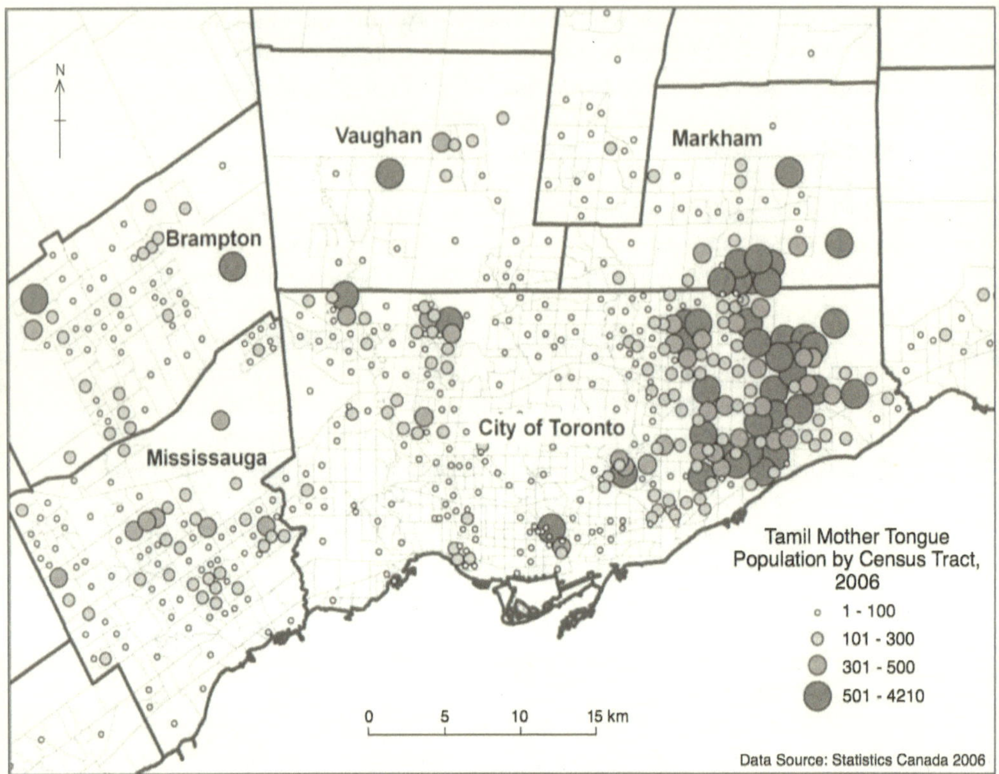


Figure 4.1 Residential locations of Sri Lankan Tamils in Toronto, created from 2006 census data.

downtown Toronto, apartments in Scarborough were relatively cheaper and larger in size, thus more affordable and suitable for Tamil composite families. Since the mid-1980s industries have relocated primarily in Scarborough and Mississauga. Tamils who sought employment in this sector preferred these areas. Finally, by the late 1990s a growing Tamil community had developed in Scarborough (e.g., in the neighbourhoods of Malvern, Agincourt, Morningside, Scarborough Village, and Dorset Park). Being institutionally complete with temples, Tamil grocery stores, and cultural centres, these areas perhaps attracted many Tamils seeking normality after several years of turmoil. Thus, the availability and cost of housing, employment opportunities, and institutional completeness may have triggered the development of Tamil clusters in the east end of Toronto, predominantly within Scarborough as well as Markham (figures 4.1 and 4.2). While "South Asians" in general are a

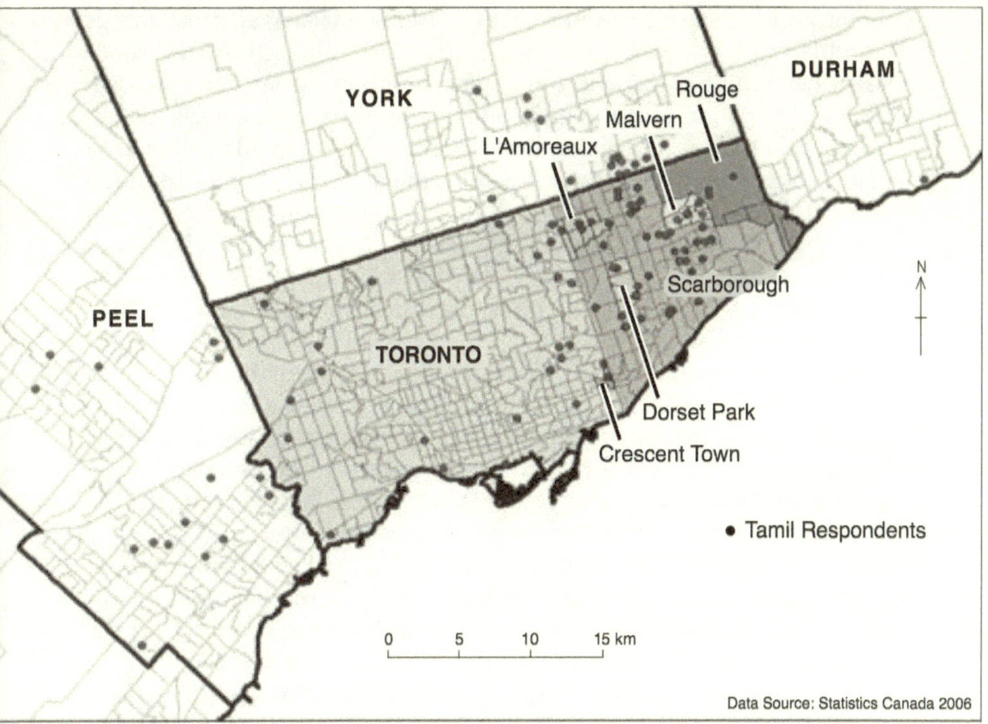


Figure 4.2 The current residential locations of Sri Lankan Tamil respondents in Toronto, 2011.

moderately segregated group in Toronto, Sri Lankan Tamils are a highly segregated group, living separately from non-"South Asians" as well as other "South Asians" (such as Indians, Pakistanis, and Bangladeshis).

*Housing Outcomes*

In this section of the chapter, the housing outcomes of the RCs and the FCMs are discussed comparatively using the interview data. First, some general information is provided on shelter use, the housing search process, and residential mobility. Next, specific aspects of housing experiences are discussed (e.g., tenure, type and structure, cost and housing satisfactions), focusing on two specific stages of the respondents housing careers: the first and the current residences.

One of the main findings of this research is that, regardless of their admissions class, economic conditions, and familial situations, none of the respondents used a public shelter. Upon their arrival, most Tamils "doubled up" with their family and friends. Although this evidence suggests a strong social network among Tamils in Toronto, and an associated expectation that the newcomers would be somewhat shielded from market forces, the reality was more complex. As will be discussed below, even though cultural obligations towards friends and extended family provided newcomers with valuable assistance, these also led to overcrowded conditions, negatively affecting the everyday rhythm of the households in diverse ways. As #Ravathy stated, "We were fine when they first came but things became difficult ... too many people in too little space you know."

During their entire housing career in Canada, none of the respondents sought the assistance of a social worker or any other non-governmental agency. This practice of non-reliance on organizations perceived as "government related" (#Anil) may have ensued from their pre-migration experiences with such organizations. In Sri Lanka, arguably, the Sinhalese government has routinely marginalized the Tamil claims to land, property, education, employment, and overall well-being for over four decades. Feelings of fear and mistrust towards various government agencies and workers seem to have been transposed to Canada, inhibiting respondents from seeking institutional assistance.

As in most "South Asian" subgroups (e.g., Bangladeshis, Sinhalese, and Punjabi Sikhs), almost all Tamils were assisted by their friends and family in the housing search process. Most homeowners sought the help of an ethnic real estate agent. Newcomers received information from their strong ties (social and familial) on specific Tamil neighbourhoods that were institutionally complete. Incidentally, although the cost of housing was relatively cheaper in these locations, the quality of housing was poor. As #Mani said, "Yes the apartment was close to other Tamil people and grocery stores but it was old. Cockroaches and insects were always crawling. Very unhealthy, not properly maintained ... But we did not say anything because we were sharing." The last point made by #Mani is particularly noteworthy. First, it is perhaps the economic needs of these newly arrived composite families that led them to poor-quality housing. Second, as #Mani says, even though they were dissatisfied with the dwelling, because they were sharing – i.e., perhaps living illegally – they were not in a position to request the building owners/supervisors to provide better living conditions.

Very few respondents were able to provide or receive any financial assistance from their social networks. This finding suggests that although most Tamil respondents are working and seem to be able to sustain themselves financially, they are not equipped to assist others in the economic realm. This may be because their housing costs are overwhelmingly high, but they may also feel the need to accumulate funding for future sponsorships. Second, seeking financial help from familial or social networks is perhaps not a culturally acceptable norm, whereas receiving assistance in kind (e.g., shelter, information on employment) is. As one respondent stated: "it is enough that they [meaning extended family members] sponsored us here … We cannot ask for money like begging" (#Bheeru).

Upon their arrival in Toronto, regardless of their admissions class, almost all Tamil households rented (91%). Most households rented from a private landlord, and only a tenth were housed in a subsidized rental unit, managed by the TCH, or Tamil Cooperative Housing (located in downtown Toronto in the Lansdowne and Bloor area). Among the few who owned, most had been living in Montreal for more than five years. Shortly after arriving in Toronto, these households purchased a high-rise condominium in a relatively low-income neighbourhood.

Like Bangladeshis, many Tamils began their housing careers in Toronto by living in high-rise apartment buildings. A significant percentage, however, also lived in a house. Often this was owned by "someone [they] knew" (#Pitija) – friends or family. Also, most Tamil households showed a relatively high level of residential mobility within the first ten years of their arrival. This was particularly noted among those arriving since the mid-1990s, renters, and large-sized households (mainly FCMs). Changes in the household structure along with issues around affordability often triggered these multiple moves.

The current housing situation of the respondents is encouraging. Unlike some "South Asian" subgroups, particularly the Bangladeshis (Ghosh 2007), most respondents seem to demonstrate high rates of homeownership. Over 80 per cent of RCs and about 75 per cent of FCMs owned their current homes. Almost all households who had arrived in Canada before 1995 had become homeowners within the first ten years of their arrival. Homeownership rates were also high among those who came after 1995 (more than 60% own). As indicated by Hiebert (2009), such high rates of ownership may be attributed to composite family structures where the number of earning members is higher than in average Canadian census families.

In terms of the structure of housing they currently occupy, Tamils seem to have achieved considerable progress as well. Regardless of tenure, most respondents were currently living in a semi-detached or a detached home. Although comparable, a slightly higher percentage of RCs (75%) than FCMs (close to 60%) were living in a semi-detached or a detached house. Among other factors, this housing "choice" also underscores the role played by composite family structures in determining the type and structure of housing occupied by Tamil households.

Cost of housing is one aspect that has remained unchanged for Tamils in Toronto. Upon their arrival, as expected, most RCs were living in an accommodation they cannot afford, thus corroborating previous research which predicted that refugees were more likely to face affordability-related issues in Toronto's competitive and tight housing market (Murdie 2010). It is important to note here, however, that the situation of the FCMs was very similar to that of the RCs. Although it is generally expected that FCMs would be protected by their social networks from facing such situations, the reality seems to be quite different. Almost all FCMs paid rent to the sponsoring family upon their arrival. Slightly more than a third of them spent more than 30 per cent and almost a fifth paid more than 50 per cent of their income on the very first house they occupied, even though they were sharing it with their sponsoring family. It was clear from the interviews as well that the newly arrived family members knew that they were expected to pay rent upon arrival. As #Bala explains: "I knew we would pay rent … [We were all] putting in a joint effort." #Sethu explained that by paying rent to the sponsoring family he was "helping those who had helped [him]." Such sentiments were echoed by many other FCMs respondents, who viewed themselves as "partners" of the sponsoring family (most of whom had come as refugees), willingly sharing the costs of resettlement.

Currently, almost all respondents (close to 90% of RCs and 84% of FCMs) were paying more than 30 per cent of their before-tax income on housing. In many cases, their struggles with affordability were the same as when they had first arrived, and in some cases they had increased. Several interconnected factors may have caused such situations. First, in order to avoid the perils of renting an accommodation in Toronto and to take advantage of the government deductions given to first-time homebuyers, many Tamils may have entered homeownership prematurely (Haan 2011). Second, significant changes could have occurred in their employment and family situations, whereby if even one person moved out of the house, the household was put under severe economic stress (also see chapters 8 and 9), as it was in the case of #Param, who

> came to Canada alone from Batticaloa as a refugee. After being in Toronto for some years Param sponsored his family members. With his brother he eventually purchased a home in Toronto. Within a short period of time, however, his brother found employment in Vancouver and moved away. Param is currently left with the responsibility of paying the mortgage and could lose the house.

Many respondents continued to live in overcrowded conditions long after the initial settlement phase (four years). Whereas upon their arrival, more than three-fourths of the RCs and almost all FCMs lived in crowded spaces, a majority in both groups were still living under overcrowded conditions. Thus, unlike other "South Asians,"[14] for this subgroup, "hidden homelessness" is not a temporary issue but a continuous housing problem. Typically, there were at least six adults and children living in a two-bedroom apartment. In order to accommodate so many persons, living rooms and corridors were also used for sleeping.

It is important to note that although residential overcrowding is a serious problem among non-white immigrants living in Canadian CMAs, this particular issue of housing has received little scholarly attention. Similar to the findings in the United States, several scholars have reported that in Canada in general and in Toronto more specifically, residential overcrowding is more prevalent among non-white refugee families, especially "South Asians" (Hiebert 2009; Preston et al. 2009; Haan 2011). Intersections of various demographic, economic, social, and cultural factors have been identified as the main causes of overcrowding. While in some cases, crowding is known to provide invaluable economic and socio-psychological benefits to new immigrants (e.g., helping them avoid the competitive and expensive rental market, providing quicker entrance into homeownership), crowding has also negatively impacted upon people's physical and mental health, their labour market success, and especially children's educational achievements. Despite these ambivalent assertions, overcrowding remains an everyday lived reality for many immigrant families – a theme that requires further investigation (Hiebert 2009; Haan 2011).

Finally, although this particular circumstance requires further investigation, it is important to note that many respondents were evicted by their own family members. Failing to pay rent or mortgage often led to evictions. Although by pooling resources the composite households were able to progress in their housing careers, many were unable to sustain their residence on their own. Also, while on the one hand even slight changes in earnings required more "partnerships" in order to keep the house, this often led to intense family problems and unwarranted evictions. It was mentioned earlier that upon arrival #Nalini's mother paid rent to her aunt, who had sponsored them. After living together as a household for some time, #Nalini's aunt suggested that they should accommodate another extended family member in the same residence so that the living costs could be minimized. When this person came to live with them, numerous conflicts between the sisters arose, and eventually #Nalini's family was asked to find a place of their own. Similarly, #Renuka sponsored her extended family (mother, married siblings, and their children) to Canada. Eventually they purchased a home together. Based on the hierarchical relationships within the family, however, access to the property was unequal. #Renuka seldom had any space of her own, even though she had purchased it. As a result, contestations over space became an everyday reality. Unable to retain the house under such circumstances, #Renuka had to sell it, and within a few days her status changed from being an owner of a house to a renter of a high-rise condominium. Even though the sponsored family had to move out, they did not always become an independent household. In many cases, due to lack of financial resources, the sponsored family moved in with another relative or friend. Consequently, as one composite family dissolved, another was formed. This situation continued – that is, moving from one place to another – until the sponsored family was able to rent a place of their own. Such a "common situation" of power differentials within Tamil households was neatly summed up by a key informant: "Families are fighting … They came here to be together in Canada, but the situation changes … Here they become renters and then landlords. They have to compete with one another, they have to ask one another to pay otherwise they will all loose … What will I say? I have the same problem in my house" (#Anu).

## Discussion and Conclusions

This study suggests that Sri Lankan Tamils may be different from the other "South Asian" subgroups in several ways. First, unlike most Indians, Pakistanis, and Bangladeshis, a majority of Tamils were forced to flee Sri Lanka in the wake of a civil war. They came to Canada predominantly as RCs with little formal education, and often carry with them traumatic experiences of living in refugee camps, which may have profound effects on the nature of their social and cultural capitals in Canada. Second, unlike some "South Asians" (e.g., Indian Bengalis), Sri Lankan Tamils seldom demonstrate close institutional attachments (weak ties). Very few households use their weak ties to come to Canada (e.g., an employment agency or immigration agency), or approach a government official/agency or a colleague from another ethnic background for information on housing. This social practice of trusting one's own for safety and defensive purposes seems to affect their residential locations and settlement patterns in Toronto, and determine their socio-economic and political integration into Canada. Third, as a result of such differences, they seem to be socio-spatially segregated from non–South Asians as well as other "South Asian" groups (e.g., Sinhalese, Pakistanis, and Indians) in Toronto.

Almost all Tamil households expressed a sense of obligation to assist not only their immediate and extended family but also village-friends in (re)settling in Canada. In this respect, therefore, forming composite families is not necessarily viewed as a "choice"; but rather as a "constrained choice." As has been mentioned earlier, the cultural practice of assisting others is perhaps interconnected with the pre-migration context in Sri Lanka, where the respondent families had witnessed the ethnic cleansing of Tamils and suffered marginalization for generations at the hands of various government machineries (civil and military). While in some respects these experiences have tightened the community bond, and increased reciprocity, they have also led to the development of a collective value whereby not assisting a fellow Tamil is not acceptable to the collective sensibility. In that sense, the "choice" to assist becomes a particular liability in the Canadian context, where the housing market is extremely competitive and the employment sector is fluctuating.

When the housing situations of the respondents are compared, it seems that, contrary to popular knowledge, RCs and FCMs may have similar experiences. Although both classes seem to have "progressed into homeownership, they were doing so at a cost – paying high prices, both monetarily and sometimes socially. Since a relatively larger percentage of RC households had sponsored their family members' move to Canada, often they needed a larger accommodation as well. On the other hand, even though the FCMs were expected to be taken care of by the sponsoring family, they had to share housing (and living) costs upon arrival. This expectation, although not resented by the FCMs, may have put many of them in a arduous situation. Furthermore, this may have also led to contestations of space between the members of the sponsoring and the sponsored household – thus pitting family members against each other in a power struggle.

Living in composite families had further perils. Changes in the housing needs and employment situations of one household affected the other, often leading to dissolution of family ties and the formation of new composite households. This study has emphasized that eviction by own family members and friends is an important problem faced by many Tamil households. Although by pooling their resources composite households were able to progress in their housing careers, many were unable to sustain a residence on their own.

Overall, this research further corroborates previous studies to argue that refugee status, in combination with other axes of social identities such as "race," creates numerous difficulties for recently arrived groups in their access to housing. By providing insights on the struggles faced by FCMs, this study highlights the importance of further research on this cohort of immigrants, who are usually "expected to do well" because of their social networks. The study attempts to make both substantive and methodological contributions to housing-career studies in Canada. Substantively, it provides timely insights on the housing careers of one of the most economically struggling "South Asian" subgroups in Toronto, which may initiate useful comparisons among other recently arrived visible minority groups in Canada, differentiated by admissions classes. The methodological approach to exploring housing careers using mixed methods (quantitative and qualitative data sources) may be applicable to the study of the settlement experiences of other groups in the Canadian urban milieu.

## NOTES

1 Refugees to Canada could ether be sponsored (by government or private organizations) or claimants. In LSIC (longitudinal data on immigrants) the refugee category only includes Government Assisted Refugees (GAR) and privately sponsored refugees.

2 "South Asians" are a relatively new addition to Toronto's social geography. At present, they are the largest visible minority group. More than one half of all "South Asians" in Canada live in Toronto.

3 "South Asian" newcomers demonstrate high rates of homeownership (58.6%).

4 In the Canadian census, immigrant groups (like "South Asians") are often arbitrarily defined, with several internally diverse subgroups under one

category merged together, the assumption being that there is some kind of commonality among them, be it "racial" (e.g., Black), regional ("West Asian"), or cultural (e.g., "Chinese"). In the context of "South Asians," such assumptions have resulted in a highly homogenized understanding of their migration and settlement experiences.

5 CMHC measures "acceptable housing" based on three indicators: adequacy (a dwelling not needing any major repairs); affordability (a dwelling that costs less than 30% of a household's before-tax income), and suitability (based on the National Occupancy Standard for housing, i.e., one bedroom for each cohabiting adult couple, unattached household member 18 years of age and over; same-sex pair of children under age 18; children of different sexes above the age of five). A household is said to be in "core need" if its housing falls below any one of these three acceptable standards.

6 PRDS data are collected by Statistics Canada for individual immigrants at the time of landing.

7 This is a part of a larger study for which a total of 201 households were interviewed. For this paper, only the experiences of the FCMs and the RCs are examined.

8 Sri Lankan Tamils have a large community in Montreal clustering in Parc Extension. This neighbourhood contains many Tamil businesses.

9 Although most Tamil-speaking persons in Canada are of Sri Lankan origin, many are also Indian, hailing from Tamil Nadu, a southern province of India.

10 In the 1950s middle- and upper-class Tamils left to seek better prospects in Europe and then North America (Imtiyaz 2008). In the 1980s, most emigrants were refugees escaping ethnic-based conflicts (Wayland 2004).

11 About a fifth of 201 respondents arrived as economic migrants; their experiences are not included in this paper.

12 The # symbol before a name indicates that it is a pseudonym.

13 Even though "South Asians" as a group have generally adhered to this pattern, the subgroups have settled in different suburbs.

14 In previous studies focussing on "South Asian" subgroups, overcrowding has not been reported as a major/continuous issue.

## REFERENCES

Carter, T., and D. Vitiello. 2012. "Immigrants, Refugees, and Housing." In *Immigrant Geographies of North American Cities*, ed. C. Teixeira, W. Li, and A. Kobayashi, 91–111. Don Mills, ON: Oxford University Press.

Caulfield, J. 1994. *City Form and Everyday Life: Toronto's Gentrification and Critical Social Practice*. Toronto: University of Toronto Press.

Chen, W., J. Myles, and G. Picot. 2012. "Why Have Poorer Neighbourhoods Stagnated Economically While the Richer Have Flourished? Neighbourhood Income Inequality in Canadian Cities." *Urban Studies* (Edinburgh, Scotland) 49 (4): 877–96. http://dx.doi.org/10.1177/0042098011408142.

CMHC. 2010. *Housing Observer*. Ottawa: Canada Mortgage and Housing Corporation.

Doucet, M. 1999. *Toronto in Transition: Demographic Change in the Late Twentieth Century*. Toronto: CERIS.

Ferdinands, S. 2001. "Sinhalese Immigrants in Toronto and Their Trajectories into Home Ownership." MA thesis, York University, Toronto.

Fiedler, R., N. Schuuman, and J. Hyndman. 2006. "Hidden Homelessness: An Indicator Based Approach for Examining the Geographies of Recent Immigrants at Risk of Homelessness in Greater Vancouver." *Cities* (London, England) 23 (3): 205–16. http://dx.doi.org/10.1016/j.cities.2006.03.004.

Fuller, S., and L. Vosko. 2008. "Temporary Employment and Social Inequality in Canada: Exploring Intersections of Gender, Race and Immigration Status." *Social Indicators Research* 88 (1): 31–50. http://dx.doi.org/10.1007/s11205-007-9201-8.

Ghosh, S. 2007. "Transnational Ties and Intraimmigrant Group Settlement Experiences: A Case Study of Indian Bengalis and Bangladeshis in Toronto." *GeoJournal* 68 (2–3): 223–42. http://link.springer.com/10.1007/s10708-007-9072-1.

Ghosh, S. 2012. "'Am I a South Asian, Really?' Constructing 'South Asians' in Canada and Being South Asian in Toronto." *South Asian Diaspora*: 1– 21.

Haan, M. 2006. "Are Immigrants Buying to Get In?" In *The Role of Ethnic Clustering on the Homeownership Propensities of 12 Toronto Immigrant Groups, 1996–2001*. 11F0019MIE no. 252. Ottawa: Statistics Canada, Analytical Studies Branch Research Chapter Series. www.statcan.gc.ca/pub/11f0019m/11f0019m2005252-eng.pdf.

Haan, M. 2011. "The Residential Crowding of Immigrants in Canada, 1971–2001." *Journal of Ethnic and Migration Studies* 37 (3): 443–65. http://dx.doi.org/10.1080/1369183X.2011.526772.

Hackworth, J. 2009. "Political Marginalisation, Misguided Nationalism, and the Destruction of Canada's Social Housing Systems." In *Where the Other Half Lives: Lower Income Housing in a Neoliberal World*, ed. S. Glynn, 257–77. London: Pluto Press.

Hiebert, D. 2009. "Newcomers in the Canadian Housing Market: A Longitudinal Study, 2001–2005." *Canadian Geographer/Géographe canadien* 53 (3): 268–87. http://dx.doi.org/10.1111/j.1541-0064.2009.00263.x.

Hulchanski, J.D. 2007. *The Three Cities within Toronto: Income Polarisation among Toronto's Neighbourhoods, 1970–2005*. Research Bulletin 41. Toronto: Centre for Urban and Community Studies, University of Toronto.

Imtiyaz, A.R.M. 2008. "Ethnic Conflict in Sri Lanka: The Dilemma of Building a Unitary State." In *Conflict and Peace in South Asia*. Contributions to Conflict Management, Peace Economics and Development, vol. 5, ed. M. Chatterji and B.M. Jain, 125–47. London: Emerald Group Publishing Limited.

Merrifield, A. 2006. *Henry Lefebvre: A Critical Introduction*. New York: Routeledge.

Murdie, R.A. 2008. "Pathways to Housing: The Experiences of Sponsored Refugees and Refugee Claimants in Accessing Permanent Housing in Toronto." *Journal of International Migration and Integration* 9 (1): 81–101. http://dx.doi.org/10.1007/s12134-008-0045-0.

Murdie, R.A. 2010. "Precarious Beginnings: The Housing Situation of Canada's Refugees." In *Newcomer's Experiences of Housing and Homelessness in Canada, Canadian Issues / Thèmes canadiens*, Fall 2010: 47–53. http://canada.metropolis.net/publications/aec_citc_fall2010_e.pdf.

Murdie, R.A., and S. Ghosh. 2010. "Does Spatial Concentration Always Mean a Lack of Integration? Exploring Ethnic Concentration and Integration in Toronto." *Journal of Ethnic and Migration Studies* 36 (2): 293–311. http://dx.doi.org/10.1080/13691830903387410.

Murdie, R.A., and J. Logan. 2011. *Precarious Housing and Hidden Homelessness among Refugees: Asylum Seekers and Immigrants. Bibliography and Review of Canadian Literature from 2005 to 2010*. Report submitted to the Homelessness Partnering Strategy (HPS) of Human Resources and Skills Development Canada. Ottawa: HRSDC.

Oliveira, L. 2004. "Housing Trajectories into Homeownership: A Case Study of Punjabi Sikh Immigrants in the Toronto CMA." Master's thesis, York University, Toronto.

Ornstein, M. 2001. *Ethno-Racial Inequality in the City of Toronto: An Analysis of the 1996 Census*. Toronto: Report of the Access and Equity Unit, Strategic and Corporate Policy Division.

Preston, V., R.A. Murdie, and A.M. Murnaghan. 2006. *The Housing Situation and Needs of Recent Immigrants in the Toronto CMA*. Ottawa: Canada Mortgage and Housing Corporation. http://publications.gc.ca/collections/collection_2011/schl-cmhc/nh18-1/NH18-1-6-4-2006-eng.pdf.

Preston, V.R., R.A. Murdie, J. Wedlock, M.J. Kwak, S. D'Addario, J. Logan, A.M. Murnaghan, S. Agrawal, and U. Anucha. 2009. "At Risk in Canada's Outer Suburbs: A Pilot Study of Immigrants and Homelessness in York Region." In *Finding Home: Policy Options for Addressing Homelessness in Canada*, ed. J.D. Hulchanski et al. Toronto: Cities Centre, University of Toronto. www.homelesshub.ca/FindingHome.

Rose, D. 2010. "The Housing Situation of Recent Immigrants to Montréal in Comparative Metropolitan Perspective: Contrasts and Convergences." In *Newcomer's Experiences of Housing and Homelessness in Canada, Canadian Issues/Thèmes canadiens*, Fall 2010: 73–8. Montreal: Association for Canadian Studies and Metropolis. http://canada.metropolis.net/publications/aec_citc_fall2010_e.pdf.

Statistics Canada. 2006. Profile of Language, Immigration, Citizenship, Mobility and Migration for Census Metropolitan Areas and Census Agglomerations. 2006 Census.

Teixeira, C. 2008. "Barriers and Outcomes in the Housing Searches of New Immigrants and Refugees: A Case Study of 'Black' Africans in Toronto's Rental Market." *Journal of Housing and the Built Environment* 23 (4): 253–76. http://dx.doi.org/10.1007/s10901-008-9118-9.

Wayland, S. 2004. "Ethno Nationalist Networks and Transnational Opportunities: The Sri Lankan Tamil Diaspora." *Review of International Studies* 30 (3): 405–26. http://dx.doi.org/10.1017/S0260210504006138.

# 5 A Two-Sided Question: The Negative and Positive Impacts of Gentrification on Portuguese Residents in West-Central Toronto[1]

ROBERT A. MURDIE AND CARLOS TEIXEIRA

Since Ruth Glass (1964) introduced the term in the 1960s, gentrification has become an integral part of the urban studies literature, and there have been numerous studies on the origin and explanations of gentrification, the nature of the process and its impacts, both negative and positive, on local neighbourhoods. Debates have emerged concerning the meaning of gentrification, but most researchers agree that gentrification is "the production of space for – and consumption by – a more affluent and very different incoming population" (Slater et al. 2004, 1145).

In spite of the extensive amount of research concerning gentrification, there has been little consideration of the intersection between ethnic groups and gentrification. As Lees (2000, 400) noted at the turn of the present century, emphasis in gentrification research has been placed much more directly on class and gender than on ethnicity or race. Seven years later, Lees (2007) reported relatively little progress in this area of research, except for Black gentrification in US cities (e.g., Boyd 2005; Freeman 2006; Moore 2009) and a few case studies on the impact of gentrification on minority groups, especially in Chicago (e.g., Boyd 2005; Nyden et al. 2006; Betancur 2009).

Research has also focused more on gentrifiers and the process of gentrification than on the experiences of non-gentrifiers living in gentrifying neighbourhoods, many of whom will likely be displaced as a result of gentrification (Rose 2004; Slater 2006; Skaburskis 2012). Displacement of low-income residents is usually identified in the literature as the major negative consequence of gentrification. However, the implications of gentrification for non-gentrifiers and the neighbourhoods in which they live are both negative and positive.

Some researchers point to negative outcomes such as residential and commercial displacement, loss of affordable housing, and the lessening of social diversity, while others emphasise positive effects such as the stabilization of declining areas, reduced vacancy rates, and increased social mix (Atkinson 2004; Lees et al. 2007; Nyden et al. 2006).

Canadian research on gentrification has focused on the extent and timing of the process in the country's three largest cities, Montreal, Toronto, and Vancouver, and the factors responsible for gentrification, concentrating particularly on the consumption side. Ley's (1996) work on the quickening pace of social status change as a marker of gentrification in inner-city neighbourhoods is especially relevant, as is Walks and Maaranen's (2008) more recent work on the timing, patterning, and forms of gentrification. Using changes in social status, artists, income, and rental tenure as indicators, Walks and Maaranen conclude that by 2001 more than one-third of pre–Second World War inner-city census tracts in Canada's three major cities had experienced complete or incomplete gentrification, with Toronto being most impacted.

There is general consensus among Canadian researchers that while artists are important catalysts in the early stages of gentrification, managers and professionals ultimately replace them as house prices increase. Both Caulfield (1994) and Ley (1996) argue that gentrifiers are part of a post-industrial middle class who reject suburban conformity and favour inner-city diversity. Ley also argues that gentrifiers are attracted to areas of older Victorian houses, urban amenities, central city employment opportunities, and proximity to existing high status enclaves. Rose (1984) has expanded this view to include "marginal gentrifiers," especially female single parents who appreciate the greater level of support services available in inner-city neighbourhoods.

Although concern has been expressed in the Canadian literature about the link between gentrification and the displacement of low-income households from inner-city neighbourhoods, precise evidence concerning the extent of displacement is not available. More generally, however, both Filion (1991) and Slater (2004) have considered the divergent impacts of gentrification on both gentrifiers and existing residents in Toronto. As Filion (1991) notes, gentrifiers benefit in terms of enhanced home equity and neighbourhood amenities, but existing residents, especially those who are displaced to less desirable neighbourhoods, are negatively impacted. Based on a case study of South

Parkdale in West Central Toronto, Slater (2004) argues that middle-class gentrification in this area, encouraged by neoliberal government policies aimed at improving the physical quality of South Parkdale, has not been beneficial to the deinstitutionalized psychiatric patients and recent immigrants who live in the area.

Despite the fact that 46 per cent of Toronto's 2006 population was foreign-born, there has been no specific study of the impact of gentrification on the city's inner-city ethnic neighbourhoods.[2] Although Toronto has experienced a substantial suburbanization of its immigrant population (see Ghosh, chapter 4, and Kataure and Walton-Roberts, chapter 6), West Central Toronto, an older, ethnically diverse residential area located immediately west of downtown Toronto still contains a large number of European migrants, especially Portuguese who immigrated in the 1960s and 1970s and a diverse group of more recently arrived immigrants. In 2006, about 45 per cent of West Central Toronto's population was foreign-born and 9 per cent had immigrated in the previous five years. During the past two decades, however, West Central Toronto has become increasingly gentrified, with important implications for the existing immigrant population (Walks and Maaranen 2008).

In this research we seek to understand the impact of gentrification on Portuguese residents living in West Central Toronto, including Little Portugal, the historic core of Toronto's Portuguese community (figures 5.1 and 5.2). Although many Portuguese have moved to other parts of the Toronto Census Metropolitan Area (CMA), Little Portugal contains most of the city's Portuguese institutions and Portuguese-oriented retailing, and Portuguese are still the largest ethnic group in West Central Toronto. The rest of the paper is divided into three major sections and a conclusion: (1) an evaluation of recent literature concerning the impact of gentrification on ethnic neighbourhoods, (2) a discussion of Portuguese settlement and the emerging pattern of gentrification in West Central Toronto, (3) an evaluation of the negative and positive impacts of gentrification on the Portuguese population in this area using key informant and focus group interviews, and (4) a conclusion in which we consider the future of immigrant settlement in West Central Toronto in the context of increased gentrification in the area and provide some general comments about the ways in which ethnic enclaves are impacted by gentrification and the extent to which the findings from this study can be generalized to other cities.

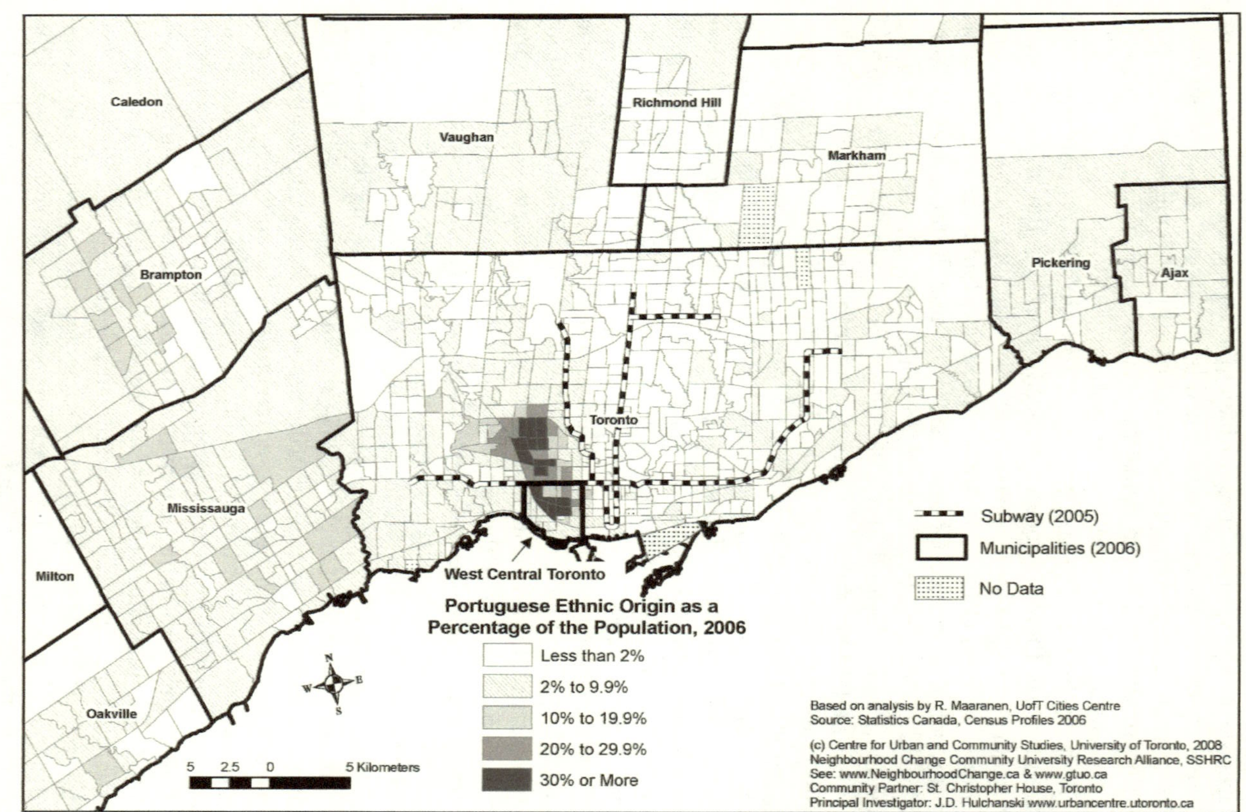


Figure 5.1 Portuguese ethnic origin by census tracts, Toronto CMA, 2006.

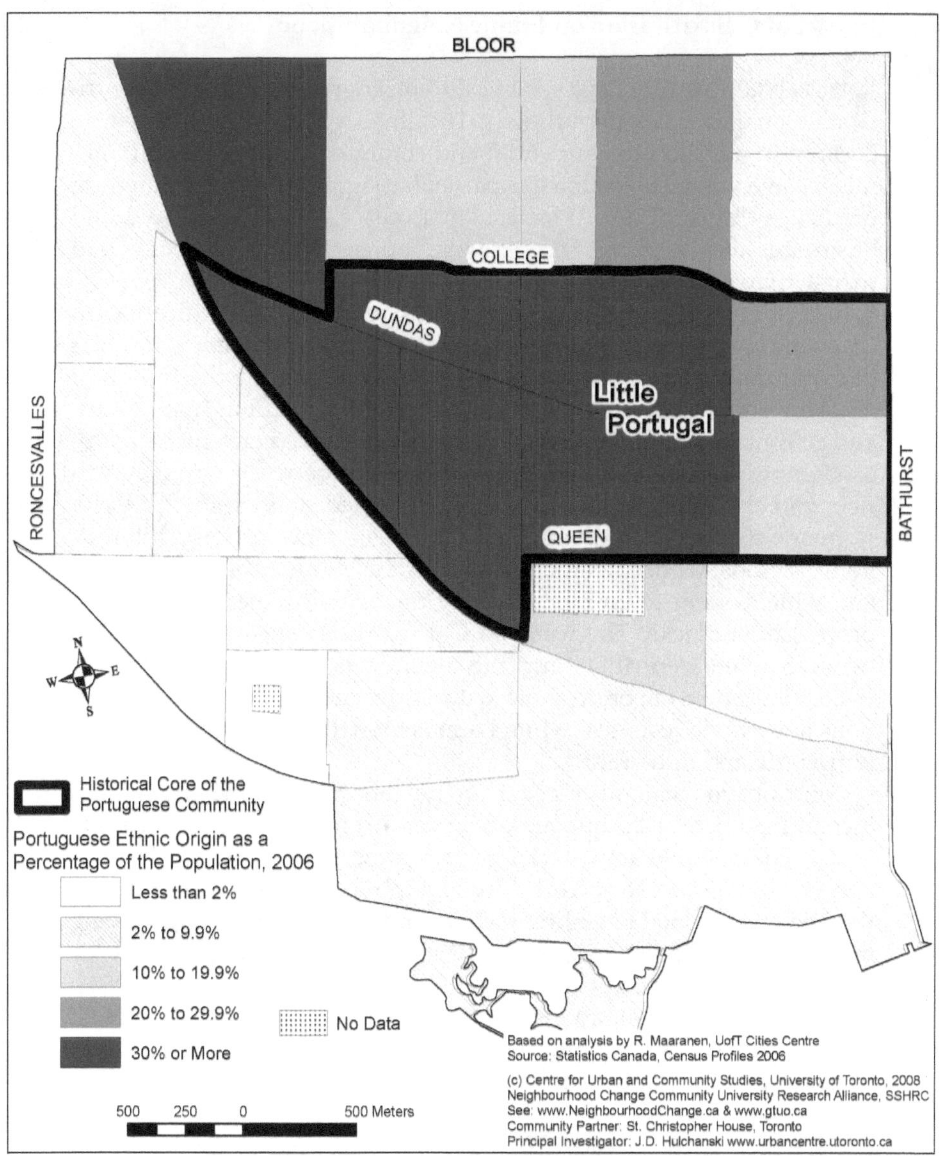


Figure 5.2 Portuguese ethnic origin by census tracts, West Central Toronto, 2006.

## Impact of Gentrification on Ethnic Neighbourhoods

Relatively few studies have specifically considered the impact of gentrification on ethnic neighbourhoods. This impact is important, however, because in addition to residential and commercial displacement, ethnic enclaves themselves have a particular significance for immigrant groups. As Krase (2005, 7) states: "Not only are local residences and businesses displaced but the symbolic representations of people and their activities are as well."

In a study of four neighbourhoods undergoing gentrification in Chicago, Nyden et al. (2006) note how gentrification effects the loss of community and ethnic/racial identity, including the disruption of neighbourhood social networks, ethnic retailing, religious institutions, and community organizations. In a highly racialized city such as Chicago gentrification also highlights differences between white gentrifiers and the minority groups being displaced, with some minority respondents characterizing gentrification as a racist process. Interestingly, Latinos, whose neighbourhoods act as a buffer between Blacks and whites, seem to be more directly impacted by the gentrification process than Blacks. The result in Chicago is an uneasy relationship between white gentrifiers and the incumbent Latinos. In particular, whites are often uncomfortable with Latino neighbourhood celebrations and ethnic festivals, while Latinos view the white gentrifiers as unfriendly and intolerant.

Gentrification can also occur in neighbourhoods where higher-income members of a group replace lower-income members of the same group. For example, intra-racial gentrification has occurred in Harlem and Clinton Hill in New York City and in the Douglas/Grand Boulevard neighbourhood of Chicago where middle-class Blacks are moving into areas occupied by lower-class Blacks, perhaps paving the way for middle- and upper-class whites to follow (Boyd 2005; Freeman 2006; Lees et al. 2007). Evaluations of intra-racial gentrification are mixed. Freeman (2006), for example, argues that displacement in Harlem has been minimal, partially because of rent-control legislation and because existing residents appreciate the improvements in amenities, services, and the physical landscape. He concedes, however, that gentrification has done little to improve employment prospects in the area. In contrast, Boyd (2005) points out that those who have encouraged Black gentrification in Chicago as a form of "racial uplift" mask the negative implications for the neighbourhood's low-income Blacks.

The few studies that have examined the impacts of gentrification on ethnic neighbourhoods have been restricted primarily to minority neighbourhoods in New York City and Chicago. The impacts range from positive to negative depending on who is impacted. As Nyden et al. (2006, 27–8) note in their Chicago study, participants who expressed ambivalence about gentrification often asked, "Who benefits?" or "Who is hurt?" These questions are at the core of this study on the impacts of gentrification on Portuguese residents in West Central Toronto.

## Portuguese Settlement and Emerging Gentrification in West Central Toronto

Portuguese immigration to Toronto began in the early 1950s and peaked in the 1970s. By 2006, 410,850 people of Portuguese ethnic background lived in Canada, almost half of whom resided in the Toronto CMA (total ethnic origin).[3] The majority of this group lived in the City of Toronto, particularly Little Portugal, adjacent areas of West Central Toronto, and an expanded corridor of Portuguese settlement to the northwest (figures 5.1 and 5.2). In 1971, people of Portuguese background (Portuguese mother tongue) accounted for about 28 per cent of the population in Little Portugal and 15 per cent in West Central Toronto (table 5.1).[4] West Central Toronto was home to almost half of Toronto's Portuguese population at that time. By 1981 Portuguese represented over half the population in Little Portugal and about one-third that in West Central Toronto.

Beginning in the 1980s Portuguese immigration to Canada declined, as did the Portuguese population in West Central Toronto, including Little Portugal. Although Little Portugal and the surrounding area of West Central Toronto are no longer the only important Portuguese residential areas in Toronto, they still contain a large number of people of Portuguese ethnic origin: more than 16,000 (total ethnic origin) in West Central Toronto, of whom almost 60 per cent live in Little Portugal. Evidence from interview studies also suggests that many Portuguese who live in other parts of the city return to Little Portugal to visit friends, take part in religious services and cultural festivals, and shop in local Portuguese stores (Teixeira 2007a).

Census data (special cross tabulations, 2001) indicate that the Portuguese in West Central Toronto exhibit the expected characteristics of an aging immigrant population. They are stayers – only about 20 per cent moved in the previous five years. About 17 per cent of the

Table 5.1 Total population and Portuguese population: Little Portugal, West Central Toronto, and the Toronto CMA, 1971, 1981, 1991, 2001, and 2006

| Area | 1971 | 1981 | 1991 | 2001 | 2006 |
|---|---|---|---|---|---|
| *Little Portugal* | | | | | |
| Total population | 44,590 | 35,140 | 33,050 | 30,940 | 28,085 |
| Portuguese | 12,285 | 19,655 | 15,945 | 12,075 | 9,040 |
| % Portuguese | 27.6 | 55.9 | 48.2 | 39.0 | 32.2 |
| *West Central Toronto* | | | | | |
| Total population | 124,355 | 103,110 | 106,276 | 108,190 | 102,375 |
| Portuguese | 18,235 | 31,645 | 27,125 | 21,450 | 16,060 |
| % Portuguese | 14.7 | 30.7 | 25.5 | 19.8 | 15.7 |
| *Toronto CMA* | | | | | |
| Total population | 2,628,125 | 2,998,947 | 3,893,046 | 4,647,955 | 5,113,149 |
| Portuguese | 39,550 | 127,635 | 124,330 | 171,545 | 188,110 |
| % Portuguese | 1.5 | 4.3 | 3.2 | 3.7 | 3.7 |
| *Portuguese population* | | | | | |
| Little Portugal/West Central Toronto | 67.4% | 62.1% | 58.8% | 56.3% | 56.3% |
| Little Portugal/ Toronto CMA | 31.1% | 16.4% | 12.8% | 7.0% | 6.9% |
| West Central Toronto/ Toronto CMA | 46.1% | 24.8% | 21.8% | 12.5% | 12.3% |

Definitions:

*Little Portugal:* Census tracts (40, 41, 42, 43, 44, 45, 46); City of Toronto neighbourhoods (Little Portugal and Trinity Bellwoods).
*West Central Toronto:* Census tracts (4, 5, 7.01, 7.02, 8, 10, 40, 41, 42, 43, 44, 45, 46, 47.01, 47.02, 48, 52, 53, 54, 55, 56, 57, 58); City of Toronto neighbourhoods (South Parkdale, Liberty Exhibition, Niagara, Trinity Bellwoods, Little Portugal, Roncesvalles, Dufferin Grove, Palmerston–Little Italy).
*Portuguese*: Portuguese mother tongue (1971); Portuguese single ethnic origin (1981, 1991); Portuguese total ethnic origin (2001, 2006).
Sources: Censuses of Canada, 1971, 1981, 1991, 2001, 2006.

Portuguese are 65 years of age and older and 27 per cent are between 45 and 64 years of age. They also have a relatively low level of educational achievement and occupational status. Almost half of Portuguese adults have no high school education and fewer than 5 per cent possess a university degree. Only about 10 per cent are engaged in high-skilled managerial or professional occupations. While average individual income for the Portuguese is relatively low, average household income is comparatively high, partially as a result of multiple earners within many Portuguese families. With regard to housing status, two-thirds of Portuguese households own their homes, more than twice the incidence of homeownership for other groups in West Central Toronto and higher than the city average. They are also more likely to occupy semi-detached and row housing. This is often the Victorian-style housing that is attractive to gentrifiers.

Currently, West Central Toronto is in transition, partly as a result of the outward movement of Portuguese, but also due to an in-movement of (1) new immigrants and refugees, including Portuguese-speaking immigrants from Brazil and Portugal's former African colonies (Angola, Mozambique, and Cape Verde Islands) and (2) an increasing number of middle-class professionals who see an opportunity to obtain relatively low-cost housing with renovation potential in close proximity to the city's downtown core (Teixeira 2007b, 2008).

Portuguese-speaking immigrants from Brazil and Africa arrived in Toronto primarily after 1980, much later than migrants from continental Portugal and the Portuguese islands of Azores and Madeira. These new groups are much less numerous than the Portuguese who arrived in the 1970s and earlier, about 6000 Brazilians and 4000 to 5000 Africans (Barbosa 2009, 219; Teixeira 2008, 259). Many of the African migrants, especially those from Cape Verde, first settled in or close to Little Portugal, while the Brazilians are more spatially dispersed. The Brazilian community in Toronto is highly differentiated by class structure, and only held together by language, cultural tradition, and soccer (Shirley 1999, 281). The African migrants are also highly differentiated. Angolans, for example, arrived primarily as refugees and lacked the social networks of the other two groups.

Concerning links between the new arrivals from Brazil and Africa and the older Portuguese community there is an ambivalence, almost a disconnect. Among the Brazilians, older immigrants and those with limited English tend to be more closely aligned with the Portuguese. Otherwise, the Brazilians, despite their internal divisions, have a strong cultural identity of their own that may dissipate over time as they become more assimilated into the Canadian mainstream (Shirley 1999, 281). Among the African migrants, there seems to be a closer relationship between the migrants from Cape Verde and the Portuguese compared to the other two groups, perhaps because of the turbulent colonial history faced by Angola and Mozambique. In part, this also relates to skin colour. Cape Verdians have a lighter skin colour than Angolans and Mozambiquans (Teixeira 2008, 262).

Gentrification in inner-city Toronto dates from the late 1960s and early 1970s and now includes much of the city's central area (e.g., Caulfield 1994; Ley 1996; Walks and Maaranen 2008). Walks and Maaranen (2008) trace the patterns of gentrification by census tract for each decade from 1961 to 2001 using a combination of variables representing changes in income, education, occupational status, housing tenure, rent, and dwelling value. They further identify areas of complete gentrification (personal income above average in 2000 from a previously below-average status in 1960) and incomplete gentrification (personal income still below average in 2000). Figure 5.3 indicates the degree to which areas of complete or incomplete gentrification were prevalent throughout inner-city Toronto in 2000. Of the twenty-three census tracts in West Central Toronto, one was completely gentrified by 2000, twelve showed incomplete gentrification, and another four were characterized as having potential for future gentrification. Walks and Maaranen (2008) indicate that gentrification has spread in a contagion fashion from the downtown area west through the eastern half of West Central Toronto and from High Park at the western edge of West Central Toronto in an easterly direction. Census tracts between these two areas exhibit either potential for gentrification or no gentrification, but if this pattern continues, it is likely that much of West Central Toronto will experience some form of gentrification in the future.

Discussion of Portuguese settlement patterns in West Central Toronto and emerging patterns of gentrification leads to a consideration of the impact of gentrification on Portuguese suburbanization. Portuguese suburbanization began in the 1970s and accelerated in the 1980s. Gentrification also began in the 1980s, but remained incomplete in most parts of West Central Toronto (Walks and Maaranen 2008). On one level it can be argued that the Portuguese simply followed the pattern established earlier by Jewish and Italian immigrants, who also settled initially in West Central Toronto and subsequently moved to the suburbs. Unlike the Portuguese, however, these groups moved before the onset

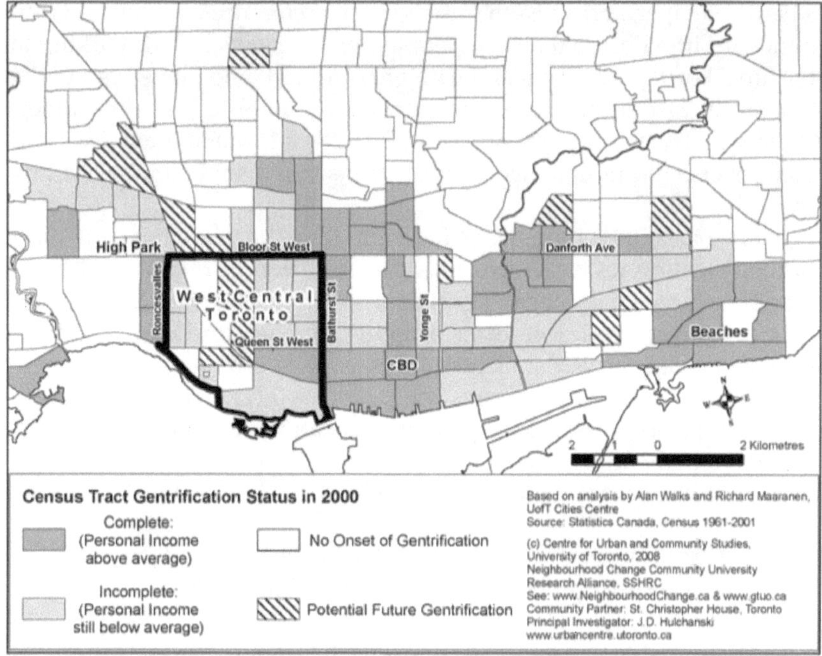


Figure 5.3 Gentrification in inner-city Toronto.

of gentrification. More likely, gentrification has accelerated Portuguese relocation to the suburbs and continues to have substantial impacts on the large number of Portuguese who still live in West Central Toronto.

## Negative and Positive Impacts of Gentrification as Identified by the Portuguese Respondents

Evidence to evaluate perceptions of neighbourhood change and the attitudes of Portuguese-speaking residents towards gentrification was obtained in 2006 from (1) informal interviews with twenty Portuguese key informants and (2) five focus groups. A total of forty-two people participated in the focus groups, thirteen Portuguese with roots in Portugal, fourteen from the Portuguese colonies, and fifteen people from other ethnic origins. The key informants come from different professions (e.g., social workers, journalists, real estate brokers, politicians, entrepreneurs, teachers, factory workers, a priest, a construction contractor, a bank manager, and a health nutritionist). Preference was given to residents of Little Portugal and to people living elsewhere but working in the area. Former residents of the neighbourhood who moved out of the area in the last decade were also interviewed in order to better understand who has left the area and why. The key informant interviews and focus group discussions were taped, transcribed, and summarized by theme.

Atkinson's (2004) framework for evaluating the impact of gentrification on existing residents was used to construct the themes. Atkinson identifies three major negatives of gentrification – (1) speculative property price increases, loss of affordable housing, and displacement, (2) commercial/industrial displacement, and (3) community resentment and conflict – and three contrasting positives – (1) increased property values, (2) stabilization of declining areas and encouragement of further development, and (3) increased social mix. In each of the three sections below, a negative is juxtaposed with a positive.

### *Speculative Property Price Increases, Loss of Affordable Housing, and Displacement versus Increased Property Values*

Loss of affordable housing, arising primarily from speculative property price increases in gentrifying areas, usually results in the displacement of existing residents, especially renters. Loss of affordable housing and displacement has been identified in numerous studies as the most negative effects of gentrification. Although measurement of displacement is difficult (Atkinson 2000) and most studies are based on incomplete evidence, there is general agreement that displacement has serious consequences, especially for low-income displacees. Not only are displacees forced to find alternative housing, but they also face the emotional impact of removal from social networks and familiar community structures. Gentrification also affects low-income residents, especially recent immigrants (see Ghosh, chapter 4), who might have otherwise found affordable housing in gentrifying neighbourhoods but are forced to look elsewhere, a phenomenon that Marcuse (1986) refers to as exclusionary displacement.

Loss of affordable housing was identified by three-quarters of the Portuguese respondents as the most important negative impact of gentrification. This is not surprising given the rapidly escalating house prices in the area. As illustrated in the following quotation many respondents blamed gentrifiers and the increased demand for housing for the escalating real estate values: "Houses in my area are listed for one price and sold later on for more than 25–30% above the asking price. The urban professionals are largely responsible for that ... They can afford."

Low-income people, including recent immigrants from Portuguese-speaking countries such as Brazil, Angola, Mozambique, and the Cape Verde Islands, are severely affected by the escalating house prices and face increased barriers in finding affordable housing in the area (Teixeira 2008). Indeed, there are already signs that some renters who have lived in the area for a long time are being forced to leave their apartments because they cannot afford the rents. One respondent even questioned the future of Little Portugal as an immigrant reception area: "I have also seen the arrival of other ethnic groups – the Asiatics, the Chinese ... but if Little Portugal is going through gentrification this area will stop being a reception area for new immigrants ... They cannot afford the rents ... Like Chinatown and Queen Street West which are in transition, Little Portugal too will follow ... it's a question of location."

Most respondents felt that smaller apartments will gradually be eliminated, and with them the ability of low-income working-class people to rent in the area. In part, this is because increased property values encourage the conversion of rental units to homeownership, thereby reducing the number of rental units in the area. Increased property values in gentrifying areas enhance the growing wealth gap between homeowners and renters (e.g., Hulchanski 2004) and make it increasingly difficult, if not impossible, for renters to enter the homeownership market in these areas of the city.

In contrast, increased property values resulting from gentrification are advantageous to homeowners who wish to capitalize on the increased equity in their house. When they sell their houses Portuguese and other homeowners stand to benefit from increased house prices, especially if they are able to or want to move to lower-cost housing in the suburbs. The increased value of housing in Little Portugal may also help the children of first-generation Portuguese achieve intergenerational class mobility: "Arrival of gentrifiers is good for Portuguese because they can sell their houses for very good prices ... They will cash in their real estate investment ... The ones who benefit the most are the sons/daughters with the sale because some of those Portuguese parents decide to live with their children. Thus, they are in a position to help financially their children." Some, however, don't want to leave

Little Portugal, and so selling their house is not an option. Regarding this dilemma, one respondent noted:

> Houses are selling for a very good price ... But what is the point? ... Most of the first generation Portuguese don't want to move ... they want to die where they spent most of their lives ... here in "Little Portugal." What's the point to sell for good bucks, cash some money and go to the suburbs ... far away from the Portuguese community? That's not what they want. So ... what's the point of having th[ese] huge housing prices here ... Who benefits? Not the Portuguese seniors, because they still need a roof to live [under].

Those homeowners who remain in Little Portugal face another dilemma. A little more than half of the respondents expressed concern about the potential displacement of existing residents due to a steady increase in property taxes and maintenance costs. Seniors on fixed incomes are the most vulnerable because many lack the necessary financial resources to maintain their houses and afford the increased property taxes. One respondent pointed to these issues but also lamented the lack of sufficient senior's housing to accommodate a growing elderly population:

> People who are older can't maintain a three storey Victorian house. Because of the very nature of their health and/or age they are not ... able to do the repairs needed in their homes ... and the high property taxes ... have gone through the roof ... Portuguese like Italians ... like to hold their houses for as long as they can ... Usually the move is due to health reasons ... Unfortunately, we don't have enough seniors' housing in our community to accommodate these people in need of a place to stay ... in a secure and comfortable place in an atmosphere where they would feel comfortable.

Seniors who decide to stay in Little Portugal often rent part of their house, usually to Portuguese-speaking immigrants, as a means of coping with escalating homeownership costs. This approach is a modern extension of a traditional Portuguese housing strategy whereby homeowners paid off their mortgages by renting out parts of their houses, often to recently arrived Portuguese immigrants. Not all seniors are comfortable renting part of their house, however, especially to non-Portuguese-speaking tenants.

A major issue is that since 1988 property taxes in Toronto have been based on current value assessment. With the inflow of urban professionals willing to pay high prices for a location near the downtown core, house prices have increased considerably during the last decade and property taxes have increased accordingly. Despite the fact that most first-generation Portuguese are homeowners and mortgage-free, they are "land rich and cash poor." Although the city offers some financial relief from property tax increases for seniors and low-income homeowners, it is not sufficient to compensate fully for high taxes and increased expenditures on maintenance and utilities. In this way, the city may be inadvertently encouraging gentrification.

More generally, many respondents viewed gentrification as a broader societal problem with implications for the well-being of the city:

> For the country itself, for the society at large, the fact that you have all those small apartments that gradually are being wiped out, that takes away the ability of the single person, the single mother, to be able to live in the centre of the city is bad. So it's good for the homeowner who owns right now to sell for a fortune … but as a city I am afraid that it makes the centre of the city completely out of reach for the working class. I don't see this area any more being an immigrant reception area as it was in the past … gone forever!

The respondents also noted that the loss of affordable housing, while of concern, is likely to be a much more serious problem in the future. Currently, the majority of urban professionals in West Central Toronto live on the periphery of Little Portugal. However, more intense gentrification in Little Portugal is just a question of time, since the process is well established east and west of Little Portugal – Little Portugal, sandwiched between the two areas, will be next. Even if some of the houses are not attractive architecturally, they can be torn down and replaced with more modern structures.

It is also unlikely, given their limited economic resources, that the Portuguese-speaking newcomers from Brazil and Africa can interrupt the progress of gentrification in West Central Toronto. Not only must they ultimately seek housing elsewhere in Toronto, but also, like many immigrant groups who came before them, this may be their preference. Similarly to many immigrants from Portugal and other European countries these groups have expressed a desire to own property, but in contrast to earlier arrivals from Portugal they wish to live in neighbourhoods without a high concentration of people from their own ethnic background and where the majority speaks English (Teixeira 2008, 270). For these groups, living in an area where Portuguese is not the dominant language is unlikely to affect their integration prospects. The Brazilians and probably the Africans are most concerned about the educational achievement of their children in the broader society.

### *Commercial and Industrial Displacement versus Stabilization of Declining Areas and Encouragement of Further Development*

In addition to residential displacement, gentrification can result in the displacement of commercial and industrial activities and the loss of working-class shopping and job opportunities. Gentrifiers, who have different tastes than the working-class population and more money to spend, demand more upscale goods and services, resulting in what Lees et al. (2007, 131) call "retail gentrification" or "boutiqueification." As Ley (1996) notes, the nature of retailing in gentrifying neighbourhoods also changes as these areas shift from the pioneer stage of gentrification to more advanced stages and the original "hippy" retailing is replaced by stores offering more luxurious goods and services.

At the same time, industrial uses either close down completely or move elsewhere, primarily because of increased land costs, obsolete factory space, and the obtrusive nature of many industrial activities that can lead to conflict and tension with middle-class professionals, resulting in a lack of political support for industrial uses. As well, many industrial buildings are ideal candidates for residential loft conversion, a housing form that is particularly attractive to gentrifiers. Industrial displacement results in the loss of good central-city manufacturing jobs that are often replaced by low-paid jobs in the informal sector. Ironically, these are often labourers who find employment in the renovation industry in gentrifying areas of the city.

In contrast to negatives associated with the displacement of commercial and industrial activities, the most positive result of gentrification is often claimed to be the stabilization of declining areas and the encouragement of further development, sometimes with public subsidy and support. Most obviously, gentrifying homeowners and retailers renovate their properties and thereby enhance the image of the neighbourhood, leading to further investment as confidence builds about positive outcomes in the area. Gentrifiers are also effective lobbyists and have the ability to direct public funds to physical and social improvements in their neighbourhoods. Business owners can also be important catalysts for enhanced residential gentrification. From a negative perspective, however, as gentrification accelerates, more low-income residents and marginal businesses upon which local residents depend are displaced from the neighbourhood.

Almost 40 per cent of the Portuguese respondents expressed concern about the future of Little Portugal's ethnic economy. Some strongly believe that the arrival of commercial gentrification and increased property taxes have the potential to displace Portuguese entrepreneurs. Portuguese businesses in Little Portugal are still doing well economically, in part because the majority of entrepreneurs own the buildings they occupy, but the number of Portuguese businesses in the area has decreased.

The reasons are varied. Some Portuguese businesses decided to follow the Portuguese who moved to the suburbs, while an increasing number of Portuguese entrepreneurs are retiring and their children show no interest in continuing the business. Other respondents said that retail gentrification is pushing out some Portuguese entrepreneurs, particularly those who rent the buildings in which their businesses are located.

Several respondents took a more positive view and argued that gentrification could promote the "stabilization of declining areas" and encourage further residential and commercial development of Little Portugal. Despite the extraordinary work by Portuguese homeowners in renovating their houses and ultimately improving the quality of life in Little Portugal, respondents recognized that some retail facilities, especially west of Dufferin St, need rehabilitation and investment. In that regard it was suggested that Portuguese entrepreneurs could cater more directly to the gentrifiers by diversifying their businesses and giving a "facelift" to the facades of their stores.

Most respondents admitted, however, that gentrifiers, who have the capital, the aesthetic taste, and the political acumen to demand high-quality municipal services, could best achieve neighbourhood rehabilitation. As one respondent, when describing the gentrifiers in positive terms, noted:

> I have a problem with a sort of boxing people as gentrifiers because – I mean – I think that people have a need for beauty and [that] shouldn't be [a] bad thing because I like to see actually Little Portugal looking a little

> bit more beautiful ... So for me, I'm feeling that there's a bad connotation to the word gentrification and I think there's also something that I appreciated about, you know, well educated people who want to live in a beautiful neighbourhood.

Another respondent noted the ability of gentrifiers to get the attention of city officials, but was a little less positive about the outcome:

> Urban professionals arrive in the area and they demand changes. Some would say positive ones ... They know how to get around ... how to get the urban planning department to do it ... how to get to the city councilors to do [it]. They have the power, they know how to do it plus they have the time and the knowledge. Because of their complaining, for example "one hour parking signs" in Little Portugal streets ... It limits the noise, the traffic ... but you can easily get a ticket when visiting the family ... Now you think twice before you come to the Portuguese feasts/festivals ... They are good in lobbying.

Overall, this group of respondents sees gentrification as an important driving force for further development, both in Little Portugal and in neighbouring areas of West Central Toronto.

### *Community Resentment and Conflict versus Increased Social Mix*

In the early stages of gentrification there will likely be resentment by existing residents, especially renters who are impacted by increased housing costs. Existing residents may also resent more general aspects of neighbourhood change, including the increased cost of upscale goods and services and an escalating loss of control over changes in their neighbourhood. At the same time, there may be a lack of tolerance by gentrifiers for existing working-class residents, often ethnic minorities, who do not share the same values and customs as themselves. On the other hand, some commentators point to the advantages of increased social mix in areas impacted by gentrification, although others are sceptical. The advantages of greater social interaction and cultural diversity are often expressed by governments as a way of reducing social segregation and the assumed negative consequences of spatially embedded disadvantage. As noted by Lees et al. (2007), governments in the United Kingdom, the Netherlands, and the United States have all used this argument as a means of promoting "positive gentrification."

Empirical evidence, however, suggests that gentrification does not necessarily result in increased social diversity or that social mix is a cure-all for the negative effects of gentrification (Lees 2008). Attitudes of gentrifiers towards the incumbent population are also mixed. Relying on interviews with a sample of residents in new non-luxury infill condominiums in Montreal's inner city, Rose (2004) found that attitudes towards social diversity and social housing run the gamut from "egalitarian" to "tolerant" to "NIMBY." Concerning issues of spatial proximity and social distance, Rose notes that in Montreal's Saint-Louis neighbourhood gentrifiers appreciate the fact that their working-class Portuguese neighbours maintain their property even though there is little communication between them, a situation that Germain and Rose (2000), in the context of multiethnic sociability in Montreal's public places, refer to as a "peaceful but fairly distant coexistence" (245).

Almost 40 per cent of the Portuguese respondents noted "community resentment and conflict" as a negative aspect of gentrification in Little Portugal. Some respondents argued that the newly arrived gentrifiers are an elitist alien group who form their own "world" (a white-collar one), while the Portuguese form another "world" (a working-class or blue-collar one). As some respondents put it, the gentrifiers come to Little Portugal in search of cultural diversity, but they end up recreating their own cultural territory, ultimately leading to social distance and in some cases tension between the Portuguese and the gentrifiers:

> Yes – exclusion ... They [urban professionals] have different uses, customs, life styles ... Also their status is different from us Portuguese. They normally spend little time at home ... They are professionals and they travel a lot. Very distinctive, even by the cars they drive you can tell them, these are people with a lot of money. They have no time and don't share interests with Portuguese immigrants ... This is another form of "ghettoization" and the Portuguese already feel insecure.

Tension between the two groups is often expressed in lifestyle or cultural differences, as indicated by the following encounter in a local park:

> They [gentrifiers] are picky. In June at the Portuguese Parade I was at the Trinity Bellwoods Park where we had the bands playing and a lot of other activities and a gentrifier complained because we were making too much noise. We have Portuguese here for 50 years and this type of party for decades ... so where is the problem? I can understand their point of view ...

> OK ... If I was a gentrifier and I paid a million bucks for the home and I am trying to surf the net and there is this noise in the Park ... It's terrible ... This is the conflict we have now ... We are very happy when we sell our houses for $750,000, but we get upset when these guys come with "decibel meters" measuring the noise ... Two worlds it seems.

On the other hand, several Portuguese respondents argued that the arrival of gentrifiers has been beneficial to the Portuguese community in general, and particularly to those who have a good knowledge of English, by helping them mix and thus integrate more broadly into Canadian society. They further argue that gentrifiers bring more cultural diversity to Little Portugal, thereby helping the Portuguese break down the closed ethnic enclave that has characterized the community since the 1960s, when the Portuguese first arrived in Toronto:

> It's positive, the arrival of gentrifiers into Little Portugal ... It destroys the "ghetto" that we had for decades. We are here highly concentrated and Portuguese didn't need to learn English because their lives were done in Portuguese within the Portuguese community ... Now our "ghetto" is diluting/disintegrating and we are integrating ourselves more into the Canadian society. There is no conflict with them ... and all of us get very happy when we sell our houses for very good money.

The gentrifiers have also been participating more in the life of Little Portugal by shopping in Portuguese businesses and taking part in Portuguese cultural events. Some respondents suggested that there are already signs of a shift in thinking among first-generation Portuguese immigrants, who are opening themselves more to people of different ethnic and socio-economic backgrounds and gradually embracing the idea of greater cultural diversity in the neighbourhood. For the majority of Portuguese, gentrifiers are also good neighbours, people who really care about their houses and the neighbourhood.

As indicated by the quotations, there is ambivalence about the arrival of gentrifiers in Little Portugal. There is general acceptance of the gentrifiers by the Portuguese, but at the same time there is a degree of social isolation between the two groups, and where cultures clash, as in the use of a park, tensions come to the surface. Regardless, the gentrifiers are well received by Portuguese homeowners who sold their property for a relatively high price.

## Conclusion

In the absence of new waves of Portuguese immigration and the further suburbanization of current Portuguese residents, West Central Toronto, and Little Portugal more specifically, has continued to lose Portuguese residents. Some respondents noted that every time a Portuguese sells a house in the area, the same house is bought by gentrifiers, speculators, or members of other immigrant groups. Thus, the number of Portuguese homeowners will decrease with time and with it some of the existing rental units as well as the informal renting that characterizes Portuguese homeowners in the area. Consequently, Little Portugal will likely decrease in importance as an institutionally complete Portuguese enclave. Indeed, this decline would be even more marked were it not for the unexpected arrival of Portuguese-speaking immigrants from former Portuguese colonies such as Brazil, Angola, and Mozambique.

A key unknown is the future of those areas in figure 5.3 that have not gentrified and are not categorized as having potential for gentrification. As noted earlier, these are areas that generally lack the spacious and architecturally appealing Victorian housing demanded by gentrifiers and suffer from environmental negatives such as proximity to rail lines and industry. On the other hand, industry is rapidly abandoning the area and there is a possibility that some of these older industrial buildings will be converted to upscale residential lofts. Alternatively, houses that are not architecturally appealing to gentrifiers may be torn down and replaced by others that are.

Almost all respondents agreed that gentrifiers are becoming the defining population in the neighbourhood. The Portuguese regard this group with mixed feelings. On the one hand, they are valued for rejuvenating the housing stock in the neighbourhood and patronizing local businesses. On the other hand, their desire to live in a multicultural neighbourhood has driven up the cost of housing in Little Portugal and the rest of West Central Toronto such that increasingly fewer Portuguese can afford to live there. While Portuguese homeowners often benefit from healthy profits upon the sale of their homes, Portuguese in the neighbourhood are under few illusions about the implications of gentrification for the long-term viability of their community. Some respondents pointed to the fate of nearby Little Italy, where gentrification has contributed to the exodus of many Italians, as the template that Little Portugal will follow in the years to come. A few of the respondents, however, remained optimistic about the future of Little Portugal and suggested that like the commercial axis of Little Italy (College St), the commercial axis of Little Portugal (Dundas St) will retain much of its Portuguese atmosphere and will continue being a magnet for Portuguese from the Toronto area.

More generally, this chapter has attempted to bring together the literature on gentrification and immigrant settlement to consider the impact of gentrification on ethnic enclaves in the inner city. The process is complex and requires further study. In Toronto, recent immigrants from a variety of origins are gradually replacing a well-established Portuguese group. At the same time, gentrification is expanding within this area of the city. We know a considerable amount about the demographic characteristics of the Portuguese and their attitude towards gentrification, but relatively little about the newly arrived immigrant groups and the gentrifiers. The different characteristics of these groups, the interaction between them, and how each group is reacting to neighbourhood change are important avenues for further research.

## NOTES

1 The final, definitive version of this paper was first published in *Urban Studies* 48 (1): 61–83, January 2011 by Sage Publications Ltd, All rights reserved. @ Urban Studies Journal Limited, 2011. It is available at http://usj.sagepub.com/. This is a revised and shortened version of the original article.
2 Unless indicated otherwise, population figures are from the Census of Canada, various years.
3 From Statistics Canada Census Profiles, 2001 and 2006. Total ethnic origin includes respondents who reported more than one ethnic origin. In 2006, 262,230 Canadians reported their ethnic origin as solely Portuguese, while 148,620 indicated Portuguese and one or more additional origins. In 2001, less than 10 per cent of the Portuguese population in West Central Toronto and Little Portugal offered more than one response to the ethnic identity question, suggesting that Portuguese identity is strong in this part of the city.
4 In 1971 the only variable available is Portuguese mother tongue (the language first learned and still understood). Given the recency of Portuguese immigration to Canada in 1971, we believe that mother tongue data are an accurate representation of the Portuguese in that year. For subsequent censuses Portuguese ethnic origin data are available. The

latter are used to document settlement patterns in 1981 and 1991 (single ethnic origin) and 2001 and 2006 (total ethnic origin). Ethnic origin figures from the 2011 National Household Survey (the replacement for the long-form census) are unlikely to be as reliable as those from previous censuses.

## REFERENCES


Atkinson, R. 2000. "Measuring Gentrification and Displacement in Greater London." *Urban Studies* (Edinburgh, Scotland) 37 (1): 149–65.

Atkinson, R. 2004. "The Evidence on the Impact of Gentrification: New Lessons for the Urban Renaissance?" *European Journal of Housing Policy* 4 (1): 107–31.

Barbosa, R. 2009. "Brazilian Immigration to Canada." *Canadian Ethnic Studies* 42 (1–2): 215–25.

Betancur, J. 2009. "Gentrification and Community Fabric in Chicago." *Urban Studies* (Edinburgh, Scotland) 48 (2): 383–406.

Boyd, M. 2005. "The Downside of Racial Uplift: The Meaning of Gentrification in an African American Neighborhood." *City & Society* 17 (2): 265–88.

Caulfield, J. 1994. *City Form and Everyday Life: Toronto's Gentrification and Critical Social Practice*. Toronto: University of Toronto Press.

Filion, P. 1991. "The Gentrification–Social Structure Dialectic: A Toronto Case Study." *International Journal of Urban and Regional Research* 15 (4): 553–74.

Freeman, L. 2006. *There Goes the 'Hood: Views of Gentrification from the Ground Up*. Philadelphia: Temple University Press.

Germain, A., and D. Rose. 2000. *Montréal: The Quest for a Metropolis*. Chichester, Eng.: Wiley.

Glass, R. 1964. "Introduction: Aspects of Change." In *London: Aspects of Change*, ed. The Centre for Urban Studies, University of London. London: MacGibbon & Kee.

Hulchanski, J.D. 2004. "A Tale of Two Canadas: Homeowners and Renters." In *Finding Room: Policy Options for a Canadian Rental Housing Strategy*, ed. J.D. Hulchanski and M. Shapcott, 81–8. Toronto: CUCS Press.

Krase, J. 2005. "Poland and Polonia: Migration, and the Re-Incorporation of Ethnic Aesthetic Practice in the Taste of Luxury." In *Gentrification in a Global Context: The New Urban Colonialism*, ed. R. Atkinson and G. Bridge, 185–208. London: Routledge.

Lees, L. 2000. "A Reappraisal of Gentrification: Towards a 'Geography' of Gentrification." *Progress in Human Geography* 24 (3): 389–408.

Lees, L. 2007. "Afterword." *Environment and Planning A* 39 (1): 228–34.

Lees, L. 2008. "Gentrification and Social Mixing: Towards an Inclusive Urban Renaissance?" *Urban Studies* (Edinburgh, Scotland) 45 (12): 2449–70.

Lees, L., T. Slater, and E. Wyly. 2007. *Gentrification*. New York, London: Routledge.

Ley, D. 1996. *The New Middle Class and the Re-Making of the Central City*. Oxford: Oxford University Press.

Marcuse, P. 1986. "Abandonment, Gentrification, and Displacement: The Linkages in New York City." In *Gentrification in the City*, ed. N. Smith and P. Williams, 153–77. Boston: Allen and Unwin.

Moore, K.S. 2009. "Gentrification in Black Face? The Return of the Black Middle Class to Urban Neighborhoods." *Urban Geography* 30 (2): 118–42.

Nyden, P., E. Edlynn, and J. Davis. 2006. *The Differential Impact of Gentrification on Communities in Chicago. Report to the City of Chicago Commission on Human Relations*. Chicago: Loyola University Chicago Centre for Urban Research and Learning.

Rose, D. 1984. "Rethinking Gentrification: Beyond the Uneven Development of Marxist Urban Theory." *Environment and Planning D: Society & Space* 2 (1): 47–74.

Rose, D. 2004. "Discourses and Experiences of Social Mix in Gentrifying Neighbourhoods: A Montréal Case Study." *Canadian Journal of Urban Research* 13 (2): 278–316.

Shirley, R. 1999. "Brazilians." In *Encyclopedia of Canada's Peoples*, ed. P.R. Magocsi, 273–82. Toronto: University of Toronto Press.

Skaburskis, A. 2012. "Gentrification and Toronto's Changing Household Characteristics and Income Distribution." *Journal of Planning Education and Research* 32 (2): 191–203.

Slater, T. 2004. "Municipally Managed Gentrification in South Parkdale, Toronto." *Canadian Geographer / Géographe canadien* 48 (3): 303–25.

Slater, T. 2006. "The Eviction of Critical Perspectives from Gentrification Research." *International Journal of Urban and Regional Research* 30 (4): 737–57.

Slater, T., W. Curran, and L. Lees. 2004. "Guest Editorial. Gentrification Research: New Directions and Critical Scholarship." *Environment and Planning A* 36 (7): 1141–50.

Teixeira, C. 2007a. "Residential Experiences and the Culture of Suburbanization: A Case Study of Portuguese Homebuyers in Mississauga." *Housing Studies* 22 (4): 495–521.

Teixeira, C. 2007b. *Toronto's Little Portugal: A Neighbourhood in Transition*. Research Bulletin 35. Toronto: Centre for Urban and Community Studies, University of Toronto.

Teixeira, C. 2008. "Barriers and Outcomes in the Housing Searches of New Immigrants and Refugees: A Case Study of 'Black' Africans in Toronto's Rental Market." *Journal of Housing and the Built Environment* 23: 253–76.

Walks, R.A., and R. Maaranen. 2008. *The Timing, Patterning and Forms of Gentrification and Neighbourhood Change in Montreal, Toronto, and Vancouver, 1961 to 2001*. Research Paper 211. Toronto: Centre for Urban and Community Studies, University of Toronto.

# 6 The Good, the Bad, and the Suburban: Tracing North American Theoretical Debates about Ethnic Enclaves, Ethnic Suburbs, and Housing Preference

VIRPAL KATAURE AND
MARGARET WALTON-ROBERTS

The implementation of more liberal immigration policies in the 1960s and increased immigration from non-European countries to Canada and the United States has led to transformations in urban settlement patterns. The diversity of immigrants has informed extensive research, most notably on the process of immigrant suburbanization. Although the settlement experiences of immigrant groups in both countries have been diverse, there seems to be a trend towards heterogeneous suburban landscapes, akin to ethnic enclaves on the periphery of major immigrant receiving cities. This chapter begins with an overview of research on immigrant settlement in suburbs and ethnic enclaves, which then leads to a discussion on how residential settlement patterns and housing trajectories in ethnic enclaves are a critical indicator of immigrant integration. We investigate the advantages and disadvantages of ethnic-enclave housing location choices of immigrants based on a recent case study of South Asians in a Canadian ethnic suburb (Kataure and Walton-Roberts 2013). The case-study research suggests that ethnic clustering in Canadian suburbs is preference based, which partly explains the strength of continued suburban ethnic concentration.

## Theoretical Grounding

Although various theories on the residential experiences of immigrants exist, no single model completely captures the diverse and complex spatial experiences of different ethno-cultural groups across North America. Nonetheless, table 6.1 provides an overview of the main settlement theories and their application to the US and Canadian contexts. Each section periodizes theories based on changing immigrant and urban settlement

Table 6.1 Evolution of ethnic urban settlement theories in the United States and Canada

| Theory / concept | Scholar(s) | Overview | United States | Canada |
|---|---|---|---|---|
| *Post-war ethnic settlement: Urban areas* | | | | |
| Human ecology, concentric zone theory, and assimilation | Burgess (1925) Park (1926) Peach (1975) | Economically underprivileged immigrants settle close together until socio-economic status improves, providing the ability to assimilate into mainstream society. Theory assumes segregation is voluntary. | This theory was based on Chicago and used principles of ecology to understand human communities and change. European immigrants eventually left the inner city by 1960s. This process is not the case for visible minorities today. | Early growth patterns of Montreal showed confirmation of this theory, where the English were segregated from the French on the periphery, and immigrants (Blacks, Italians, and Chinese) were in the central business district (Dawson and Gettys 1929). |
| Social class division | Duncan (1957) | Lower classes (including immigrants) occupy the old, cheap housing in downtowns and higher-status groups (such as Northern Europeans) move to the suburbs. | Increases in income have shown only moderate influence over residential segregation, especially for Blacks, where segregation persists (Massey and Denton 1993; Logan, Stults, and Farley 2004). | Social rank, family status, and ethnicity factors were examined in Winnipeg and found characteristics of social class factors (Driedger 1991). |
| Multiple-nuclei theory | Hoyt (1939) Harris and Ullman (1945) | Nuclei settlements shape the degree of interaction between group members in enclaves. Socio-economic status, ethnicity, history, topography, economy, politics, and culture influence residential location. | This was evident in a study of Los Angeles, where social rank, urbanization, and segregation accounted for varitations between census tracts (Shevky and Bell 1955). | Canada's suburban landscapes developed in the late 20th century, creating decentralized suburban nodes. A study analysing the social ecology of Toronto between 1951 and 1961 resembled this model (Murdie 1969). |

Table 6.1 (*continued*)

| Theory / concept | Scholar(s) | Overview | United States | Canada |
|---|---|---|---|---|
| *Effects of racialization* | | | | |
| Ghetto | Philpott (1978)<br>Massey and Denton (1993)<br>Peach (1996) | Formed through exclusion and discrimination from housing and labour markets by the host society regardless of personal resources or preferences, and difficult to escape (Walks and Bourne 2006, 276). Can contain extreme poverty rates greater than 40%, which is often the threshold to identify a ghetto. | The arrival of racial minorities stimulated negative attitudes, stereotyping, and racist views (Harwood 1986), which led to harsh living conditions of immigrant groups in specific neighbourhoods (Massey 1985). Chicago in the 1930s had over 90% of the Black population living in areas forming over 80% of the population (Philpott 1978). Today, Black populations still continue to face ghettoization in the United States (Ellis, Wright, and Parks 2004). | Historically were created as a result of discriminatory laws that prevented land ownership and professional employment for racialized immigrants (Anderson 1987). Vancouver's Chinatown demonstrates how legislation and restrictions formed by the British Columbia legislature in the late 19th and early 20th century created a population outside of mainstream society. There is less evidence today to support ghetto formation in the Canadian context (Walks and Bourne 2006). |
| Ethnic enclave | Massey (1985)<br>Breton (1964)<br>Anderson (1987)<br>Logan, Alba, and Zhang (2002) | Area of high-single ethnic-density neighbourhoods surrounded by zones of lower ethnic concentrations. Settlement is seen as reflecting group's cultural values and symbols. | Segregation is considered natural, but only temporary given immigrants' market resources and ethnically bound cultural and social capital (Massey 1985). Those with more financial resources and mainstream jobs avoid ethnic zones, and the second generation also moves away (Logan, Alba, and Zhang 2002). | Segregation is considered a matter of choice rather than outcome of discrimination. Members of the second-plus generation continue to settle in ethnic enclaves to maintain cultural heritage, language, norms, etc. (Qadeer, Agrawal, and Lovell 2010). |

| | | | | |
|---|---|---|---|---|
| *"New" immigration and globalization: Suburbs* | | | | |
| Social mosaic hypothesis and dispersed city | Bourne (1989) Doucet (1999) | Changing socio-economic status of newly arrived immigrants represents increased socio-economic capital. Homogeneous suburbs become immigrant reception zones, creating heterogeneous communities. | Post-1965, wealthier immigrants left concentrated ethnic communities for more affluent diverse neighbourhoods in suburbs, such those of New York and Los Angeles (Logan, Alba, and Zhang 2002). | The increase in highly skilled immigrants from diverse origins via the "points system," allowed immigrants with greater capital to move directly to the suburbs. Studies confirm this in Vancouver (Hiebert 1994), Montreal (Archambault et al. 1999), and Toronto (Doucet 1999). |
| Ethnoburb | Li (1998) | Ethnic clusters found in suburban residential areas with cultural values and symbols reflecting community's ethnic character. Can be multi-ethnic communities where newer immigrants with various origins and statuses settle directly to the suburbs. | In the 1960s, more upwardly mobile Chinese and higher-capital immigrants moved out of Chinatowns to the suburbs for better housing and neighbourhoods (Li 1998). Ethnoburbs coexist with traditional ethnic ghettos in inner cities, but the local ethnic economy in the ethnoburbs play a larger role in the global economy. | Immigrants moved to the suburbs as a form of integration, but not necessarily assimilation, since ethnic identity and clustering persists (Zucchi 2007). Jewish immigrants in Toronto and South Asian immigrants in Vancouver show resegregation occurring in the suburbs through a desire for new homes and communal focus on religious and cultural institutions (Balakrishnan, Maxim, and Jurdi, 2005; Qadeer, Agrawal, and Lovell 2010). |

Table 6.1 (*continued*)

| Theory / concept | Scholar(s) | Overview | United States | Canada |
|---|---|---|---|---|
| Segmented assimilation theory | Gans (1992)<br>Portes and Zhou (1993)<br>Perlmann and Waldinger (1997)<br>Boyd (2002) | Three possible relationships for the second generation: (1) traditional upward mobility and assimilation; (2) opposite downward mobility and ethnic segregation creating second generation "underclass"; (3) the New Assimiliation Theory: upward mobility while sustaining ethnically homogeneous immigrant communities. | The native-born population of African Americans and Puerto Ricans often constitute the underclass (Portes and Rumbaut 2001). Others have argued that the second generation will experience upward mobility, since parents begin at the bottom of the ladder (Farley and Alba 2002). | Canadian research does not show visible minorities contributing to the underclass, given their higher educational attainment levels (Boyd 2002). Immigrant groups maintain ethnic enclaves longer than traditional theories suggest, forming more durable ethnic communities, giving rise to positive and growing enclave effects (Balakrishnan, Maxim, and Jurdi 2005; Boyd 2002; Kalbach 1990; Zhou 1997). |

patterns. The first wave of these theories emerged in the post-war period, focusing on the process of assimilation for European immigrants mainly settling in urban centres. The second period examines how, with the arrival of more visible-minority immigrants, settlement changed from models based mostly on exclusion to those incorporating elements of voluntary clustering. The third period considers the arrival of immigrants with higher capital moving to the suburbs, and the experience of the second generation. The section following provides a more focused analysis that charts the chronological evolution of these main theorizations, leading into a discussion on how suburban ethnic-enclave immigrant settlement can have positive and negative outcomes.

### *Early Settlement Patterns and Assimilation*

Research on immigrant settlement patterns originated from the Chicago School of Urban Sociology, where Burgess (1925) and Park (1926), using the Spatial Assimilation theory, introduced a human ecology model known as the Concentric Zone theory. This theory suggested the integration of immigrants into the host society as they moved away from the clustered inner-city ethnic community out towards suburban settlements. Duncan (1957) added to this and argued that lower classes, such as immigrants, live in cheap, "slum" housing downtown, and by the second generation, as their social class improves, the group would move to the suburbs. In these models, inner-city ethnic housing concentrations are theorized as temporary areas of residence inhabited during the early phases of settlement, with end-point settlement in affluent suburban communities displaying successful assimilation (Massey 1985; Massey and Denton 1987). "Traditional" enclaves were created before the Second World War due to deliberate housing discrimination, and laws resulting in institutions and internal labour markets that acted as a form of protection and survival for ethnic communities (Li 1998). Subsequent urban models employed the Multiple Nuclei approach, which extended the enclave perspective to suburban regions as well as the inner city.

### *Immigrant Suburbanization in North America*

The movement of immigrants to the suburbs began as early as the 1900s. The increased use of automobiles by the 1930s allowed immigrants to relocate to the rural-urban fringe following the decentralization of jobs, specifically manufacturing (Harris and Lewis 2001). This movement, however, did not necessarily imply social or residential integration, since suburban ethnic enclaves formed through the re-establishment of immigrant clusters (Harris and Lewis 2001).

After the Second World War, suburbs, which were previously regarded as strictly serving residential functions, began to serve commercial functions as well and saw labour market growth, leading to a general population shift to the periphery. After 1965, traditional enclaves became saturated, and "new" enclaves formed through the presence of newly arrived immigrant and refugee populations (Li 1998). The improved socio-economic status of immigrants arriving from new source regions such as Asia, Africa, the Caribbean, the Middle East, and Central/South America, allowed immigrants to bypass the traditional urban-core spatial assimilation process, and move directly to the urban fringe. This "internationalization" of immigration has changed the socio-demographic profile of North America (Teixeira 2007, 496–7). Affluent business immigrants, mostly from Asia, have settled directly in single detached dwellings in the suburbs; however, refugees and other poor immigrants often live in low-rent and low-quality apartments or housing (Murdie and Teixeira 2004).

Continued suburban segregation is not always the case for all immigrant groups, since some disperse after improvements in socio-economic position, while others were never heavily clustered to begin with (Balakrishnan and Hou 1999; Teixeira 2007).[1] Overall, suburban ethnic concentration can be classified as generally heterogeneous across different national or ethnic groups, and at times segregated by class (Zucchi 2007, 20). This has resulted in complex and unique suburban settlement forms that reflect diverse immigrant populations (Murdie and Teixeira 2004, 1). Today, the suburban geographies of both Canada and the United States reflect sprawling mixed-use suburban zones that are automobile-dependent, highway-oriented, and generally autonomous from central cities (Fiedler and Addie 2008, 2).

### *Origins of the Suburban Ethnic Enclave*

Ethnic enclaves can be defined as residential areas where a single ethnic group is prominent. Three factors are necessary for ethnic enclave development: a high volume of immigration from a given area with a sustained flow of immigrants over time, and residential concentration (Boyd 2002). These clusters often emerge through chain migration (such as family reunification and sponsorship policies), changing labour market processes (from urban to suburban cities), and the development of ethnic businesses and community organizations (Li 1998; Murdie and Skop 2012; Teixeira 2007). They are accompanied by various levels of "institutional completeness" (Breton 1964); economic activities, local services, organizations, or institutions that address the specific needs and reflect the culture of the ethnic group. Settlement geographies are not merely the product of existing economic conditions, but are also shaped by socio-spatial context and the power of collective cultural identity.

The economic function of the suburbs can be as important as the residential one for non-traditional immigrants. Using the case of ethnic Chinese immigrant groups Li (1998) explores "ethnoburbs," suburban ethnic settlements characterized by integrated ethnic economies specific to one ethnic group and their ties to the global economy. Li's (1998) study examined the demographic profiles and socio-economic characteristics of ethnic Chinese in American metropolitan areas, specifically Los Angeles. Ethnoburbs include multiethnic communities where one visible minority group is significantly concentrated complete with businesses, institutions (such as churches, clubs, associations, etc.), social networks, and goods or services reflecting the community's character, although not necessarily constituting a majority of the population. She argues that the ethnoburb formed through profound changes in the global economy and industry, changes in national immigration policies, foreign direct investment, the growth in multinational corporations, and geopolitical change in the sending region (482).

The settlement choices and experiences of immigrants can also affect the housing outcomes of their children, known as the second generation. The Segmented Assimilation theory proposed by Portes and Zhou (1993) was among the first to propose diverse residential patterns for immigrant offspring based on social class and economic and racial stratification, rather than just the assumption of a linear assimilation pattern. Portes and Zhou (1993), among other scholars (Boyd 2002; Gans 1992; Perlmann and Waldinger 1997), discuss three possible relationships the second generation could exhibit. The first is traditional upward mobility and linear assimilation to the host society. The second is opposite, or downward mobility due to racialization experiences, also known as second generation "decline," "revolt," or "underclass." Lastly, the New Assimilation theory proposes the possibility of upward mobility, while still sustaining ethnically homogeneous immigrant communities rather than spatial assimilation (Logan, Alba, and Zhang 2002). In addition, there is the possibility of spillover, where the existing enclave is no longer able to accommodate all members, and thus settlement into adjacent non-ethnic neighbourhoods occurs (Massey and Denton 1987).

The concept of the "underclass" has become tangled between academic, media, and political debates, resulting in an unclear understanding of who constitutes the underclass, what criteria are used, and where they are applied, especially when examining differences between Canada and the United States (Waldinger and Feliciano 2003, 11). The underclass scenario of the Segmented Assimilation model uses indicators that (a) are often inconsistently applied between studies, (b) assume low levels of parental and community resources, and (c) assume highly racialized populations (particularly Black populations) facing structural barriers that differ from the majority (Miles 1989). The historical context that led to the development of institutional barriers and the factors influencing race relations differ between Canada and the United States, since Canada did not experience succession over slavery (see the 'Important Distinctions' section below), and the presence of an underclass in Canada is more strongly linked to adverse socio-economic factors rather than visible-minority status (Mendez 2009, 94; Walks and Bourne 2006). There currently remains a limited Canadian reference group for the second-generation underclass model, since many minority groups are not large enough in size, and immigrants in Canada tend to improve upon their residential location outcomes over time, although discrimination and racialization still exist (Mendez 2009). These issues are important to consider when interpreting the findings below.

### *Segregation: Advantages and Disadvantages*

The main ethnic enclave debate is whether they represent external structural inequality or internal social cohesion, and concerns the nature of the actual and perceived consequences associated with clustering (Agrawal 2006; Hiebert 2000, 31; Lo and Wang 1997; Ray 1999, 82). Below we examine immigrant experiences in the ethnic enclave with a specific focus on homeownership and housing experiences, services and the ethnic economy, occupation and income, social networks, and education.

### HOMEOWNERSHIP AND HOUSING EXPERIENCES

Immigrants, especially visible minorities, may be discriminated against in the housing market in both the renting and buying of housing (see chapters 2 and 3 above). Since immigrant households tend to be larger and have a lower income than the average Canadian household, immigrant groups may be susceptible to overcrowded housing conditions, leading to discrimination from landlords (Hiebert 2000; Qadeer 2003; Teixeira and Murdie 2004). Affordability problems can become exacerbated when housing availability is reduced due to increased demand and restricted supply, especially in non-market or assisted housing (Wayland 2007). Finally, reduced assimilation can occur when ethnic enclaves only rent or sell to individuals of the same ethnic group or when majority populations try to maintain the exclusivity of a neighbourhood. Clustering, by contrast, allows minority groups to defend against racially motivated bias and provides social support within their co-ethnic group (Bauder and Sharpe 2002). Alternatively, majority groups may feel justified in perpetuating stereotypes and prejudices that would otherwise be considered inappropriate (Balakrishnan and Hou 1999, 202; Qadeer 2003).

Limited access to affordable housing might be mitigated when immigrants seek housing within ethnic enclaves (Bauder and Sharpe 2002; Papillon 2002). Scholars have pointed out that immigrants' strong desire for homeownership allows them to become homeowners within the first generation (Kim and Boyd 2009, 12), and makes them more likely to own their homes than the Canadian-born population (Hiebert 2000). Most studies confirm that Chinese and South Asian populations in Canada and the United States tend to have high homeownership rates, whereas Mexican and Black ethnic groups perform poorly in the housing market, displaying higher tendencies to enter the American "underclass" (but this is less evident among Black Canadians) (Boyd and Grieco 1998; Farley and Alba 2002; Haan 2005, 435; Li 1998). Evidence suggests that immigrants spend more of their income on housing than non-immigrant populations, raising concerns about poverty (Murdie and Teixeira 2004), although poverty does not necessarily correlate with racial concentrations in Canada (Wayland 2007).

### SERVICES AND THE ETHNIC ECONOMY

The degree of institutional completeness present in a neighbourhood depends upon the number of services the community can provide to its co-ethnic members (Breton 1964), and plays an important role in settlement experiences. Ethnic enclave neighbourhoods allow newcomers with limited skills to have access to more culturally sensitive skills training and education programs (Papillon 2002). Religious, educational, and welfare institutions also create and maintain social interaction between minority groups, which act as a form of support for continuing services that may otherwise not be available (Balakrishnan and Hou 1999). The opportunities generated by ethnic businesses can allow for minority groups to succeed by offering familiar and essential goods and services in their own language (Fong 1996; Teixeira and Murdie 2004).

Conversely, ethnic institutions may present unequal access to socioeconomic resources such as education, health, and other public facilities (Balakrishnan and Hou 1999). This is often attributable to unequal budget allocation, poor school facilities, school segregation, and increased dropout rates that can drastically extend the initial integration phase (Balakrishnan and Hou 1999). Ethnic enclaves may also restrict immigrant access to employment, experience exploitation, and delay language attainment (Qadeer 2003; Wayland 2007). Ethnic enclaves may also be perceived as an economic and political threat to the host community, resulting in discriminatory attitudes (Balakrishnan and Hou 1999).

### OCCUPATION AND INCOME

Occupational and income success of immigrants has been an area of concern (see chapters 8 and 9 below). Bauder and Sharpe (2002) explain that income disparities among immigrant and Canadian-born populations reflect differences in human capital as a result of the under-evaluation of skills. Newcomers do fare as well as their equivalent Canadian-born counterparts in terms of employment and income, even though immigrants are often more educated and skilled (Wayland 2007). In extreme cases, poverty may become evident and persist over generations due to ethnic segregation, as seen in the United States (Farley and Alba 2002; Portes and Zhou 1993). It is this concern that differences in immigrant income might be transferred to the second generation which has animated a great deal of North American research. Gans (1992) and Portes and Zhou (1993) believe that there is a growing gap between the achievements of the "old" second generation (European descendents of immigrants who arrived in the early 1900s) and the "new" second generation (whose parents arrived after 1965), where occupational segmentation results in more limited opportunities for upward mobility. However, Canadian research on the second generation has shown that the occupational and educational achievements of immigrant offspring outperform third and higher generations (Reitz, Zhang, and Hawkins 2011; Picot and Hou 2011). Nevertheless, other studies have found that ethnic-minority Canadian and foreign-born immigrants (second and first generations) have lower earnings than the overall White population (Pendakur and Pendakur 2011).

#### SOCIAL NETWORKS

The importance of social networks for the immigration process is evident in both the migration and settlement phase (Massey and Aysa-Lastra 2011). Researchers have considered how immigrants use different types of networks or links, and to what degree these assist them in integration, especially for women, children, seniors, and those not fluent in English (Rose et al. 1998). There is an overwhelming motive to join family or friends due to the economic and psychological support they offer (Qadeer 2003; Wayland 2007).

Studies have also shown that immigrant offspring's educational achievement may even be improved through continued settlement in ethnic enclaves, since tight-knit communities may encourage young immigrants to pursue higher education through the sharing of similar values and obligations, social support systems, a strong sense of family, and the monitoring of children's activities within the community, which supports the Segmented Assimilation theory (Coleman 1990; Portes and Schauffler 1994; Zhou 1997). However, children who do not have exposure to diverse communities could have difficulty improving their education attainment and socio-economic status, and limit cross-cultural understanding (Balakrishnan and Hou 1999; Borjas 2006; Papillon 2002). The importance of social networks in migration and settlement is now widespread in the literature, a recognition that settlement geographies are not merely the product of economic conditions, but are also shaped by socio-spatial context and the power of cultural identity (Ley and Germain 2000; Putnam 2001; Reitz and Sklar 1997).

#### EDUCATION

In Canada, conventional wisdom suggests that the clustering of immigrants is voluntary, and therefore the process offers some *optimistic* outcomes for immigrant settlement (Logan, Alba, and Zhang 2002). Allen (2006) argues that immigrants who bring capital are able to enter the middle class and potentially avoid underclass scenarios. Canadian evidence shows little reason to believe that continued ethnic attachment would limit human capital success, or limit education, especially among South Asian and Chinese populations (Coleman 1990; Qadeer, Agrawal, and Lovell 2010; Reitz and Sklar 1997).[2] Conversely, the American melting pot or assimilation perspective supports the idea that cultural differences should diminish over time to eventually conform to the dominant culture, demonstrating successful integration.[3]

In general, compared to Canada, ethnic enclaves in the United States have traditionally been linked to *pessimistic* views. Some suggest that ethnic segregation increases discrimination in the housing market, leading to ghetto formation and possibly increase prejudice against the group (Galster 1989). After the 1960s Civil Rights Movement and subsequent anti-discrimination legislation in the United States, there was a fundamental shift in how the residential segregation of minorities was perceived; where it was once seen as a result of discriminatory housing markets, it became seen as a preference-based decision (Clark 2009, 339).[4] Even after housing legislation change in the United States, residential concentration still persists (Massey and Denton 1993). Massey and Denton argue that residential segregation can lead to a concentration of low-achieving students, while others suggest this may limit English-language attainment, eventually affecting upward intergenerational mobility (Portes and Zhou 1993).

### *Important Distinctions*

While there is no clear indication as to whether settlement patterns and immigration policies are great predictors of immigrant outcomes, they are important to bear in mind, especially when there are distinct differences between immigration perspectives in both countries (Aydemir and Sweetman 2006). The US case is somewhat differentiated from the Canadian one by its distinctive racial history. In the United States, the experience of the African American population is entrenched in a history of slavery, disenfranchisement, and legislated and de facto racial segregation, although such forms of historical segregation are no longer legislatively upheld (see Darden and Wyly 2010). The ethnic composition of visible minorities also differs, since the United States has a greater percentage of Black and Hispanic population versus Canada, which has a greater South Asian and Chinese population. Similarly, most Blacks in the United States are native born, which is not necessarily the case for

Black communities in Canada. The socio-economic and demographic characteristics of visible minority populations also vary in each country leading to unique settlement patterns, which differ most evidently between Black and Asian populations (Portes and Macleod 1999). The implementation of the 1967 points-based immigration system in Canada led to a more highly educated immigrant population, often more educated than the native-born Canadian population (Boyd 2002; Boyd and Grieco 1998), whereas the percentage of undocumented migrants compared to documented residents (citizens or visa holders) is much higher in the United States than in Canada (Allen 2006). Overall, substantial racial residential separation seems to still remain among groups in Canada and the United States, even when controlling for socio-economic status (Farley et al. 1993; Fong 1996; Kalbach 1990; Massey and Denton 1993). Future immigrants may experience divergent outcomes as circumstancs change in terms of their human and social capital, their location in urban space, and their ethnic background (Farley and Alba 2002).

## Case Study: Brampton's Ethnic Enclave

While discussing the general literature on ethnic enclaves and suburbs in North America is beset by problems of generality, this section examines one Canadian suburb in order to assess these settlement theories in more detail. There has been limited analysis of ethnic enclaves specific to suburban cities, even though such ethnic concentrations are central to outer suburban development in Canada (see chapter 7 for the US Chinese experience). South Asians constitute the largest visible minority group in Canada, and most are of Indian national origin (Statistics Canada 2006). Nonetheless, little research has contributed towards understanding settlement patterns specific to the South Asian community (Qadeer 2003; Ray 1999). Available research has focused on major city centres – Toronto, Vancouver, and Montreal (Agrawal 2006; Fong 1996; Ray 1999; Myles and Hou 2003; Walks and Bourne 2006) – even though visible minorities are increasingly clustering in suburban neighbourhoods (Hiebert 2000; Qadeer, Agrawal, and Lovell 2010; Walks and Bourne 2006).

### CASE STUDY: BRAMPTON

Brampton grew by 20.8 per cent between 2006 and 2011, the highest rate of population growth in Canada. In 2011, over half (53%) of Canada's South Asian population lived in the Toronto Census Metropolitan Area

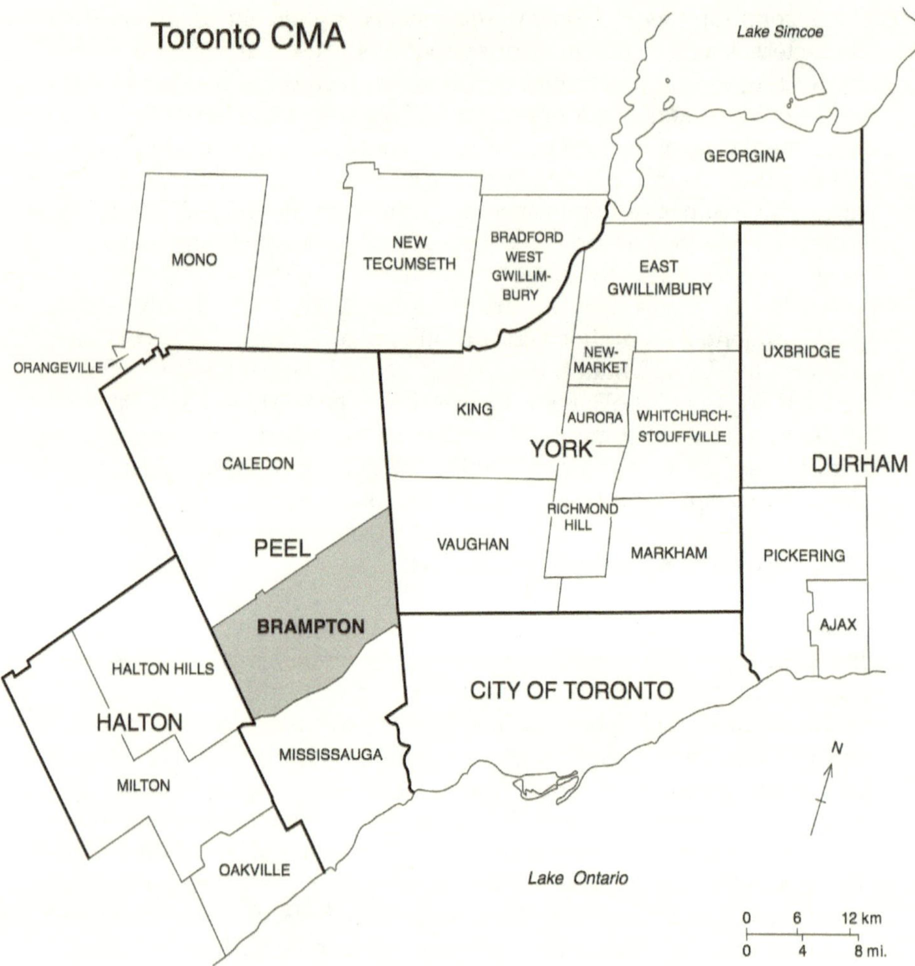


Figure 6.1 The Greater Toronto Area, CMA and municipalities
Source: Metropolitan Toronto Council 1995.

(CMA), which includes the City of Brampton (figure 6.1). Brampton has the largest concentration of South Asians in the Toronto CMA, equal to 38 per cent of Brampton's total population, and 66 per cent of the city's total visible minority population. According to census self-identification questions, 89 per cent of the total South Asian population is of "East

Indian" or "Punjabi" heritage and 57 per cent reported Punjabi as their mother tongue. The number of South Asians living in ethnic enclaves in Brampton increased between 2001 and 2006 from 30 to 49 per cent (Qadeer, Agrawal, and Lovell 2010).

*Methods*

This section reports on aspects of research conducted on the city-suburb of Brampton (Kataure 2012). Three methods were used for this case study exploring ethnic settlement preference: telephone surveys, focus groups, and key informant interviews. Ethnic enclaves were identified as census tracts (CTs) that included a South Asian population equal to or above 50 per cent of the total population (based on Statistics Canada 2006 data). Five census tracts with the highest percentage of residents of South Asian origin were surveyed. Phone numbers for the telephone survey were obtained from an online directory, and households with last names of South Asian origin were contacted. Respondents were selected and asked to complete a survey consisting of fourteen mostly closed-ended questions if they were over twenty years of age, of South Asian heritage, and living within the CT. An average of 767 phone calls was made per CT, with an average response rate of 2.3 per cent. A total of 3930 phone calls were connected, yielding 103 completed surveys. Almost all participants (92%) were of Indian heritage, with an average age of 24.5.

Three focus groups were also conducted with a total of fifteen second-generation South Asian participants using the same eligibility criteria as the telephone survey and employing the snowball method. The average age of participants was 22.8 and all were of Indian herti-age. The purpose of the focus groups was to gather detailed information on second-generation settlement preferences, perceptions of their existing neighbourhoods, and ideas and insights about future suburban or urban neighbourhoods (see Kataure and Walton-Roberts 2013). Three key informant semi-structured interviews were also conducted with a real estate agent, a regional-level planner, and a local private-sector planner. Interviews were aimed at understanding the reasons behind South-Asian Canadians' housing preferences and existing consumption trends, and to further contextualize the research findings. While this research was explicitly focused on the second generation, the interviews and survey findings also provided information on the general experiences of living in an ethnic suburb (since many of the second-generational respondents lived with their first-generation immigrant parents). We report some findings below that inform our earlier discussion on the formation and function of suburban ethnic enclaves.

### *Research Findings*

Survey and focus group respondents saw the ethnic enclave as beneficial in promoting the successful settlement of immigrants due to the affordability of housing, proximity to family and social networks, the preference of being in proximity to the co-ethnic community, and access to services in their own language. The ethnic enclave was also viewed as promoting the successful upbringing of offspring through community ties, the instilling strong cultural values and norms, and the development of multigenerational households. However, issues of concern included discrimination and stereotyping by the majority group, reduced assimilation of immigrants and offspring, and a lack of privacy and diversity.

#### UNDERSTANDING IMMIGRANT HOUSING PREFERENCES

The telephone survey results indicated a split between those who wanted to continue residing in Brampton (49.5%) and those wanted to move away (50.5%). The popualtion who wanted to continue to reside in Brampton was slightly older (25.1%), with the overwhelming justification being proximity to family and friends (66%). Those who wanted to move away were younger (23.9) and wanted a lifestyle change (41%). Of the total participants surveyed, 68 per cent wanted to reside in a suburban city, while only 23 per cent wanted to move to an urban centre. This supports Sigelman and Henig's (2001) claim that some ethnic groups view suburban living more positively.

In an interview with a private-sector planner, the fact that South Asians prefer larger suburban homes was related to the need to accommodate extended family members. The suburbs provide such dwellings at affordable prices. Much of Brampton's South Asian community is from Punjab, India, where higher-status individuals reside on large lots of land. Residing in cities in a more densely built environment and renting was seen as a lower-class lifestyle prone to higher crime and related risk. This perspective appears to have been transfered to Canada, and is reflected in the desire of South Asian families to reside in large suburban homes on larger detached lots. It seems that traditional assimilation theories and the movement to low density suburban-style homes communicate socio-economic success and integration for South Asians in reference to both the Canadian and the pre-migratory Indian norm (see Haan and Yu, chapter 3, detailing the homeownership of immigrant groups). This transnational confirmation of status achievement has not been examined in terms of immigrant settlement theories, which focus more closely on the post-migration residential context.

Housing trajectories for South Asian immigrants are also multigenerational projects. In an interview with a real estate agent, she indicated that individuals were willing to reside with the in-laws or parents within the ethnic enclave until they could afford a house on their own. Even when a young couple move out of their parents' home, they often seek to remain in close proximity to family and friends. For example, most of the focus group participants enjoyed their experience living in the suburbs, and felt that they benefited from being raised in an ethnic enclave. Family was highly valued, and continued to be important for them, as one focus group participant stated: "Yeah growing up here, I loved it ... [I had] a good relationship with [my] grandparents and realize[d] how important your elders are in your life and ... that's one of the things I love about our culture, and I definitely want to impart that on [my children]" (Kataure and Walton-Roberts 2013).

Participants also argued that the adaptation process was enhanced because of the large social network of South Asians within the ethnic enclave. This clustering reduced, though it did not completely erase, the feeling of social isolation for the elderly. As one participant states: "My grandmother tells me all the time, she's home alone ... because everybody's gone to work. She's so bored but when she was in India, when everybody left, it was fine because her neighbours would come or someone in the village would come over ... Brampton is like the closest they can pretty much get to it [the village] in Canada."

Conversely, there were some concerns about how elders were adapting to Canadian culture in the ethnic enclave. Many participants agreed that living in a South Asian–dominated community hindered the language development of immigrants, stunted immigrant integration into society, and possibly reduced their acceptance of the Canadian culture. Participants saw a difference between themselves and their grandparents, who immigrated thirty or forty years ago, and who were, in a way, forced to learn English in order to settle successfully. In contrast, participants argued that grandparents who immigrated in the past ten years had not learned any English, since they were able to get by given the large number of South Asians able to provide translations. As one participant stated: "I think by clustering South Asians in an area, it's sort of forming a resistance to Canadian culture. And it's dumbing the older generation."

There were also some perceptions of racism, where individuals heard discriminatory remarks from others outside the ethnic community, and felt that negative stereotypes were being associated with housing in the area (such as poor maintenance of the house's exterior, people sitting in garages, and driving large vehicles around town displaying specific cultural symbols). Ethnic enclave clustering may lead to adjacent or majority populations developing stereotypical views about the ethnic enclave population based on essentialized practices.

Still, many participants talked about how they appreciated the underlying sense of trust and the level of comfort felt among the South Asian community. Participants gave examples of giving co-ethnic elders a ride from the temple or the grocery store on their way home because of the sense of communal trust built among South Asians living in the neighbourhood. As one respondent stated: "There's so many Indians there and if … they all of a sudden need a ride home, they can like just ask somebody because they're Indian. They will trust them."

When it comes to grandparents, it is traditional in the South Asian culture for their offspring to care for them, since placing parents or grandparents in nursing homes is viewed as culturally unacceptable. Participants showed a sense of obligation to support elders and a reluctance to place parents in nursing homes (see also Walton-Roberts and Pratt 2005, 183). The real estate agent commented on a trend she was seeing among older homebuyers, who wanted to maintain multigenerational households in order to care for their elderly parents. One would expect that with assimilation, such traditional views would be abandoned or modified, although these findings show continuation and strong support for these values.

### OFFSPRING

Many participants felt that the ethnic enclave helped ensure the successful upbringing of immigrant offspring regarding acceptable behaviours, educational attainment, and cultural ties. Even the small number of survey participants who wanted to reside in cities (23%) said they would move back to the suburban centres to raise their families. This suggests a form of segmented assimilation in that successful integration still retains a housing trajectory that reflects cultural norms shaped overwhelming by the centrality of multigenerational households. This aspect of segemented assimilation is also evident with regard to the upbringing of the second generation. Suburban ethnic settlement clearly allows the community to monitor youth activities. One individual stated: "There's [a] sense of community and if they're watching you do something bad, the whole community will know about it and shun you … in society." Another participant states: "[It's] how we grew up. We have morals right. We were taught … respect and reputation … In a sense that's what keeps us grounded … We stopped ourselves from doing a lot of stuff because in our community and in our culture it's like frowned upon."

Such settlement patterns may promote positive behaviours and achievements, but this kind of community influence is also highly gendered. Females felt that they were being treated unfairly in the ethnic enclave when it came to monitoring their activities. They felt that there was no sense of privacy when it came to their personal lives, and some male participants also mentioned this as an area of concern. Some individuals felt that living in the ethnic community led to pressures to do well in school and go to university, not college. Second-generation educational outcomes were interpreted by some respondents as a form of competition, which fuelled parents to push their children to succeed educationally in order to effectively compete with their neighbours. One individual states: "They [parents] are really up for that competition."

The research revealed how residing close to family was seen as ensuring an appropriate upbringing of children (similarly to Sodhi 2008). The nature of that upbringing was seen in direct contrast to spatial assimilation, as one second-generation South Asian participant blatantly says: "I just don't want my kids to be 'White washed.'"[5] However, other participants said they would not raise their children in a highly concentrated ethnic enclave because they wanted them exposed to more diverse environments. Place plays a major role in the socialization of children because their social interactions are limited to the neighbourhood (Filion, Bunting, and Warriner 1999, 1320). Younger generations may become more closed-minded, especially if their social circle consists mostly of only other South Asians, as one respondent commented: "It's good to be in touch with your roots, but you should also … be willing to accept and understand other people." Another participant states: "You want to teach your kids [to have] diverse friends. You [have] to be friends with everybody. You can't just be friends with one kind." There seems to be a struggle between residing near family and friends so that children can be in touch with their roots and cultural values, but at the same time not wanting to raise children in an ethnically homogeneous enclave.

Overall, the focus groups and interview transcripts suggest that the South Asian population in Brampton clusters out of preference. The ethnic enclave was viewed positively due to the closeness to family and friends, the affordability of suburban locations, achieving homeownership, access to ethnic networks, feelings of comfort and trust, and the unique cultural community that reinforced living-arrangement traditions, and values and norms, among offspring. However, there were some emerging concerns regarding discrimination and the development of stereotypes by the host society, lack of English-language acquisition, lack of privacy, a sense of cultural competition, lack of intercultural mixing, and closed-mindedness.

The results of this case study indicate that the ethnoburbs will remain an important contributor to suburban growth in Canada, and ethnic enclaves will likely continue to increase in concentration and remain attractive to the second generation, especially considering the undeniable influence of hegemonic cultural norms and social class. The characteristics of survey participants did not indicate an "underclass" second-generation population. Instead, the qualitative findings support the theory of New Assimilation, whereby second generations are, for the most part, satisfied with their experience within ethnic enclaves, and their continued residence there is deemed to have little or no negative impact on their level of socio-economic integration. However, other negative impacts (outside of socio-economic achievements which have been documented) regarding the New Assimilation theory remain undefined, and have only been identified as possiblilities due a lack of study areas and populations. Future research on the actual housing choices of immigrant offspring living in enclaves, the perceptions of the host society, the real estate environment in the neighbourhood, and intra-group experiences would better assess the outcomes of the New Assimilation theory.

## Conclusion

The growth of immigrant communities, especially in suburban locations, represents one of the most important demographic changes occurring across the North American urban landscape. The composition of more recent immigration has changed the social, physical, and cultural landscape, leading to diverse urban forms and housing patterns. This chapter has outlined the historical evolution of theories that explain the movement of immigrant populations to the suburbs, and how this has been made possible by technological, labour, geopolitical, and socio-economic changes. This chapter has indicated that housing experiences differ in the US and Canadian context based on processes and distinct histories of immigration. This complexity has generated substantial research that explores the potential negative and positive outcomes of ethnic clustering for immigrant housing and integration experiences. Research has evolved over time from explanations of residential clustering caused by external discriminatory processes to more recent explanations that highlight an internal socio-cultural preference for clustering.

Based on research exploring the largest visible minority community in Canada (South Asians), in one of the fastest growing suburbs of Canada (Brampton), we suggest that suburban living in an ethnic enclave represents a choice rooted in socio-cultural preference. The case study findings suggest that although there are some drawbacks to ethnic clustering, the benefits outweigh the potential negative impacts, which supports the New Assimilation theory. This was highlighted by the sample's research participants, who appreciated the positive qualities of their neighbourhood. However, research on whether or not these communities will remain attractive and how they will impact the achievements of the third-plus generation remains limited. In addition, we have highlighted the specific cultural features of this particular South-Asian group; the experiences of other groups (such as the Sri Lankan population Ghosh explores in chapter 4) may be markedly different. We maintain that this area of research is important, since South Asian immigrants are a growing component of the Canadian population, and suburban residential development will continue to play an increasingly important role in urban development in Canada and North America.

As immigrants continue to settle in North America and the second generation continues to grow, we will need to better understand integration processes, economic outcomes, living arrangements, and settlement patterns in order to determine whether planners and local governments are adequately meeting their needs. Housing attainment is one of the largest monetary purchases one can make, and among immigrant groups, purchasing a home in the suburbs is one important reflection of their successful integration into society. Assessing the consequences of continued clustering on homeownership, the occupational and educational trajectories of immigrants, and their offspring in Canada and the United States is central to any interpretation of future prospects in terms of housing experiences and settlement patterns across North America.

## NOTES

1 Unlike earlier European immigrant groups, when visible minorities suburbanize it does not necessarily accompany a decline in residential concentration in inner cities, as the ethnic enclave in the urban centre can continue to provide commercial and retail services (Hiebert and Ley 2003; Murdie and Teixeira 2004).

2 A distinct element of this perspective is the acceptance of parallel and autonomous social formations that represent other forms of social organization in the host society – Canada would be defined by diversity and equity, rather than a set of common practices or viewpoints (Hiebert and Ley 2005, 17). Furthermore, diversity and equity entitles immigrants to preserve their heritage, language, religion, customs, and culture via the formation of enclaves (Qadeer 2003). Although Canadian ethnic enclaves are seen as positive, some scholars express concern for the low-density housing style, which may contribute to spatial and social isolation, although the dispersion of economic and social activities to the suburbs allows for the growth of local ethnic services and businesses (Fong, Chen, and Luk 2007, 121).

3 In this view, retention of old country pride and identity would interfere with the process of assimilation (Allen 2006). This comes with the normative assumption that there is a unified core of Americans to which immigrants are expected to assimilate, regardless of origin, characteristics, background, etc. (Zhou 1997).

4 The US government also implemented the Fair Housing Act 1968 and 1988, and the Housing and Community Development Act 1974 to prevent discrimination in housing.

5 This refers to becoming "Westernized" by adopting and assimilating with the white population, and is a term often used in a negative tone, indicating that one has discarded their ethnic culture or values.

## REFERENCES

Agrawal, S.K. 2006. "Housing Adaptation: A Study of Asian Indian Immigrant Homes in Toronto." *Canadian Ethnic Studies* 38 (1): 117–30.

Allen, J.P. 2006. "How Successful Are Recent Immigrants to the United States and Their Children?" *Association of Pacific Coast Geographers* 68 (1): 9–32. http://dx.doi.org/10.1353/pcg.2006.0002.

Anderson, K.J. 1987. "The Idea of Chinatown: The Power of Place and Institutional Practice in the Making of a Social Category." *Annals of the Association of American Geographers* 77 (4): 580–98. http://dx.doi.org/10.1111/j.1467-8306.1987.tb00182.x.

Archambault, J., B. Ray, D. Rose, and A.M. Séguin. 1999. *Atlas – Immigration et Métropoles.* http://www.im.metropolis.net/research-policy/researcg_content/atlas/index.html.

Aydemir, A., and A. Sweetman. 2006. "First and Second Generation Immigrant Educational Attainment and Labour Market Outcomes: A Comparison of the United States and Canada." Institute for the Study of Labor Discussion, no. 2298.

Balakrishnan, T., and F. Hou. 1999. "Socioeconomic Integration and Spatial Residential Patterns of Immigrant Groups in Canada." *Population Research and Policy Review* 18 (3): 201–17. http://dx.doi.org/10.1023/A:1006118502121.

Balakrishnan, T.R., P. Maxim, and R. Jurdi. 2005. "Social Class versus Cultural Identity as Factors in the Residential Segregation of Ethnic Groups in Toronto, Montreal and Vancouver for 2001." *Canadian Studies in Population* 32 (2): 203–27.

Bauder, H., and B. Sharpe. 2002. "Residential Segregation of Visible Minorities in Canada's Gateway Cities." *Canadian Geographer / Géographe canadien* 46 (3): 204–22. http://dx.doi.org/10.1111/j.1541-0064.2002.tb00741.x.

Borjas, G. 2006. "Making It in America: Social Mobility in the Immigrant Population." *Opportunity in America* 16 (2): 55–71.

Bourne, L.S. 1989. "Are New Urban Forms Emerging? Empirical Tests for Canadian Urban Areas." *Canadian Geographer / Géographe canadien* 33 (4): 312–28. http://dx.doi.org/10.1111/j.1541-0064.1989.tb00918.x.

Boyd, M. 2002. "Educational Attainments of Immigrant Offspring: Success or Segmented Assimilation?" *International Migration Review* 36 (4): 1037–60. http://dx.doi.org/10.1111/j.1747-7379.2002.tb00117.x.

Boyd, M., and E.M. Grieco. 1998. "Socioeconomic Achievements of the Second Generation in Canada." *International Migration Review* 32 (4): 853–76. http://dx.doi.org/10.2307/2547663.

Breton, R. 1964. "Institutional Completeness of Ethnic Communities and the Personal Relations of Immigrants." *American Journal of Sociology* 70 (2): 193–205. http://dx.doi.org/10.1086/223793.

Burgess, E. 1925. "The Growth of the City: An Introduction to a Research Project." In *The City*, ed. R.E. Park, W. Burgess, and R.D. McKenzie, 47–62. Chicago: University of Chicago Press.

Clark, W.A.V. 2009. "Changing Residential Preferences Across Income, Education, and Age: Findings from the Multi-City Study of Urban Inequality." *Urban Affairs Review* 44 (3): 334–55. http://dx.doi.org/10.1177/1078087408321497.

Coleman, J. 1990. *Foundations of Social Theory*. Cambridge, MA: Belknap Press of Harvard University Press.

Darden, J., and E. Wyly. 2010. "Cartographic Editorial: Mapping the Racial/Ethnic Topography of Subprime Inequality in Urban America." *Urban Geography* 31 (4): 425–33. http://dx.doi.org/10.2747/0272-3638.31.4.425.

Dawson, C.A., and W.E. Gettys. 1929. *Introduction to Sociology*. New York: Ronald Press.

Doucet, M. 1999. "Toronto in Transition: Demographic Change in the Late Twentieth Century." Toronto: CERIS.

Driedger, L. 1991. *The Urban Factor: Sociology of Canadian Cities*. Toronto: Oxford University Press.

Duncan, O.D. 1957. *The Negro Population of Chicago: A Study of Residential Succession*. Chicago: University of Chicago Press.

Ellis, M., R. Wright, and V. Parks. 2004. "Work Together, Live Apart? Geographies of Racial and Ethnic Segregation at Home and at Work." *Annals of the Association of American Geographers* 94 (3): 620–37. http://dx.doi.org/10.1111/j.1467-8306.2004.00417.x.

Farley, R., and R. Alba. 2002. "The New Second Generation in the United States." *International Migration Review* 36 (2): 669–701.

Farley, R., C. Steeh, J. Jackson, M. Kryson, and K. Reeves. 1993. "Continued Racial Residential Segregation in Detroit: A Chocolate City, Vanilla Suburbs." *Journal of Housing Research* 4 (1): 1–38.

Fiedler, R., and J. Addie. 2008. "Canadian Cities on the Edge: Reassessing the Canadian Suburb." City Institute at York University, Occasional paper.

Filion, P., T. Bunting, and K. Warriner. 1999. "The Entrenchment of Urban Dispersion: Residential Preferences and Location Patterns in the Dispersed City." *Urban Studies* (Edinburgh, Scotland) 36 (8): 1317–47. http://dx.doi.org/10.1080/0042098993015.

Fong, E. 1996. "A Comparative Perspective on Racial Residential Segregation: American and Canadian Experiences." *Sociological Quarterly* 37 (2): 199–226. http://dx.doi.org/10.1111/j.1533-8525.1996.tb01746.x.

Fong, E., W. Chen, and C. Luk. 2007. "A Comparison of Ethnic Businesses in Suburbs and City." *City & Community* 6 (2): 119–36. http://dx.doi.org/10.1111/j.1540-6040.2007.00204.x.

Galster, G. 1989. "Residential Segregation in American Cities: A Further Response to Clark." *Population Research and Policy Review* 8 (2): 181–92. http://dx.doi.org/10.1007/BF00126732.

Gans, H. 1992. "Second Generation Decline: Scenarios for the Economic and Ethnic Futures of the Post-1965 American Immigrants." *Ethnic and Racial Studies* 15 (2): 173–92. http://dx.doi.org/10.1080/01419870.1992.9993740.

Haan, M. 2005. *The Decline of the Immigrant Homeownership Advantage: Life-Cycle, Declining Fortunes and Changing Housing Careers in Montreal, Toronto and Vancouver, 1981–2001.* Ottawa: Minster of Industry.

Harris, R., and R. Lewis. 2001. "The Geography of North American Cities and Suburbs, 1900–1950: A New Synthesis." *Journal of Urban History* 27 (3): 262–92. http://dx.doi.org/10.1177/009614420102700302.

Harris, C., and E.L. Ullman. 1945. "The Nature of Cities." *Annals of the Academy of Political and Social Science* 242 (1): 7–17. http://dx.doi.org/10.1177/000271624524200103.

Harwood, E. 1986. "American Public Opinion and U.S. Immigration Policy." *Annals of the American Academy of Political and Social Science* 487 (1): 201–12. http://dx.doi.org/10.1177/0002716286487001013.

Hiebert, D. 1994. "Immigration and the Changing Social Geography of Greater Vancouver." *BC Studies* 121: 35–82.

Hiebert, D. 2000. "Immigration and the Changing Canadian City." *Canadian Geographer / Géographe canadien* 44 (1): 25–43. http://dx.doi.org/10.1111/j.1541-0064.2000.tb00691.x.

Hiebert, D., and D. Ley. 2003. "Assimilation, Cultural Pluralism, and Social Exclusion among Ethnocultural Groups in Vancouver." *Urban Geography* 24 (1): 16–44. http://dx.doi.org/10.2747/0272-3638.24.1.16.

Hoyt, H. 1939. *The Structure and Growth of Residential Neighborhoods in American Cities.* Washington: Federal Housing Administration.

Kalbach, W.E. 1990. "Ethnic Residential Segregation and Its Significance for the Individual in an Urban Setting." In *Ethnic Identity and Equality*, ed. R. Breton et al., 92–134. Toronto: University of Toronto Press.

Kataure, V. 2012. "A Suburban State of Mind: The Housing Preferences and Location Choices of Second Generation South Asians." MA thesis, Wilfrid Laurier University, Waterloo.

Kataure, V., and M. Walton-Roberts. 2013. "The Housing Preferences and Location Choices of Second-Generation South Asians Living in Ethnic Enclaves." *South Asian Diaspora* 5 (1): 57–76. http://dx.doi.org/10.1080/19438192.2013.722385.

Kim, A., and M. Boyd. 2009. "Housing Tenure and Condos: Ownership by Immigrant Generations and the Timing of Arrival." *Canadian Journal of Urban Research* 18 (1): 47–73.

Ley, D., and Germain, A. 2000. "Immigration and the Changing Social Geography of Large Canadian Cities." *Plan Canada* 40 (4).

Li, W. 1998. "Anatomy of a New Ethnic Settlement: The Chinese Ethnoburb in Los Angeles." *Urban Studies* (Edinburgh, Scotland) 35 (3): 479–501. http://dx.doi.org/10.1080/0042098984871.

Lo, L., and S. Wang. 1997. "Settlement Patterns of Toronto's Chinese Immigrants: Convergence or Divergence?" *Canadian Journal of Regional Science* 20 (1): 49–72.

Logan, J.R., R.D. Alba, and W. Zhang. 2002. "Immigrant Enclaves and Ethnic Communities in New York and Los Angeles." *American Sociological Review* 67 (2): 299–322. http://dx.doi.org/10.2307/3088897.

Logan, J., B. Stults, and R. Farley. 2004. "Segregation of Minorities in the Metropolis: Two Decades of Change." *Demography* 41 (1): 1–22. http://dx.doi.org/10.1353/dem.2004.0007.

Massey, D.S. 1985. "Ethnic Residential Segregation: A Theoretical Synthesis and Empirical Review." *Social Research* 69: 315–50.

Massey, D.S., and N.A. Denton. 1987. "Trends in the Residential Segregation of Blacks, Hispanics, and Asians: 1970–1980." *American Sociological Review* 52 (6): 802–25. http://dx.doi.org/10.2307/2095836.

Massey, D.S., and N.A. Denton. 1993. *American Apartheid: Segregation and the Making of the Underclass*. Cambridge, MA: Harvard University Press.

Massey, D.S., and M. Aysa-Lastra. 2011. "Social Capital and International Migration from Latin America." *International Journal of Population Research* 2011: 1–18. http://dx.doi.org/10.1155/2011/834145.

Mendez, P. 2009. "Immigrant Residential Geographies and the 'Spatial Assimilation' Debate in Canada, 1997–2007." *Journal of International Migration and Integration* 10 (1): 89–108. http://dx.doi.org/10.1007/s12134-008-0090-8.

Metropolitan Toronto Council. 1995. "There's No Turning Back: A Proposal for Change. Submission to the Greater Toronto Area Task Force." Municipality of Metropolitan Toronto, Chief Administrator's Office, Corporate Planning Division.

Miles, R. 1989. *Racism*. London: Routledge.

Murdie, R. 1969. *Factorial Ecology of Metropolitan Toronto, 1951–1961*. Chicago: University of Chicago Press.

Murdie, R., and E. Skop. 2012. "Immigration and Urban and Suburban Settlements." In *Immigrant Geographies of North American Cities*, ed. C.W. Teixeira, 48–68. Don Mills, ON: Oxford University Press.

Murdie, R.A., and C. Teixeira. 2004. "Towards a Comfortable Neighbourhood and Appropriate Housing: Immigrant Experiences in Toronto." *Metropolis*. CERIS Working paper 10, 1–31.

Myles, J., and F. Hou. 2003. "Neighbourhood Attainment and Residential Segregation among Toronto's Visible Minorities." Statistics Canada, Business and Labour Market Analysis Division 24, no. 206. http://dx.doi.org/10.2139/ssrn.486146.

Papillon, M. 2002. *Immigration, Diversity and Social Inclusion in Canada's Cities*. Discussion Paper F/27, Family Network, Canadian Policy Research Network.

Park, R.E. 1926. "The Urban Community as a Spatial Pattern and a Moral Order." In *The Urban Community*, ed. E.W. Burgess, 3–18. Chicago: University of Chicago Press.

Peach, C. 1975. *Urban Social Segregation*. London: Longman.

Peach, C. 1996. "Good Segregation, Bad Segregation." *Planning Perspectives* 11 (4): 379–98. http://dx.doi.org/10.1080/026654396364817.

Pendakur, K., and K. Pendakur. 2011. *Colour by Numbers: Minority Earnings in Canada 1996–2006*. Vancouver: Metropolis British Columbia – CERIS.

Perlmann, J., and R. Waldinger. 1997. "Second Generation Decline? Children of Immigrants, Past and Present – A Reconsideration." *International Migration Review* 31 (4): 893–922. http://dx.doi.org/10.2307/2547418.

Philpott, T.L. 1978. *The Slum and the Ghetto: Neighbourhood Deterioration and Middle Class Reform, Chicago, 1880–1930*. New York: Oxford University Press.

Picot, G., and F. Hou. 2011. *Preparing for Success in Canada and the United States: The Determinants of Educational Attainment among the Children of Immigrants*. Statistics Canada: Social Analysis Division.

Portes, A., and D. Macleod. 1999. "Educating the Second Generation: Determinants of Academic Achievement among Children of Immigrants in the United States." *Journal of Ethnic and Migration Studies* 25 (3): 373–96. http://dx.doi.org/10.1080/1369183X.1999.9976693.

Portes, A., and R. Rumbaut. 2001. *Legacies: The Story of the Immigrant Second Generation*. Berkeley, New York: University of California Press and Russell Sage Foundation.

Portes, A., and R. Schauffler. 1994. "Language and the Second Generation: Bilingualism Yesterday and Today." *International Migration Review* 28 (4): 640–61. http://dx.doi.org/10.2307/2547152.

Portes, A., and M. Zhou. 1993. "The New Second Generation: Segmented Assimilation and Its Variants." *Annals of the American Academy of Political and Social Science* 530 (1): 74–96. http://dx.doi.org/10.1177/0002716293530001006.

Putnam, R.D. 2001. "Social Capital: Measurement and Consequences." *Isuma: Canadian Journal of Policy Research* 2 (1): 41–51.

Qadeer, M. 2003. "Ethnic Segregation in a Multicultural City: The Case of Toronto, Canada." Metropolis Toronto CERIS – Working paper.

Qadeer, M., S.K. Agrawal, and Alexander Lovell. 2010. "Evolution of Ethnic Enclaves in the Toronto Metropolitan Area, 2001–2006." *Journal of International Migration and Integration* 11 (3): 315–39. http://dx.doi.org/10.1007/s12134-010-0142-8.

Ray, B. 1999. "Plural Geographies in Canadian Cities: Interpreting Immigrant Residential Spaces in Toronto and Montreal." *Canadian Journal of Regional Science* 2 (Spring/Summer): 65–86.

Reitz, J.G., and S.M. Sklar. 1997. "Culture, Race, and the Economic Assimilation of Immigrants." *Sociological Forum* 12 (2): 233–77. http://dx.doi.org/10.1023/A:1024649916361.

Reitz, J.G., H. Zhang, and N. Hawkins. 2011. "Comparisons of the Success of Racial Minority Immigrant Offspring in the United States, Canada and Australia." *Social Science Research* 40 (4): 1051–66. http://dx.doi.org/10.1016/j.ssresearch.2011.03.009.

Rose, D., P. Carrasco, and J. Charbonneau. 1998. *The Role of "Weak Ties" in the Settlement Experiences of Immigrant Women with Young Children: The Case of Central Americans in Montréal.* Joint Centre of Excellence for Research on Immigration and Settlement.

Shevky, E., and W. Bell. 1955. *Social Area Analysis.* Connecticut: Greenwood Press.

Sigelman, L., and J. Henig. 2001. "Crossing the Great Divide: Race and Preference for Living in the City versus the Suburbs." *Urban Affairs Review* 37 (1): 3–18. http://dx.doi.org/10.1177/10780870122185163.

Sodhi, P. 2008. "Bicultural Identity Formation of Second-Generation Indo-Canadians." *Canadian Ethnic Studies* 40 (2): 187–99. http://dx.doi.org/10.1353/ces.2010.0005.

Statistics Canada. 2006. 2006 Community Profile: Brampton. http://www12.statcan.gc.ca/census-recensement/2006/dp-pd/prof/92-591/details/search-recherche/frm_res.cfm?Lang=E.

Teixeira, C. 2007. "Residential Experiences and the Culture of Suburbanization: A Case Study of Portuguese Homebuyers in Mississauga." *Housing Studies* 22 (4): 495–521. http://dx.doi.org/10.1080/02673030701387622.

Teixeira, C., and R. Murdie. 2004. "Towards a Comfortable Neighbourhood and Appropriate Housing: Immigrant Experiences in Toronto." Centre of Excellence for Research on Immigration and Settlement (CERIS),Working paper no. 10. http://ceris.metropolis.net/Virtual%20Library/housing_neighbourhoods/murdie/murdieteixeira1.html.

Waldinger, R.D., and C. Feliciano. 2003. "Will the New Second Generation Experience 'Downward Assimilation'? Segmented Assimilation Re-Assessed." Department of Sociology, UCLA: eScholarship.

Walks, A., and L. Bourne. 2006. "Ghettos in Canada's Cities? Racial Segregation, Ethnic Enclaves and Poverty Concentration in Canadian Urban Areas." *Canadian Geographer/Le Géographe canadien* 50 (3): 273–97. http://dx.doi.org/10.1111/j.1541-0064.2006.00142.x.

Walton-Roberts, M., and G. Pratt. 2005. "Mobile Modernities: A South Asian Family Negotiates Immigration, Gender and Class in Canada." *Gender, Place and Culture* 12 (2): 173–95. http://dx.doi.org/10.1080/09663690500094823.

Wayland, S. 2007. "The Housing Needs of Immigrants and Refugees in Canada." *Canadian Housing and Renewal Association*: 1–54.

Zhou, M. 1997. "Growing Up American: The Challenge Confronting Immigrant Children and Children of Immigrants." *Annual Review of Sociology* 23 (1): 63–95. http://dx.doi.org/10.1146/annurev.soc.23.1.63.

Zucchi, J. 2007. *A History of Ethnic Enclaves*. Ottawa: Canadian Historical Association.

# 7 Housing Experiences and Trajectories among Ethnoburban Chinese in Los Angeles: Achieving Chinese Immigrants' American Dream

WAN YU

Obtaining a home in their receiving country is always a remarkable step for immigrant individuals and families, as housing represents a fundamental aspect of immigrants' living situations, their socio-economic status, their assimilation/integration process, and their communities (Alba and Logan 1992; Hiebert et al. 2008; Myers and Lee 1996). As contemporary globalization greatly accelerates and diversifies the immigrant compositions of migrant-receiving countries like the United States, analysis of immigrant housing experiences becomes an important way to understand their realities and issues.

Chinese immigrant housing experiences in US metropolitan areas have been an interesting and hotly debated area of study for contemporary immigration scholars. Despite early Chinese immigrants' long history in Chinatowns (McKeown 2001), the Chinese who lived in these inner-city enclaves did not begin their suburbanization process until the second half of the twentieth century (Fan 2002; Fong 1994). Many of these well-to-do Chinese manifest a distinct type of suburban immigrant housing experience, representing one of the Chinese immigrant American dreams ("to own a home, to be my own boss, and to send my children to the Ivy League") (Zhou 2004, 35). Li (1997) coined the term "ethnoburb" to describe this new type of suburban Chinese community, presenting the Chinese ethnic suburbs in the San Gabriel Valley of Los Angeles as typical examples of this kind of Chinese suburban community. Since the 1990s, when globalization increased the interdependence and interaction between the United States and Mainland China, an increasing number of immigrants from China have settled in American suburbs, intent on achieving their "American Dream." These new Chinese immigrants – ranging from labour migrants to highly skilled professionals – exhibit different housing characteristics and trajectories, which consequently impact the landscapes of the suburban communities they live in and result in housing disparities among Chinese ethnoburbs.

In this context, this chapter focuses on comparing the housing characteristics and trajectories of ethnoburban Chinese in order to examine their diverse housing experiences. It first revisits theoretical debates about contemporary immigrant housing and settlements, and then briefly explains the methodologies and case studies of this chapter. After examining housing disparities among ethnoburban Chinese in terms of demographics and housing characteristics between two ethnoburbs, it compares their housing types in relation to their distinct housing characteristics. Moreover, the differences in housing characteristics among ethnoburban Chinese and the housing types between two ethnoburbs result in diverse housing trajectories (moving from the "old" ethnoburb to the "new" ethnoburb, or remaining in the "old" ethnoburb) among Chinese immigrant subgroups. These diverse housing trajectories substantially impact the community landscape of Chinese ethnoburbs, influence the housing experiences of ethnoburban Chinese, and also reflect a path to Chinese immigrants for achieving their "American Dream."

## Conceptual Frameworks

To analyse Chinese immigrants' housing experiences and trajectories, it is necessary to revisit theoretical debates surrounding immigrant and ethnic housing as it pertains to minority groups' access to the housing market, their homeownership, and other housing issues. Discussions on immigrant settlements and integration, and particularly Chinese immigrant settlements in the United States, are also addressed.

### *Immigrant and Ethnic Housing*

Research on immigrant housing sometimes is intertwined with the discussion of ethnic minority housing. Recent scholarly debates on immigrant and ethnic housing have extended beyond urban settings and are actively involved in suburban and rural geography (Li and Teixeira in chapter 1 and Kataure and Walton-Roberts in chapter 6). One visible variable of ethnic housing is homeownership rate, as it relates closely to ethnic integration, socio-economic status, educational attainment, as well as social and public benefits (Darden in chapter 2; Myers, Megbolugbe, and Lee 1998; Rossi and Weber 1996). Access to the housing market and homeownership can sometimes provide opportunities for upward social and economic mobility for ethnic minority groups (Hiebert et al. 2008). There has been much research examining ethnic minorities' lack of access to the housing market as a result of their socio-economic status (Myers and Wolch 1995; Schill et al. 1998), or due to structural and institutional barriers such as race, nationality, and religion (Ahmed and Hammarstedt 2008; Ray and Preston 2009). These barriers could be systemic institutional forces such as access to housing mortgage loans (Munnell et al. 1992), could be evident in terms of prejudice and discrimination in the housing market (Alba and Logan 1992; Darden and Kamel 2000), or could take the form of risks such as homelessness and exclusion (Fiedler et al. 2006). Another interesting variable would be housing quality, reflected by issues such as overcrowding and "hidden homelessness," which are usually more prevalent among immigrant newcomers (Fiedler et al. 2006; Krivo 1995; Myers and Lee 1996; Myers, Baer, and Choi 1996).

Thus, immigrant residential distribution can be viewed as the collective spatial outcome of individual immigrants' housing strategies coping with their constraints on accessing better housing in the market, such as their socio-economic status or barriers keeping them from affordable housing, and their preferences regarding home quality or community environment, such as a preference for a particular type of housing and co-ethnic environment (Carter and Vitiello 2012). Particularly in regard to the housing experiences of immigrants, their housing considerations and trajectories are also correlated to immigrant characteristics such as indefinite immigration status and lack of citizenship or permanent residency (Coulson 1999; Darden and Kamel 2000; Teixeira 2011), English attainment (Alba and Logan 1992), length of stay in their receiving country (Myers, Megbolugbe, and Lee 1998), the process of assimilation to the host societies (Friedman and Rosenbaum 2004; Myers and Lee 1996), return migration intention (Murdie 2002; Owusu 1998), the effects of living in ethnic enclaves (Borjas 2002), and, most important, the receiving country's policies on immigrant housing. A common perspective among recent studies is that housing conditions can be negatively impacted by immigration status, as immigrant newcomers usually have lower homeownership rates than their native-born counterparts (Carter and Vitiello 2012; Hiebert et al. 2008; Saiz 2007), but this negative effect gradually diminishes as immigrants extend their length of stay and integrate to the host society (Myers and Lee 1996;

Myers, Megbolugbe, and Lee 1998). Spatially concentrated residential clusters in urban ethnic enclaves as well as compact suburban ethnic communities hence demonstrate lower homeownership rates among ethnic minority groups (Borjas 2002; Myers and Lee 1996; Painter et al. 2004; Saiz 2007). Several studies have examined immigrant housing experiences across generations and argue that homeownership rates remain stable after immigrants' length of stay reaches a certain point (see also Kataure and Walton-Roberts in chapter 6). And interestingly, homeownership could even drop for later generations of immigrants (Andersen 2012; Kim and Boyd 2009). Moreover, some studies also indicate that, instead of declining or delaying homeownership, many immigrants adopt strategies to cope with their housing conditions, such as subleasing part of their houses (Teixeira 2011) or moving their homes outside ethnic enclaves (Saiz 2007).

Previous literature explicates immigrant housing conditions and issues in receiving countries, yet contemporary diversifying immigrant demographic and socio-economic characteristics and the consequent impact on their housing experiences and trajectories have been under-studied. Some work distinguishes different "housing appetites" and preferences that exist between "old" immigrants and "new" immigrants in terms of their length of stay in the receiving countries and immigrant backgrounds, seeking to explain their diverse housing characteristics and trajectories (Haan 2005; Kim and Boyd 2009). Yet only a little research has focused on the recent affluent immigrant newcomers in North America and transnational migrants whose socio-economic statuses and immigration background drives them to select different housing trajectories from their expatriates (Hiebert et al. 2008).

### *Immigrant Settlement and Integration to the United States*

Despite many studies confirming a positive correlation between immigrants' housing conditions and their assimilation process, some research shows immigrants' residential settlement is also tightly related to their integration to the receiving country (Borjas 2002; Haan 2005). Many immigrants, especially newcomers, choose to stay in neighbourhoods with large co-ethnic populations, and with convenient ethnic social networks, such as urban enclaves or ethnic suburbs where specific housing types are prevalent (Zavodney 1998; Zhou 1992). This housing preference, combining with residential concentration, consequently affects the immigrant integration process. A large body of literature on immigrant settlement is related to ethnic segregation – from considering it an outcome of social and economic competition among ethnic groups, or external characteristics, as in chapter 2, to believing it is a result of individual preferences, or internal characteristics, as in chapter 2. From an urban ecologist's perspective, the invasion and succession model illustrates ethnic groups' diverse residential settlements and their integration processes (Park et al. 1967), whereas some scholars consider residential segregation as always detrimental to social integration (Sundstrom 2004).

Specifically, Chinese settlements in the United States have experienced the establishment of inner-city enclaves, to the prosperity of both urban enclaves and suburban communities. Since the 1950s, an increasing amount of literature has analysed Chinese immigrants in inner-city enclaves and the function of these segregated communities in urban areas (Barth 1964). In the 1980s, Chinese immigration to the United States was significantly accelerated as economic globalization and China's "open door" policy increased the interaction between China and the United States (Fan 2002; Li 1998). Meanwhile, the suburbanization process of Chinese Americans has led to emerging studies on a new type of Chinese communities, Chinese suburbs. Fong (1994) first described the rapidly growing suburban population in Monterey Park, a city in Los Angeles County, and defined it as the "first suburban Chinatown" to reflect its large Chinese population. Instead of conforming to Fong's "suburban Chinatown" model, Li introduces the *ethnoburb* concept to describe "suburban ethnic clusters of residential areas and business districts in large metropolitan areas. They are multiethnic communities, in which one ethnic minority group has a significant concentration, but does not necessarily comprise a majority" (Li 1997, 2). An ethnoburb can be viewed as a spatial outcome of immigrant groups being integrated to the receiving society (Li 1998). Yet other scholars test the validity of the ethnoburb concept, arguing about ethnoburbs' nature as ethnic enclaves, whose residents' assimilation process is still slow (Lin and Robinson 2005).

In conclusion, previous literature on immigrants' housing and settlement in the United States mostly focuses on their housing issues and the manifestation of immigrant settlements. Little research has focused on distinguishing the diversifying immigrant populations nowadays and their consequently shifting housing choices and trajectories. Analysis of housing dynamics and trajectories within one type of ethnic settlement – especially inside suburban communities – and investigations of the diverse housing experiences within single immigrant groups – especially Chinese immigrants, due to their large numbers in recent decades – are also rare. Positing the research purpose of filling this gap, this chapter examines the different housing types underlying the ethnoburb's conceptual framework, as well as the housing trajectories within the Chinese immigrant group, in order to better understand the diverse housing experiences among ethnoburban Chinese in suburban Los Angeles.

## Data and Methods

In order to analyse the housing disparities among ethnoburban Chinese, as well as among different ethnoburbs, two Chinese ethnoburbs which emerged at different times are selected to conduct a comparative analysis. With the largest suburban Chinese population in the United States, Los Angeles is considered to have the most diverse Chinese communities of any metropolitan area, and the selection of case study areas in this study derives from the changing residential clusters of Chinese according to 1990 and 2010 Los Angeles County census data. According to figure 7.1, by 1990 in Los Angeles County, the suburban community

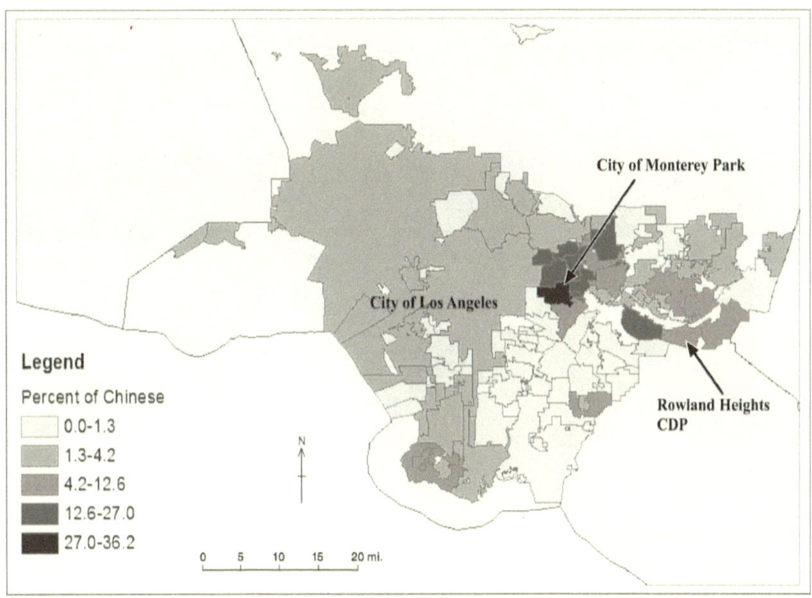


Figure 7.1 Percentage of Chinese among suburbs in Southern Los Angeles County, 1990

Note: Rowland Heights CDP stands for Rowland Heights Census Designated Place.
Source: US Census Bureau, Census 1990, Summary file 1, matrices P1 and PCT5.

with the highest percentage of Chinese was Monterey Park, with 36.2 per cent. However, by 2010 (figure 7.2), coinciding with the Chinese concentration in Monterey Park gradually expanding north and eastward, there was a new Chinese population cluster emerging on the east side of the San Gabriel Valley, centring in the community of Rowland Heights. Therefore, these ethnoburbs, each established at different times, serve as the case studies of my research.

This study uses decennial census data and multi-year American Community Survey (ACS) data to obtain the demographic and housing characteristics of Chinese in Monterey Park and Rowland Heights. In addition, qualitative data include twenty-six semi-structural interviews conducted in the summer of 2008, with thirteen in Monterey Park and thirteen in Rowland Heights. All interviewees were approached and selected on the streets or in each community's public spaces. Street locations are selected from four residential neighbourhoods in each ethnoburb, with one neighbourhood from each

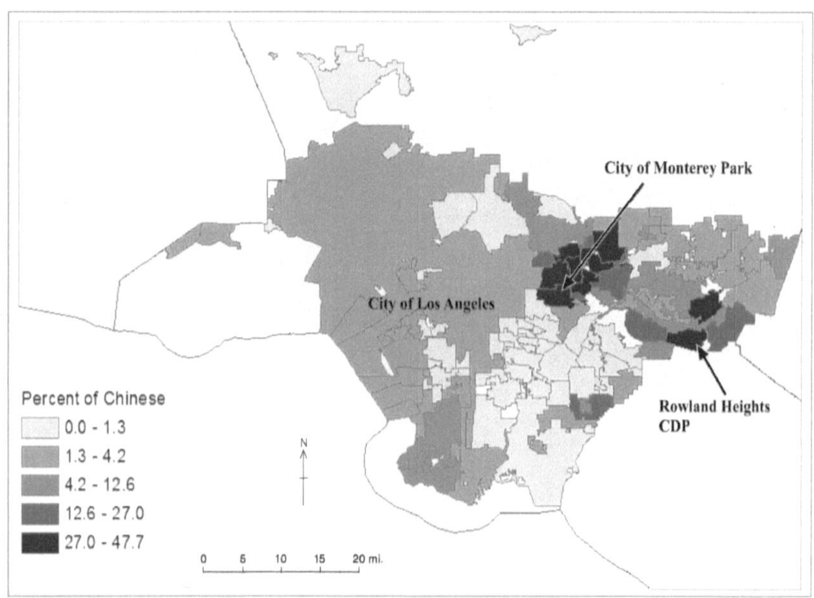


Figure 7.2 Percentage of Chinese in Southern Los Angeles County, 2010.

Note: Rowland Heights CDP stands for Rowland Heights Census Designated Place.
Source: US Census Bureau, Census 2010, SF-1.

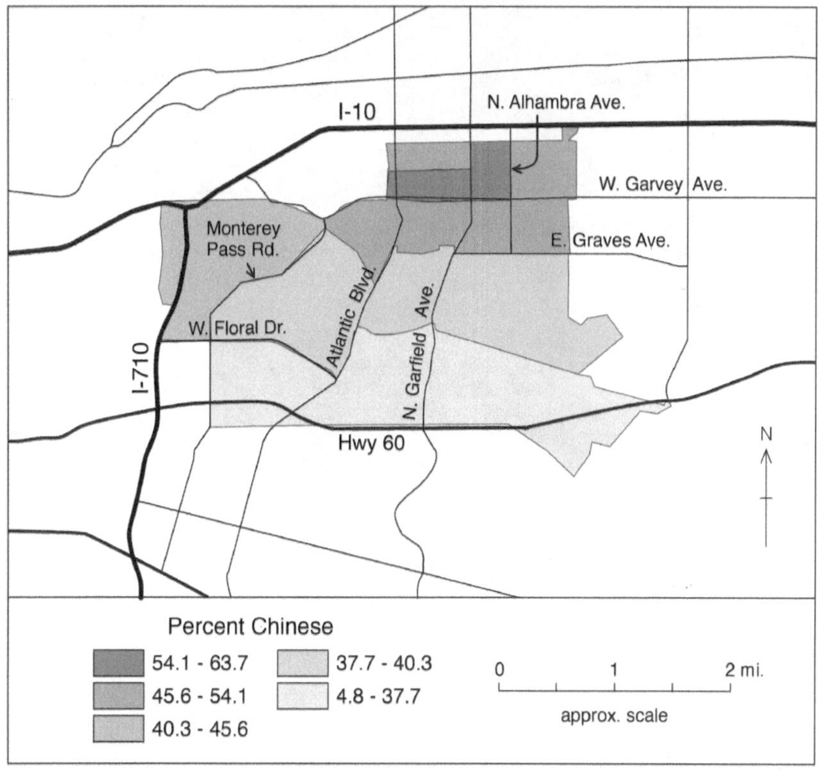


Figure 7.3 Percentage of Chinese in Monterey Park by census tract, 2000

Note: Map downloaded through Thematic Map Module from Census 2000 at www.census.gov.

Source: US Census Bureau, Census 2000 Summary file 1, matrices P1 and PCT7.

of the highest four categories of percentage of Chinese in figure 7.3 and figure 7.4. This method of selection excludes major commercial and business areas, in order to eliminate the possibility of approaching Chinese who don't live in the relevant ethnoburb (because both ethnoburbs in this case study function as commercial and cultural hubs for Chinese in the region, their customer base is beyond the city boundaries). Moreover, by selecting one location in each category of the residential neighbourhoods with 37.7 per cent or higher Chinese in Monterey Park and 17.9 per cent or higher Chinese in Rowland Heights (because the total population of both lowest categories of

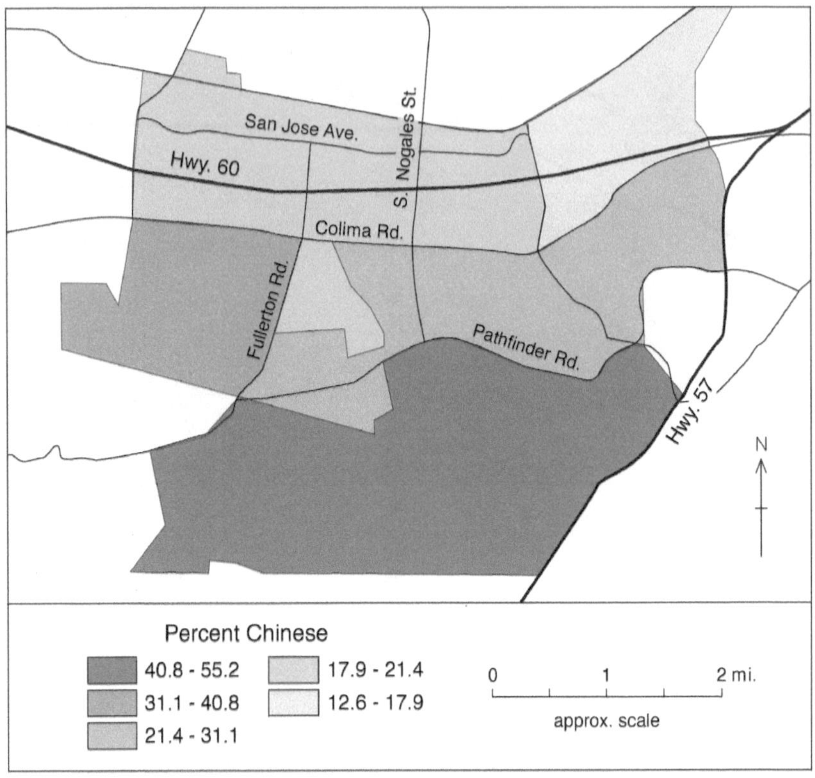


Figure 7.4 Percentage of Chinese in Rowland Heights by census tract, 2000

Note: Map downloaded through Thematic Map Module from Census 2000 at www.census.gov.

Source: US Census Bureau, Census 2000, Summary file 1, matrices P1 and PCT8.

figure 7.3 and figure 7.4 is relatively low), this method aims to lower the bias of one particular housing type becoming over-represented from the interview data. Among the interviewees, ten are male, with six in Monterey Park and four in Rowland Heights; sixteen are female, with seven in Monterey Park and nine in Rowland Heights. All interviewees had been living in their respective community for more than one year. The interview questions cover Chinese residents' migration backgrounds, their conceptions of the community, their length of stay in the community, and their housing experiences.

## Housing Characteristics of Ethnoburban Chinese in Monterey Park and Rowland Heights

### *Ethnoburban Chinese in Monterey Park and Rowland Heights*

For the past two decades both Monterey Park and Rowland Heights have witnessed booming Chinese populations stimulated by increasing economic globalization and transnational activities. According to table 7.1, in the past three decades, the Chinese population in Monterey Park grew steadily, from 36.2 per cent of the total population in 1990 to 42.7 per cent in 2000, and to 47.7 per cent in 2010, a bit lower than that of Los Angeles County. More drastically, the Chinese population in Rowland Heights increased from 10.9 per cent of the total population in 1990 to 29.9 per cent in 2000, and to 36.2 per cent in 2010.

In terms of Chinese immigrants' socio-economic status, generally, Chinese in Rowland Heights have a relatively higher socio-economic status than those living in Monterey Park. According to figure 7.5, Monterey Park has a higher percentage than Rowland Heights of Chinese families earning less than $75,000 per year, and even higher than Chinese in Los Angeles County.

Rowland Heights has a higher percentage than Monterey Park of Chinese families with incomes of at least $75,000. Both Monterey Park and Rowland Heights have a significantly lower percentage than Los Angeles County of Chinese families with household incomes *above* $200,000. This income disparity indicates that Monterey Park has more working-class Chinese and Chinese in poverty. And instead of attracting wealthy Chinese families, Rowland Heights is the ethnic suburb where most middle-class Chinese families prefer to live.

Table 7.1 Percentage of Chinese among total population in Monterey Park and Rowland Heights, 1990–2010

| Year | Monterey Park | Rowland Heights | Los Angeles County |
|---|---|---|---|
| 1990 | 36.2 | 10.9 | 2.63 |
| 2000 | 42.7 | 29.9 | 3.46 |
| 2010 | 47.7 | 36.2 | 4.01 |

Source: Decennial Census, 1990, 2000, 2010, Summary File 1.

Figure 7.5 Chinese household income distribution in Monterey Park, Rowland Heights, and Los Angeles County, 2010

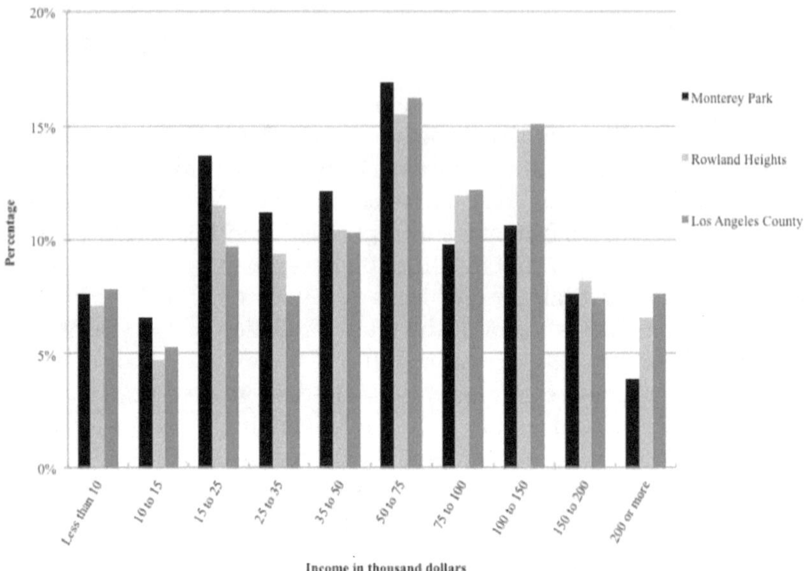


Source: American Community Survey, 2006–2010 five-year estimates.

On the aspect of English-speaking ability (table 7.2), Rowland Heights has a much higher percentage of Chinese who can speak English very well or only speak English, whereas Monterey Park has more than 60 per cent of Chinese who speak English less than very well. Similar disparities also exist in educational attainment among Chinese in these two ethnoburbs: Monterey Park's percentage (25.9%) of Chinese whose degree is lower than high school is more than twice than that of Rowland Heights (12.6%), whereas its percentage of Chinese with a bachelor's degree and above is much lower than Rowland Heights'.

In terms of occupational types, Rowland Heights has more Chinese with sales, office, management, and professional occupations, while Monterey Park has a higher percentage of Chinese working in the production, transportation, and service industries. These occupational differences suggest that Rowland Heights has a much larger population of Chinese professionals, whereas Monterey Park has more working-class Chinese (table 7.2).

Table 7.2 Selected demographic and socio-economic characteristics of Chinese in Monterey Park and Rowland Heights, 2010

| | Monterey Park (%) | Rowland Heights (%) |
|---|---|---|
| *Speaking English, population 5 years and over* | | |
| Very well or English only | 37.0 | 44.8 |
| Less than very well | 63.0 | 55.2 |
| *Educational attainment, population 25 years and over* | | |
| Percentage with less than high school | 25.9 | 12.6 |
| Percentage with bachelor's degree and above | 31.7 | 45.2 |
| *Occupation by employed civilian population, 16 years and over* | | |
| Total | 100 | 100 |
| Service | 20.3 | 11.1 |
| Production/transportation | 9.4 | 6.5 |
| Construction/maintenance | 3.7 | 3.5 |
| Sales/office | 30.7 | 37.9 |
| Management/professional | 35.9 | 40.1 |
| *Age groups* | | |
| Under 18 years old | 15.1 | 17.1 |
| 18 to 24 years old | 8.4 | 8.2 |
| 25 to 44 years old | 25.3 | 27.3 |
| 45 to 64 years old | 32.1 | 35.1 |
| 65 years old and over | 19.1 | 13.3 |

Source: American Community Survey, 2006–2010 five-year estimates.

## *Subgroups of Chinese in Monterey Park and Rowland Heights*

Based on the demographic comparison above, ethnoburban Chinese in Monterey Park can be mainly categorized into four subgroups by their distinct demographic and socio-economic status: Chinese professionals such as academics, technological engineers, and managers, who mostly came to the United States after the 1980s and have a relatively higher socio-economic status; Chinese entrepreneurs who live and own businesses in Monterey Park; working-class Chinese, usually cooks, waiters, and waitresses, who frequently cannot speak English well and have lower levels of education; and elderly Chinese who have lived in Los Angeles for a much longer time than those in other subgroups. Chinese immigrants in Rowland Heights can also be viewed within similar subgroups: Chinese professionals, working-class Chinese, and the Chinese entrepreneurs. These three subgroups constitute the majority of the Chinese population in Rowland Heights, with elderly Chinese accounting for a much smaller share (13.3%, according to table 7.2).Thus, ethnoburban Chinese in Monterey Park and Rowland Heights do not vary in terms of types of migrants, but in the relative composition of each subgroup. This demographic disparity consequently impacts their differential housing characteristics.

### *Housing Characteristics of Ethnoburban Chinese in Monterey Park and Rowland Heights*

Despite both ethnoburbs representing large concentrations of Chinese, the spatial distributions of Chinese neighbourhoods are drastically different between the two ethnbourbs. Figure 7.3 indicates that the densest concentrations of Chinese residents occur within several blocks in the northern part of Monterey Park. According to the City's zoning plan, the two main land-use categories in this area are High Density Residential (HDR) and Medium Density Residential (MDR) land use, indicating the high housing density of ethnoburban Chinese in Monterey Park. Contrastingly, in Rowland Heights the residential cluster of Chinese is located in the southern part (figure 7.4), where low-density residential land use is coded according to the Community General Plan. The area is located on mountainous terrain, mainly occupied by light agricultural land use (in the form of ranch farms), with several residential planned-development land use neighbourhoods.

### *Tenure*

Notable housing disparities also exist between ethnoburban Chinese in the two ethnoburbs. Table 7.3 indicates that Rowland Heights has a much higher percentage of Chinese who own their home, whereas Monterey Park has a higher percentage of Chinese renters. This higher rate of homeownership among Chinese in Rowland Heights, as mentioned before, brings them one closer step to their "American Dream."

Table 7.3 Selected housing characteristics of Chinese in Monterey Park and Rowland Heights, 2010

| | Monterey Park (%) | Rowland Heights (%) |
|---|---|---|
| *Tenure* | | |
| Owner-occupied housing units | 54.1 | 71.0 |
| Renter-occupied housing units | 45.9 | 29.0 |
| *Units in structure* | | |
| OWNER OCCUPIED | | |
| 1, detached or attached | 84.4 | 96.2 |
| 2 to 4 | 3.8 | 1.0 |
| 5 or more | 11.7 | 0.2 |
| Other | 0.1 | 2.7 |
| RENTER OCCUPIED | | |
| 1, detached or attached | 35.8 | 35.4 |
| 2 to 4 | 20.5 | 18.6 |
| 5 or more | 43.3 | 43.4 |
| Other | 0.4 | 2.6 |
| *Year householder moved into unit* | | |
| 2005 or later | 27.4 | 32.2 |
| 2000–4 | 31.0 | 31.0 |
| 1990–9 | 22.0 | 27.2 |
| 1980–9 | 12.3 | 9.3 |
| 1970 or earlier | 7.2 | 0.3 |
| *Vehicles available* | | |
| No vehicles | 13.3 | 5.9 |
| 1 vehicle | 25.7 | 18.2 |
| 2 or more vehicles | 61.1 | 75.9 |

Source: Census Bureau, Census 2010 Summary file 1; American Community Survey, 2006–2010 five-year estimates.

### *Units in Structure*

Chinese in Rowland Heights are not only more likely to own their homes, but they also tend to live in single-family houses. Table 7.3 shows that Chinese homeowners in Rowland Heights overwhelmingly reside in single-family houses (96.2%), a higher percentage than in Monterey Park (84.4%). However, such disparity does not reflect on Chinese renters, who exhibit similar housing types in the two ethnoburbs.

*Move-in Year of Householder*

According to table 7.3, the majority of Chinese householders in Rowland Heights moved into their homes within the last two decades (90.4%), which coincides with the Chinese population boom in that community. Yet, the number in Monterey Park is comparatively lower, with only 80.5 per cent of Chinese moving into their homes between 1990 and 2010, and 19.5 per cent moving in during earlier decades. This housing disparity can be explained by the different year-of-entry to the United States among Chinese immigrants. According to 2006–10 ACS five-year estimates, 48.2 per cent of the foreign-born Chinese in Rowland Heights came to the United States during the 1980s, much more than that of Monterey Park, at 34.3 per cent; conversely, Rowland Heights has a much lower percentage (11.8%) of Chinese immigrants coming to the United States before the 1980s than Monterey Park (23.7%). This indicates that Chinese immigrants in Rowland Heights are mainly recent immigrants who came to the United States during and after the 1980s, and who tended to settle elsewhere before moving in Rowland Heights, whereas Monterey Park has a higher percentage of early Chinese immigrants.

*Vehicles per Household*

As shown in table 7.3, Rowland Heights has a much higher percentage (94.1%) of households that own at least one vehicle, while Monterey Park has larger share of households without any vehicles (13.3%). Monterey Park's unusually high percentage of households without a vehicle can be explained by the high housing density of the neighbourhood and its large population of elderly Chinese.

In conclusion, ethnoburban Chinese in Monterey Park and Rowland Heights vary by groups due to their demographic and socio-economic statuses, and these group disparities also lead to their identifiable housing characteristics. Monterey Park hosts a larger Chinese population in a much denser neighbourhood with longer-term residents. Rowland

Heights, by contrast, accommodates more recent Chinese homeowners living in a more dispersed neighbourhood.

## Housing Types of Chinese Neighbourhoods in Monterey Park and Rowland Heights

### *City of Monterey Park*

Located in the west San Gabriel Valley, the City of Monterey Park experienced a series of demographic, economic, and social changes in the past several decades. Before the 1960s, the majority of Monterey Park's population was white. It was, in fact, once known as "one of the whitest spots in Southern California" in the first half of twentieth century (Fong 1994, 28). After the passage of the US Immigration and Nationality Act in 1965, the city, regarded as a liberal and ethnic-friendly community, had become popular among middle-class immigrants. At the end of the 1980s, Monterey Park was first reported as a "majority minority" city, with an only 25 per cent white population. Yet, in the past two decades, its ethnic composition has become more diverse, due in part to the recent booming population of Chinese immigrants to the United States.

The present housing types of Monterey Park are outcomes of the residential transformations of the 1970s, when the city started to adjust to a large influx of Chinese immigrants and financial capital. After the arrival of wealthy Chinese immigrants – who preferred to purchase houses with bags of cash – the area's residential property prices skyrocketed. With many white families moving out, Chinese investors tore down the former single-family homes and established multiple-unit complexes for new immigrants (Monterey Park Historical Heritage Commission 1991). With Chinese investors converting one piece of property into two or even more pieces, more houses were constructed on smaller lots to accommodate the increasing Chinese population and expanding demand in the housing market. This transformation of housing types was considered real estate investment and profited many early Chinese immigrants at that time (Monterey Park Historical Heritage Commission 1991). As the population density increased, property units were cut into smaller and smaller pieces. The spatial outcome of this process gradually reshaped the community's landscape, with increasing residential density, a booming ethnic economy, and shrinking public spaces.

Figure 7.6 Condominiums in Monterey Park
Source: Author's fieldwork, 2013.

Figure 7.3 shows that the highest percentage of Chinese population is located in the northern part of Monterey Park, where residential density is relatively high. Within this Chinese residential cluster, condominiums, apartments, and townhouses are the most common housing types (figure 7.6). One female interviewee, who moved in Monterey Park in the late 1980s, now rents three houses to tenants, all of them located in the Chinese residential cluster:

> There is a [land use] restriction on what kind of houses you can build. I used to own a house on the other side [of Garvey Avenue], but it wasn't as profitable as houses here, so I sold it out in 1999 and bought this new one … The houses [in this neighborhood] had just built up and [were] open for sale when I came here. People rushed to buy them. All of my friends recommended me to put my money on these houses as an investment. I finally bought two sets of them and rent them out after decoration … Most of the tenants are new Chinese immigrants who just came to the US …

Figure 7.7 Single-family homes in Monterey Park
Source: Author's fieldwork, 2013.

> Chinese people, especially the immigrants who just landed in the US, prefer to live with other Chinese.

When being asked about how she felt about the compact housing type, she answered:

> To most newcomers, it is already spacious enough. Many of them came from places in China even denser than here. Besides, this is just their first stop. Many of them moved out after two years.

There are Chinese families living in single-family detached houses as well. Despite their single-family home type in terms of land use, some houses occupy almost all of their respective lots, leaving little space for a front or back yard. Even if yard space is left out, there is usually an open garage or storage structure (figure 7.7). These "monster houses" are counted among the single-family houses in Monterey Park.

A housewife who lives with her husband and father-in-law in Monterey Park said:

> My husband and I bought this house when he got a job in Santa Monica … because it has a good Chinese neighborhood for me to make friends and for his dad to go around … Most of the single-family houses are like this, with small backyards or gardens … [A] lawn is of little use for us and it is hard to take care of. We don't have time or money to keep it [up].

### *Rowland Heights CDP*

The dramatic growth of the Chinese population in Rowland Heights started in the late 1980s, when middle- and upper-class Taiwanese and South Koreans moved in. The first group of Chinese, though only a couple of households, settled down in Rowland Heights in the mid-1970s when the government implemented a program that enabled veterans of the Second World War to purchase houses at a lower interest rate. At that time (and until the 1980s) Rowland Heights was still a predominately middle-class white neighbourhood. The second wave of Chinese in Rowland Heights emerged in the late 1980s after the massive influx of Taiwanese and South Koreans. The former Asian population in Rowland Heights had built a fundamentally diverse community environment as well as convenient facilities, which favoured the Chinese newcomers settling there.

According to figure 7.4, the residential cluster of Chinese in Rowland Heights is located in the southern part of that community, where housing types are mainly in the form of single-family detached homes. Most of these single-family detached houses are occupied by Chinese homeowners who prefer bigger houses for family living. Houses on the hilly terrain have bigger lots, more spacious rooms and good views from the hill, which attracts a lot of Chinese to settle in the lower-density single-family houses in this area (figure 7.8).

As one of the homeowners in Rowland Heights said:

> [Here,] with the same price, I can buy a house twice as big as the one I could have bought in the east [San Gabriel Valley]. Now, my house can [accomodate] my [own] family, my parents, my parents' in-laws … We have a huge backyard. My wife and my mother in-law made a small organic garden [in the backyard], [with] tomatoes, leek, cucumbers … It may not even compensate the utility bill, but what the heck, it is for fun.

Figure 7.8 Single-family homes in Rowland Heights
Source: Author's fieldwork, 2008.

Figure 7.9 Apartments on Colima Road, Rowland Heights
Source: Author's fieldwork, 2008.

Some Chinese also reside in the western part of Rowland Heights, which in the Rowland Heights General Plan is defined as a Residential Planned Development Area, a relatively complex residential area with low-density and high-density land use combined, and with much lower housing prices for condominiums and apartments (figure 7.9). Due to the high fluidity of occupations in service industries, especially in Chinese ethnic businesses, many working-class Chinese prefer to reside in such places in order to be closer to work and to have flexible housing choices.

Compared with the prominent multi-unit housing types in Monterey Park, the Chinese housing in Rowland Heights is relatively sparse. This is due to a relatively short history of Chinese settlement and to spatially dispersed settings. Thus, with similar metropolitan backgrounds, the housing types of Chinese communities in Monterey Park and Rowland Heights vary significantly.

## Housing Trajectories of Chinese Subgroups in Monterey Park and Rowland Heights

From different housing characteristics of ethnoburban Chinese and different housing types of Chinese neighbourhoods in Monterey Park and Rowland Heights, it is evident that different housing trajectories exist among Chinese subgroups residing in different ethnoburbs.

### *Chinese Professionals*

To Chinese professionals, Monterey Park functions as their residential, economic, and cultural centre in the San Gabriel Valley. However, high population density, poor living conditions, and an isolated environment also drive them to move out. One of the interviewees in Monterey Park mentioned: "[The reasons to move] for me are the housing price and bad environment. With the same money, you can buy bigger houses in many good neighborhoods with fewer Chinese, like El Monte, Temple City, [and East Los Angeles] ... Those neighbourhoods are cleaner and nicer than here."

The increasing housing density and changing housing types in Monterey Park also cause decreased quality of life, as one long-term Chinese resident in Monterey Park mentioned: "There are newcomers moving in every day ... [Did you] see that house (the house next door)? The house owner shares their house with three other guys. Many Chinese do the same thing in this neighbourhood. They rent the spare bedrooms to newcomers and use that income to pay the loan ... The neighbourhood is changing. I don't know my neighbours anymore."

Besides changes in housing prices and housing types, diminishing open spaces and aggregated sanitation problems also prompt Chinese professionals to move out of the community. As one housewife in Monterey Park mentioned: "It is not the best place to raise children, no yards, no playground, and no diversity. I want my child grown up in a diverse neighbourhood or some places with more contact to the mainstream society ... If we were rich, we definitely wouldn't live here. [We would] move to Arcadia or some place to the east ... Monterey Park is the first not the last stop of our life ... [Moving out is] just [a] matter[] of time."

On the other hand, Rowland Heights, emerging as a new and thriving Chinese community, represents a new middle-class Chinese ethnoburb – the "American Dream" lifestyle. The business expansion in adjacent cities – City of Industry and Diamond Bar – attracts many Chinese professionals, who get jobs or open businesses there to seek nearby residential communities to settle their homes. The convenient transportation to the adjacent communities, the lower housing density, and better school districts attract many Chinese professionals. As one engineer (whose job is in Diamond Bar) said: "[The reason for moving in] is just because I got a job in Diamond Bar. I got my job in a software company in 1995 so I moved here with my family ... [Why I chose this community is because] it's near to my work and some of my Chinese colleagues live here and recommended me this place ... The housing price is not that high so I can buy bigger houses for my kids and my parents."

Consequently, Chinese professionals who can afford a better life are the first group to move out of the "old" ethnoburbs to "new" ethnoburbs to change their suburban housing.

### *Chinese Entrepreneurs*

Most of the Chinese entrepreneurs move their houses to follow the large Chinese population. After Chinese professionals move out of the "old" ethnoburbs to the "new" ethnoburbs, the Chinese entrepreneurs start to move their businesses and homes to the new ethnoburbs, to seek more market opportunities. As one Chinese handbag shop owner in Rowland Heights indicated: "[The reason I moved here is] because my friends told me the rent [for a business] here was much cheaper and the housing prices [were] low. That's why I came, as well as my family."

### *Working-class Chinese*

The business and housing changes of Chinese entrepreneurs create a large demand for Chinese labourers in "new" ethnoburbs, which triggers working-class Chinese to move their homes from "old" ethnoburbs to "new" ethnoburbs. To many working-class Chinese, Monterey Park is a place mixed with opportunities and dilemmas: while the large Chinese population provides an ethnic-friendly environment for living and a booming economy for working, the increasing amount of labour from China also accelerates competition in the job market and reduces the wages of Chinese workers. With declining wages on one side and increasing housing prices on the other, working-class Chinese in Monterey Park find themselves trapped at the bottom. One Chinese healthcare worker in Monterey Park said:

> Every day is about struggling to survive. I live in Alhambra with my friends. Four of us share a one-bedroom apartment for $1,000 rent a month … The rent in Monterey Park is even higher, I can't afford it. I take the bus to work every day. My work is to take care of the patient ten hours a day, six days a week, and the salary is $1,100 a month. This barely affords the basic living expense here … I don't have health insurance. If I am sick, the only way for me is to go to the church for help.

Compared to Monterey Park, Rowland Heights, for most of the working-class Chinese, is a community with more job opportunities and lower housing costs. With some level of language ability, working-class Chinese can receive better salaries, and rent more affordable houses or apartments. Thus, the working class who have developed themselves start to move their homes to new ethnoburbs. As one male maintenance person in Rowland Heights mentioned: "My daughters are in high school here … We don't have that much money … But life is easier here than in other places. We [he and his wife] probably wouldn't both find jobs in Chinatown or somewhere like that."

### *Elderly Chinese*

With an average age over 65 by 2010, most of the early Chinese immigrants are out of the labour force and living with their families (Zhou 1992). Their low mobility requires them to reside in a walkable and spatially dense neighbourhood, such as Monterey Park. As one elderly

Chinese lady said, "I don't know where else I could live if I left this place." Public facilities, such as churches, libraries, parks, and hospitals, in a walkable distance are also a motivation for elderly Chinese to keep their homes in Monterey Park. As one 64-year-old woman said:

> Everything here is easy to get ... I don't have a car but I do grocery shopping for my family ... I usually leave home in the morning when my son and daughter-in-law go to work. Then I go to the library, chat with my friends, go to the community Mah-jong centre to play a while, and buy groceries on my way home. After dinner I always go to the park [Garvey Ranch Park] with my family ... I can see my friends and neighbours there too. If I am sick, I can even go to the hospital (Garfield Medical Center) by myself. It just takes me 10 minutes to walk there ... In Monterey Park, you don't need a car to survive.

Monterey Park is also equipped with well-developed senior housing for elderly Chinese. One male interviewee living in Temple City (another emerging ethnoburb in the East San Gabriel Valley), said he visited his parents in Monterey Park twice a week. His parents chose to stay at a seniors' condo in Monterey Park when he moved out to Temple City.

> They are more comfortable here [in Monterey Park] and feel more secure. They stayed with me at my home (in Temple City) for two weeks, just two weeks, and they told me they were not comfortable there and wanted to move back. Here [at the seniors' condo] they have friends and places to kill time ... [The seniors' condo] have nurses and physiotherapists to take care of them. I feel safe to put them there.

In summary, with "old" ethnoburbs' housing changing due to a large inflow of newcomers, Chinese professionals start moving out of old ethnoburbs and into new ethnoburbs. Chinese entrepreneurs then find better opportunities for their businesses, and new housing in new ethnoburbs, contributing to the prosperity of Chinese businesses in the new ethnoburbs. This entices working-class Chinese immigrants to move from the "old" to the "new" ethnoburbs. Thus, through different housing trajectories among Chinese subgroups, Chinese immigrants gradually settle their new homes and create their ethnic communities in new ethnoburbs.

With elderly Chinese staying still and new immigrants coming from China, old ethnoburbs remain high-density neighbourhoods with multi-unit housing types. They function as the residential, commercial, and cultural centre of their suburban area, as well as the "immigration gateway" for new immigrants from China. Besides changes made in old ethnoburbs, the housing in new ethnoburbs changes as well. With more Chinese immigrants settling down, new ethnoburbs gradually become multi-ethnic communities with considerable Chinese populations, and attract more well-established Chinese. The Chinese population soon booms. With the increasing Chinese population at a higher socio-economic level, more properties are needed in the real estate market. This large demand on the housing market would soon elevate housing prices, shrink open spaces, and increase community density, which eventually drives affluent Chinese families out to even newer and less dense suburbs, which consequently starts a new cycle of immigrant housing changes among ethnoburban Chinese and begins the emergence of yet newer ethnoburbs. Thus, the diverse housing trajectories of ethnoburban Chinese are highly intertwined with their housing characteristics and the dynamic housing types of Chinese ethnoburbs.

## Conclusion

Using comparative analysis between two Chinese "ethnoburbs," this chapter provides empirical insights for a contemporary theoretical discussion of immigrant housing and settlement in North America. By interviewing different Chinese subgroups in each ethnoburb, this chapter tries to depict a comprehensive picture of recent Chinese immigrants' diverse housing experiences and trajectories.

Through a comparison of the demographics and housing characteristics of Chinese immigrants and housing types in two representative Chinese neighbourhoods, this chapter first reveals the existence of demographic and housing disparities between ethnoburbs that emerged at different times. In terms of housing characteristics, "old" ethnoburbs are mostly occupied by elderly Chinese and Chinese newcomers, most of whom do not own homes, while Chinese residents in "new" ethnoburbs are mostly professionals, entrepreneurs, and working-class individuals, the majority of whom obtain ownership of their homes. In regard to immigrant housing types, "old" ethnoburbs are spatially dense and compact, usually with multi-unit housing types, whereas "new" ethnoburbs are geographically dispersed and scattered, and have a greater percentage of single-family houses. Housing characteristics among ethnoburban Chinese, associated with the different housing types among ethnoburbs, lead to diverse housing trajectories among ethnoburban Chinese.

Starting with Chinese professionals moving in, and ending with functioning as new "ports of entry," ethnoburbs evolve over time, representing a dynamic model of immigrant settlement pattern in the American suburbia. The fundamental reason for this dynamic settlement pattern is the diversifying housing choices among contemporary Chinese immigrants. And the underlying forces for such a diverse and prospering immigrant population comes from the constant inflow of immigrant newcomers from China. As increasing numbers of new immigrants from China settle in old ethnoburbs, more well-established Chinese will move out to new ethnoburbs, which in turn results in increasing housing density in new ethnoburbs and a new wave of Chinese seeking better housing in even newer ethnoburbs. As long as the globalizing economy continues to provide an influx of new Chinese immigrants landing at suburban "ports of entry," the housing shift of suburban Chinese will continue, and its impact on the local community landscape will continue. Thus, the finding of this chapter suggests a more profound understanding of geography in migration movements at different scales. Changing immigrant housing choices and trajectories, the dynamic immigrant settlement landscapes, and their diverse assimilation and integration levels all compose an interconnected relationship between geography and human: immigrant individual characteristics in shaping their housing experiences, at the household level; collective immigrant housing trajectories in shaping their neighbourhood and settlement landscape, at the community level; spatial transformation of immigrant settlement in shaping regional internal migration and the housing market, at the metropolitan level; and regional and national demographic and economic contexts in shaping transnational migration flows, at the international level. All are mirrored in the basic assimilation path of Chinese immigrants – to integrate to US society and achieve their "American Dream."

## REFERENCES

Ahmed, A.M., and M. Hammarstedt. 2008. "Discrimination in the Rental Housing Market: A Field Experiment on the Internet." *Journal of Urban Economics* 64 (2): 362–72. http://dx.doi.org/10.1016/j.jue.2008.02.004.

Alba, R.D., and J.R. Logan. 1992. "Assimilation and Stratification in the Homeownership Patterns of Racial and Ethnic Groups." *International Migration Review* 26 (4): 1314–41. http://dx.doi.org/10.2307/2546885.

Andersen, H.S. 2012. "Immigrants' Moves into Homeownership in Copenhagen: A Longitudinal Study." Paper presented at Boligforsker Seminar 2012, Denmark.

Barth, G. 1964. *Bitter Strength: A History of the Chinese in the United States, 1850–1870*. Cambridge, MA: Harvard University Press.

Borjas, G. 2002. "Homeownership in the Immigrant Population." *Journal of Urban Economics* 52 (3): 448–76. http://dx.doi.org/10.1016/S0094-1190(02)00529-6.

Carter, T., and D. Vitiello. 2012. "Immigrants, Refugees, and Housing." In *Immigrant Geographies of North American Cities*, ed. C. Teixeira, W. Li, and A. Kobayashi, 91–111. Don Mills, ON: Oxford University Press.

Coulson, E. 1999. "Why Are Hispanic- and Asian-American Homeownership Rates So Low? Immigration and Other Factors." *Journal of Urban Economics* 45 (2): 209–27. http://dx.doi.org/10.1006/juec.1998.2094.

Darden, J., and S. Kamel. 2000. "Black Residential Segregation in the City and Suburbs of Detroit: Does Socioeconomic Status Matter?" *Journal of Urban Affairs* 22 (1): 1–13. http://dx.doi.org/10.1111/0735-2166.00036.

Fan, C. 2002. "Chinese Americans: Immigration, Settlement, and Social Geography." In *The Chinese Diaspora: Space, Place, Mobility, and Identity*, ed. L. Ma and C. Cartier, 261–91. Lanham, MD: Rowman and Littlefield.

Fiedler, R., N. Schuurman, and J. Hyndman. 2006. "Hidden Homelessness: An Indicator-based Approach for Examining the Geographies of Recent Immigrants At-risk of Homelessness in Greater Vancouver." *Cities* (London, England) 23 (3): 205–16. http://dx.doi.org/10.1016/j.cities.2006.03.004.

Fong, T. 1994. *The First Suburban Chinatown: The Remaking of Monterey Park, California*. Philadelphia: Temple University Press.

Friedman, S., and E. Rosenbaum. 2004. "Nativity Status and Racial/Ethnic Differences in Access to Quality Housing: Does Homeownership Bring Greater Parity?" *Housing Policy Debate* 15 (4): 865–901. http://www.tandfonline.com/doi/abs/10.1080/10511482.2004.9521525#.VGpvpjTF8l8.

Haan, M. 2005. "The Decline of the Immigrant Home-Ownership Advantage: Life-Cycle, Declining Fortunes and Changing Housing Careers in Montreal, Toronto and Vancouver, 1981–2001." *Urban Studies* (Edinburgh, Scotland) 42 (12): 2191–212. http://dx.doi.org/10.1080/00420980500331983.

Hiebert, D., P. Mendez, and E. Wyly. 2008. "The Housing Situation and Needs of Recent Immigrants in the Vancouver Metropolitan Area." Working paper no. 08-01. Centre of Excellence for Research on Immigration and Diversity.

Kim, A.H., and M. Boyd. 2009. "Housing Tenure and Condos: Ownership by Immigrant Generations and the Timing of Arrival." *Canadian Journal of Urban Research* 18 (1): 47–73.

Krivo, L.J. 1995. "Immigrant Characteristics and Hispanic-Anglo Housing Inequality." *Demography* 32 (4): 599–615. http://dx.doi.org/10.2307/2061677.

Li, W. 1997. "Ethnoburb versus Chinatown: Two Types of Urban Ethnic Communities in Los Angeles." *Cybergeo*, 11 December: 1–11.

Li, W. 1998. "Anatomy of a New Ethnic Settlement: The Chinese Ethnoburb in Los Angeles." *Urban Studies* (Edinburgh, Scotland) 35 (3): 479–501. http://dx.doi.org/10.1080/0042098984871.

Lin, J., and P. Robinson. 2005. "Spatial Disparities in the Expansion of the Chinese Ethnoburb of Los Angeles." *GeoJournal* 64 (1): 51–61. http://dx.doi.org/10.1007/s10708-005-3923-4.

McKeown, A. 2001. *Chinese Migrant Networks and Cultural Change: Peru, Chicago, Hawaii, 1900–1936*. Chicago: University of Chicago Press.

Monterey Park Historical Heritage Commission. 1991. *Monterey Park Oral History Project, Interview transcripts*. Monterey Park, CA: Bruggemeyer Memorial Library.

Munnell, A., H.L.E. Browne, J. McEneaney, and G.M.B. Tootell. 1992. "Mortgage Lending in Boston: Interpreting HMDA Data." Working paper no. 92-7. Boston: Federal Reserve Bank of Boston.

Murdie, R.A. 2002. "The Housing Careers of Polish and Somali Newcomers in Toronto's Rental Market." *Housing Studies* 17 (3): 423–43. http://dx.doi.org/10.1080/02673030220134935.

Myers, D., W. Baer, and S.Y. Choi. 1996. "The Changing Problem of Overcrowded Housing." *Journal of the American Planning Association* 62 (1): 66–84. http://dx.doi.org/10.1080/01944369608975671.

Myers, D., and S.W. Lee. 1996. "Immigration Cohorts and Residential Overcrowding in Southern California." *Demography* 33 (1): 51–65. http://dx.doi.org/10.2307/2061713.

Myers, D., I. Megbolugbe, and S. Lee. 1998. "Cohort Estimation of Homeownership Attainment among Native-born and Immigrant Populations." *Journal of Housing Research* 9: 237–69.

Myers, D., and J. Wolch. 1995. "The Polarization of Housing Status." In *State of the Union: America in the 1990s*, vol. 1, *Economic Trends*, ed. R. Farley, 269–334. New York: Russell Sage Foundation.

Owusu, T.Y. 1998. "To Buy or Not to Buy: Determinants of Home Ownership among Ghanaian Immigrants in Toronto." *Canadian Geographer/Le Géographe canadien* 42 (1): 40–52. http://dx.doi.org/10.1111/j.1541-0064.1998.tb01551.x.

Painter, G., L. Yang, and Z. Yu. 2004. "Homeownership Determinants for Chinese Americans: Assimilation, Ethnic Concentration and Nativity." *Real Estate Economics* 32 (3): 509–39. http://dx.doi.org/10.1111/j.1080-8620.2004.00101.x.

Park, R.E., E.W. Burgess, and R. McKenzie, eds. 1967. *The City* (1925). Chicago: University of Chicago Press.

Ray, B., and V. Preston. 2009. "Geographies of Discrimination: Variations in Perceived Discomfort and Discrimination in Canada's Gateway Cities." *Journal of Immigrant & Refugee Studies* 7: 228–49.

Rossi, P.H., and E. Weber. 1996. "The Social Benefits of Homeownership: Empirical Evidence from National Surveys." *Housing Policy Debate* 7 (1): 1–35. http://www.tandfonline.com/doi/abs/10.1080/10511482.1996.9521212#.VGpwNjTF8l8.

Saiz, A. 2007. "Immigration and Housing Rents in American Cities." *Journal of Urban Economics* 61 (2): 345–71. http://dx.doi.org/10.1016/j.jue.2006.07.004.

Schill, M.H., S. Friedman, and E. Rosenbaum. 1998. "The Housing Conditions of Immigrants in New York City." *Journal of Housing Research* 9 (2): 201–35.

Sundstrom, R. 2004. "Racial Politics in Residential Segregation Studies." *Philosophy and Geography* 7 (1): 61–78. http://dx.doi.org/10.1080/1090377042000196029.

Teixeira, C. 2011. "Finding a Home of Their Own: Immigrant Housing Experiences in Central Okanagan, British Columbia, and Policy Recommendations for Change." *Journal of International Migration and Integration* 12 (2): 173–97. http://dx.doi.org/10.1007/s12134-011-0181-9.

Zavodney, M. 1998. "Welfare and the Locational Choices of New Immigrants." *Federal Reserve Bank of Dallas Economic Review*, 2nd quarter, 1997.

Zhou, M. 1992. *Chinatown: The Socioeconomic Potential of an Urban Enclave.* Philadelphia: Temple University Press.

Zhou, M. 2004. "Are Asian Americans Becoming White?" *Contexts* 3 (1): 29–37. http://dx.doi.org/10.1525/ctx.2004.3.1.29.

# PART TWO

# The Economic Experiences of Immigrants

# Introduction to Part Two: The Economic Experiences of Immigrants in Canada and the United States

JOHN R. MIRON

The notion of agency – the idea that people can and do act purposefully to improve the conditions in which they find themselves[1] – is an important foundation for this book. Given the focus of the second half of this book on the economic experiences of immigrants, a clear understanding of motivation and behaviour is essential because how immigrants experience and cope with life in their new country reflects at least in part what they were trying to achieve by immigrating and the strategies they used. My focus in introducing the five chapters on the economic experience of immigrants is on the agency of individuals: how, when, and why they cope with the problems and barriers associated with immigration. In the vein of agency, the act of immigration – and the strategy underlying it – can be seen as a "choice" and migrants as mindful of the consequences of alternative choices.[2] Therefore, a focus on agency here helps us think about and interpret the economic experiences of immigrants. My purpose is to situate the chapters in the second half of this book within the literature on agency in migration.

The five chapters in this section have different approaches and methods: chapters 8 and 9 analyse large national data sets to tease out trends and experiences of immigrant wage-earners in labour markets in Canada and the United States respectively, whereas chapters 10–12 present empirical case studies on the various aspects of immigrant entrepreneurship. All these chapters use an inductive approach – process tracing – wherein authors use empirical evidence to build a consistent story that explains the experience of immigrants. My approach here in this introduction is deductive and is based on the assumption that individuals make rational choices. A deductive perspective in introducing chapters that are largely inductive gives us a useful triangulation, a kind of ground truthing or cross-checking that helps us to understand better both the strengths of these chapters as well as areas and questions that need more work.

## Structure and Agency

In thinking about the human condition, social scientists often distinguish between agency and structure. In everyday language, "structure" refers to (1) an arrangement of – or relation between – the parts or elements of something complex or (2) the quality of being organized. In both cases, "structure" can be seen as a characterization (schema) imposed by the observer in order to simplify and to highlight the operation of a process under study. From an agency perspective, structure is a context – often envisaged as a set of rules and resources – thought to shape the outcomes experienced by individuals and thereby their choices from among alternatives.[3] That is my approach in this introduction. From a structural perspective, the important questions would concern how and why notions of culture (identity), class, political and/or network affiliation, age, gender, social status, religious beliefs, ethnicity, sexual orientation, or immigrant status help us think about the outcomes that people experience. I do not pursue a structural perspective here. At the same time, the distinctions between agency and structure are easily blurred. It can be argued, for example, that culture is fragmented across groups and inconsistent in manifestation. This is culture as complex rule-like structures – resources that an individual can put to strategic use.[4] In this respect, agency and culture (indeed any aspect of structure) may not be distinct perspectives.

A different way of casting the argument about agency versus structure is to think about the implications of having a great variation in human condition and experience: that is, human diversity. When we ask what immigrants were hoping to achieve by immigrating, what strategies they used, and how they experienced and coped with life in their new country, we are averaging over outcomes that vary considerably from person to person. Empirically, we often structure (disaggregate) immigrants by characteristics so that we can estimate how the outcomes differ from group to group. Used one way, this is structure exogenously constraining the outcomes on immigrants: for example, when the labour force earnings expected by male immigrants are higher on average than those expected by their female peers. Used differently, this is structure with which immigrants interact to improve their well-being. An example here might be where success in the job market is affected by whether immigrants received their degrees from a Canadian university versus one abroad. An immigrant might therefore choose to return to school to improve his or her job prospects.

## Markets, Strategy, Allocation, and Experience

We can imagine that immigrants act in various ways – make choices – that then give rise to economic experiences. Their actions – taken in combination as a "strategy" – include choices they make as producers (be they workers supplying labour or entrepreneurs supplying products, capital, or services) and as consumers of an array of goods and services across various markets.

The notion of "economic experience" covers what might be characterized as three distinct sets of ideas. One set includes outcomes as reflected in returns and costs experienced by the immigrant (both pecuniary and non-pecuniary) as well as the risks. A second set includes outcomes as reflected in education and skill acquisition (human capital) experienced by the immigrant. A third set includes outcomes with respect to savings, investment, and financial capital experienced by the immigrant. These outcomes may well differ from what the individual might have anticipated before immigrating.

Much of this economic experience happens in the context of the operation of local markets: for instance, markets for labour, markets for capital, and markets for products and services. The notion of a market is that price is used to allocate products and services: for example, the allocation of labour by households to firms and the allocation of product by firms to consumers. Important here too is the operation of markets in the presence of heterogeneity. Sometimes, markets readily adjust to variations in quality; other times, they do not. An immigrant perceived to have language difficulties, an unfamiliar or different education or skill training, or a difference in culture or disposition may risk discrimination, unemployment, or underemployment as a result.

However, not all allocation is based on price. A household uses its labour in part for various household production activities. A firm also makes allocations within the firm without relying on market prices. To fully appreciate the economic experience of immigrants, we need to take into account also the production and allocation of goods and services that occur outside a price mechanism. Particularly important here is the role of social networks, be they a household, community, firm, or industry, in shaping outcomes that help form the economic experience of immigrants.[5]

## Motivation in Immigration

In a seminal study, Ravenstein (1885, 1889) interpreted internal migration in the United Kingdom as evidenced in the 1881 census. Ravenstein also saw people are purposeful in their actions; to him, migration was a response to industrialization and the growth of jobs in urban areas. What did Ravenstein discover about the strategies of migrants? He concluded that (1) most migrants move a short distance and in the direction of cities; (2) the likelihood of migration declines with distance; (3) the process of dispersion is the inverse of the process of absorption; (4) each main current of migration produces a compensating countercurrent of return migration; (5) migrants proceeding long distances go to a city; (6) residents of towns are less likely to migrate than rural residents; (7) females are more likely to migrate than are males. His analysis covered Ireland, Scotland, England, and Wales: areas with differences in language, ethnicity, and custom. Presumably, these differences had some effect on an individual's decision to migrate and on their subsequent economic experience. However, none of this made its way directly into Ravenstein's seven laws. Instead, geographic distance can be thought to proxy such differences. Put differently, Ravenstein's conclusions evidence a strategy for migration to the extent that migrants are reluctant to move long distances.

Extending Ravenstein, purposeful people might be seen to immigrate for a variety of reasons, each of which instances an improvement in their condition but that may vary greatly in terms of subsequent economic experience. Sometimes, it is an act of the heart, as when family or friends who have already left entreat you to join them abroad. Sometimes, it is the attraction of local amenities (e.g., stable democracy, social tolerance, good infrastructure, social safety net, clean air, potable water, mild temperatures, beaches or even nearby ski slopes) compared to your current home. Other times, the notion of immigration is attractive because you seek the adventure – the challenge, excitement, and novelty (new life) – associated with living abroad. Still other times, you want to immigrate elsewhere because adverse conditions or risks at home – be they, for example, religious, social, political, cultural, or related to personal security – make living or working elsewhere necessary or more attractive. These reasons may have little to do with economic opportunities at the destination and may well lead to immigration despite job prospects.

Other times, an employer offers a good job opportunity in the form of a transfer to another country. Some times, you have a business, trade, or occupation that might be struggling locally but that you think might do better at abroad. Alternatively, you may have a business of your own that develops in a way that makes relocation to another country attractive. In all these cases, you might immigrate fully expecting an economic experience that is better than what you had before.

## Impacts of Immigration

As with any other demographic change, immigration has impacts on local markets that affect the experiences of immigrants and non-immigrants alike. In an important sense, these impacts are imponderable. One immigrant with a grand vision, a will and ability to succeed, a network of backers and risk-takers, and a fortunate sequence of circumstances may be able to reshape singlehandedly the urban economy and the commercial prospects of the destination city. Along the way, local markets of all kinds (e.g., labour, capital, land, real estate, and exports) get reshaped or even reinvented.[6]

What I can say something about is the ordinary operation of competitive markets. This book focuses on two markets in particular: the market for housing and the market for labour. Here I can use the comparative-statics analysis of economists to think about outcomes. In either market, we can imagine equilibrium first in the absence of immigration and then in the presence of immigration. This inherently involves some counterfactual speculation: the world that could have been versus the world we see. The "effect" of immigration is the differences between these two outcomes. Methodologically, we need to be clear about how we can know the world that could have been.

In terms of the market for labour, immigration has two sets of consequences. One is in terms of final demands for the output of industries. More immigration means a larger market locally. That means more demand for goods and services. Individual firms can now operate at a larger scale. This in turns raises the possibility of a greater division of labour. In this way, immigration can produce greater profit, higher wages, and lower prices.[7] The other is in terms of the supply of labour. Immigration means more workers with particular skills and education. In general, this means lower wages, at least initially, in those labour market sectors where the number of immigrant workers is the largest. The resolution of these competing effects is at the heart of the methodology in our empirical analysis.

What then does a focus on agency tell us about how we should look at the economic experience of immigrants? First, the variety of reasons for immigrating suggests that the economic experiences of immigrants may vary broadly. Second, it is helpful to look at experience in terms of costs and returns. Third, we need to look at the risks facing immigrants. Fourth, we need to look at the role of social networks (be these household, family, or community) in affecting economic experience. Fifth, we need to look at savings behaviour and its role in the immigrant experience. Let me conclude with comments on each of these.

## Costs and Returns to Immigration and Economic Experience

In deciding whether to immigrate, our perspective and strategy can range from short term to longer term. When our perspective is short term, the act of immigration looks more like an "adventure in trade." We immigrate in the hope of immediate gain, knowledgeable of the risks, and our strategy is to move on again (or return home) soon if that better opportunity is not quickly realized. At this end of the scale, I would include migrant (seasonal) workers. Also, one might include here immigrants who are permanent residents but see themselves as members of a family or other social group that spans two or more nations (i.e., are transnationals) and use a strategy of remitting to, or otherwise benefitting, members abroad. At the other end of the scale, our perspective and strategy may be longer term: e.g., based more on the benefits that immigration may bring for our children and future generations.

Whatever the perspective and our strategy in moving, there are benefits and costs associated with the choice to immigrate. I use the term "benefit" here to mean any change in our condition that makes us better off; in contrast, a "cost" is any change that leaves us worse off or that requires an expenditure of money or time and effort to overcome. A new country means new sets of laws and citizenship or residency requirements and might mean a new language and new institutions, customs, and practices. Indeed, some of these changes are among the benefits that we sought by moving. However, other changes, including some unforeseen consequences (risks), may be more like costs.

Purposeful people might be expected to use strategies that allow them to receive the benefits of immigration while keeping costs as low as possible. In this sense, people can be thought to manage risks and returns. In this respect, the act of immigration is similar to a financial investment. Along these lines, Sjaastad (1962) blazed a new approach to the study of migration. He developed ideas about important costs and returns to migration: including both monetary and non-pecuniary amounts. The novelty of his analysis is that it treats migration explicitly as an investment (in human capital) which requires resources and which generates a return. Further, he argues that migration cannot be viewed in isolation; complementary investments in human capital are as important as migration itself. Sjaastad's ideas have been widely used to study interregional migration across large market-oriented national economies such as those of the United States,[8] Canada,[9] and the European Union.[10] More importantly for this book, Sjaastad has also been an important influence in major work in the area of international migration: for example, Borjas 1987, Massey et al. 1993, Massey and Espinosa 1997, and Chiquiar and Hanson 2005.

For Ravenstein and his intellectual descendants, the distance between origin and destination place can be thought to be a rough proxy for the costs of migration. This led naturally to a focus on the gravity model[11] and the intervening opportunity model[12] to explain flows of migrants. However, a focus on distance alone does not help us much when we are trying to characterize the economic experience of immigrants.

One shortcoming of Sjaastad is that emigration and immigration are simply two sides of the same coin. However, Ravenstein and more recent scholars argue that the factors that are important in determining the level of emigration from one country are different from the factors important in determining the level of immigration to another. Lowry (1966) exemplifies this perspective, arguing (p. 95) that the volume of emigration (out-migration) from any place is demographically driven (i.e., related to the size and composition of its population), while the volume of immigration to a prosperous place is driven by favourable labour market conditions. Let me cast that argument a different way. Young adults are the most mobile. Emigration depletes the local population of young adults. In the receiving country, immigration adds to the population of young adults and hence increases outmigration downstream. That emigration and immigration are two distinct processes also helps us to begin to understand why outmigration from one country subsequently leads to heightened levels of return and onward migration.

In a sense, some benefits and costs can be directly translated into economic experiences that are correspondingly positive or negative and readily measured. For example, a better real income can be measured as the difference in earnings or incomes that migration makes possible. In other senses, the benefits and costs may be more difficult to measure. Costs might be incurred, for example, in (1) becoming fluent in a new language, (2) acquiring information about and finding employment in a regulated occupation or profession or employment appropriate to one's education or training, (3) adapting to local customs, habits, and practices, (4) coping with bias and discrimination, and (5) keeping one's family or group together – including traditions and identity – in the face of competing attractions (and possibly disapproval) in the new country.

In chapter 8, Pendakur and Pendakur argue that minorities in Canada suffered a growing earnings disadvantage between 1990 and 2005, especially in the early 1990s. The decline in relative earnings was about 10 percentage points for Canadian-born visible minorities. This pattern is observable for both men and women, in each of Canada's three largest CMAs, and for most visible minorities. The increasing disparity faced by minorities born in Canada is troubling. These individuals are raised and educated in Canada, and in general do not face the same barriers as immigrants with respect to language knowledge, recognition of credentials, accent penalties, or absence of social networks. Importantly for this book, the same pattern is also observed broadly among immigrants. The decline in relative earnings was about 20 percentage points for both white and visible-minority immigrants. On this basis, Pendakur and Pendakur suggest that the relative labour market outcomes of recent immigrants have worsened substantially in the past two decades. That the immigrant earnings disparity has increased over time is troubling in light of Canada's large intake of immigrants and steady increase in ethnic diversity. Canada's cities are seen worldwide as roadmaps to cohesive diversity; growing immigrant and visible minority labour market disparity threatens this.

## Risks in Immigration and Economic Experience

The 1970s saw two major extensions of the ideas of Sjaastad and Lowry. One development was the Harris-Todaro model, which emerged as a principal economic explanation for rural-to-urban migration in less developed countries (LDCs).[13] To take into account the high levels of urban unemployment in LDCs, the Harris-Todaro model saw migration as a response to differences in expected incomes after the uncertainty associated with unemployment is taken into account.[14] Put differently, in deciding whether to migrate to the city, rural residents traded off a known low income there against a better possible income in the city that came with a high risk of unemployment. In deciding whether to migrate, what is important here are attitudes to risk.[15]

The second development was in the work of David (1974), Miron (1978), Katz and Stark (1986), and others that make use of job search theory. Unlike the Harris-Todaro model, in which a prospective migrant is deterred by high unemployment, a job search perspective imagines that the migrant collects multiple job offers and has to decide when to stop searching. Because the migrant chooses the job offer with the highest wage, the variability of the wage distribution is as important as its expected value. This leads to what David calls "fortune": the prospect of a wage much higher than expected for that market.

Considerations of risk and fortune imply that not everyone will view the prospects for migration the same. To the extent that individuals differ in their willingness to take risks, immigrants will self-select. Borjas (1987) argues that the earnings of immigrants differ from those of the native-born because of this. Looking at the experiences of immigrants from forty-one countries from the 1970 and 1980 US censuses, Borjas finds that immigrants from countries that have a high GNP, low level of income inequality, and a competitive political system tend to have high earnings relative to native-born individuals at the same skill level.

In chapter 9, Wang and Lysanko offer another perspective on the risks in immigration in the United States. They define a person as underemployed when he or she possesses more years of education than the job requires. They first calculate the average educational attainment (measured by the years in school) for each occupation; employees are considered underemployed when their years in school are more than one standard deviation above the mean for their occupation. Not surprisingly, immigrants with low level of educational attainment (e.g., from Mexico, Guatemala, and El Salvador) have a low rate of underemployment. Chinese, Filipinos, and Koreans, by contrast, are 1.5 times more likely to be underemployed than the US-born whites. This suggests that the better-educated foreign-born have difficulty in transferring their education credential and human capital obtained in their country of origin to the US labour market.

### Household, Family, and Community and Economic Experience

Another stream of thought emerged in work on the economic nature of social groups or networks including the household. I define a household here as a set of individuals sharing a private dwelling (e.g., a detached house, flat, or apartment). Here, we must be careful not to think of a household as exogenously defined. The very formation of a household itself (persons with whom we share accommodation, from lodger to extended family) may be part of the coping strategy used by individuals and nuclear families. Starting from the work that largely began with Becker (1981), the household – to take one case of a social group or network – is seen as a non-market institution for the production and allocation of goods and services. There are different ways of thinking about decision-making within the household: ranging from altruism to bargaining. However, an essential element is that the household provides a shared home in the event of unemployment or illness: in effect, it insures its members against adverse outcomes in the labour market. Expressed differently, the household supplies labour when the market demands more workers and absorbs labour back into itself when the market demands less (see Wallerstein 1984).

Suppose instead we start to think about the social group or network as an extended family: a group of people related by blood or marriage. Unlike a household, the members of a family need not reside together. The popular press often reports on the large volume of international remittances that some immigrants send to family members in the home country. Why do some immigrants do this? The common presumption is that these transfers represent altruistic payments by the immigrant. Hoddinott (1994), in contrast, envisages a rural nuclear family consisting only of a prospective migrant and his parents: all assumed purposeful. Hoddinott imagines an implicit contract between parents and son that specifies how the costs and benefits of migration are to be shared. Some of this sharing may be in the form of cash: typically as gifts or loans. Remittances by the migrant son give the parents an additional source of income or capital. In turn, the son gets psychic and perhaps financial support while getting established in the urban area. Further, the parents provide implicit insurance against unemployment: a home to return to should the son be laid off. The parents may also provide a bequest in the form of a family home (investment) to return to when the son retires. These ideas are tested in de la Brière et al. (2002) in the case of immigrants to the United States from the Dominican Republic. Further, Amuedo-Dorantes and Pozo (2006) argue that remittances provide a way for immigrants to insure a venture into a risky labour market abroad that they might not otherwise attempt.

As immigrants start to flow from a given origin country, a community of recent immigrants ("expats") starts to emerge in the destination country. Word-of-mouth expands the information available to a nuclear family in the origin country. Community migrant networks add to the information and assistance available to new immigrants and may make the nuclear family less important as a consequence.[16] Fafchamps and Gubert (2007) find that geographic proximity, age, and wealth differences are major determinants of mutual insurance links among immigrants and that gifts respond to shocks and to differences in health status. Variations in the density of community networks across the destination country come to affect where individuals choose to migrate. The result, "chain migration," helps lead to the development of ethnic enclaves.

Earlier in this book, Kataure and Walton-Roberts (chapter 6) argue that suburban immigrant communities represent an important demographic change in the structure of Canadian and American cities. They see suburban living in an ethnic enclave as a choice by immigrants that is rooted in social and cultural preferences. Despite the drawbacks to ethnic clustering, the ethnic neighbourhood has served a positive role for first-generation immigrants. Immigrants also widely see the purchase of a home in these suburbs as reflecting their integration into the adopted country. However, it is less clear whether these ethnic enclaves will also be attractive to the next generation. How, for example, will the ethnic enclave play into the ways that adult offspring of immigrants measure their own success (e.g., in terms of economic outcomes and life courses) and integration?

## Savings Behaviour and Risk Mitigation in Immigration

I now turn to the role of capital accumulation and savings over the life cycle of an immigrant. Immigrants can arrive with substantially different levels of capital: from penniless refugee to wealthy investor. Further, their capital on hand at the time of immigration may take the form of net worth or a loan. Either way, they will purposefully plan to increase their level of net worth: repaying loans and accumulating assets. Their need for capital might be driven by, for example, a new business they want to start, plans for retirement or a child's education, precautionary savings (anticipation of a rainy day), plans to purchase a family home, or plans to purchase other consumers durables (e.g., an automobile). Dustmann (1997), modelling the effect of uncertainty on precautionary savings for immigrants who plan one day to return home, argues that whether precautionary savings of migrants are above or below those of natives depends on the risk in host- and home-country labour markets and on the correlation of labour market shocks between these markets. As with any other household, immigrants must strike a balance among these factors, taking into account the risks and returns of each.

This perspective on risks and returns has implications too for the housing experience of immigrants, the subject matter of the first half of this book. An investment in housing – in the form of homeownership – carries its own special level of risks. Housing is widely seen to provide a hedge against inflation. However, as an investment, housing historically has a low rate of return; households that overinvest in housing fall behind other households. As well, housing poses a risk of serious loss that can linger over a period of several years at a time. Henley (1998), among others, argues that when house prices drop, households with substantial mortgages (ranging up to negative equity) become less likely to move and therefore become less likely to be able to take up better-paying jobs elsewhere. As Darden (chapter 2) notes earlier in this book, this can be particularly important to immigrants with meagre savings; an investment in owner-occupied housing may be just too risky for them.

At the same time, we must be mindful of the role played by suppliers of rental housing. For a household to "choose" to rent, there must be an investor willing to be a landlord. The landlord who takes out a mortgage in order to purchase a rental apartment building takes on a price risk in borrowing "long" to lend "short." Put differently, the investor assumes a long-term debt obligation in a market where market rent (and occupancy) may be volatile in the short run. In markets where price risk is higher (often taken to be local markets that are smaller and hence less liquid), investors will be less interested in rental housing as an investment. In such cases, we might expect to see immigrant nuclear families opt instead for homeownership (despite the risk), either living alone or in shared accommodation.

Be they immigrants or native-born, Miron (2004) argues, consumers choose living arrangement, tenure, and housing expenditure on the basis of price, income, wealth, and tastes. As it is costly to alter one's housing conditions, consumers employ strategies to cope with labour market risks and expectations about future earnings. These strategies link household savings with housing career. Miron shows how individuals and families in Canada responded to a decline in the real earnings of men in the 1980s and 1990s by use of coping strategies that included doubling up, substandard housing, and "overspending" on rental housing (i.e., spending more than 30% of household income on rent). Recent immigrants (those who immigrated to Canada in the last eight years) were more likely to live in shared accommodation, more likely to be renters, and less likely to overspend on rent (see also chapter 4 of this book). These are all suggestive of the consumer who – while not yet a homeowner – has a strategy of saving towards homeownership.

Finally, I now turn to the relationships between self-employment and immigration and the former's ties to savings behaviour. I define two categories of self-employment.

A. This includes persons with the abilities, motivation, skills, and presence to establish their own successful enterprises and hire employees. A good education, extensive work experience, a good professional reputation, and sufficient capital funding are commonly thought to be among the requirements for success in this kind of self-employment. These are the self-employed as "boss" and they may well have a need for substantial capital equipment and working capital in their enterprise.

B. This includes persons who are self-employed because of structural constraints they face. They are characterized by (a) an absence of full-time employment, (b) a category of pseudo-employment such as "associate," "intern," "representative," or "consultant," or (c) having no employees of their own. The capital equipment and working capital needs of these self-employed are typically modest.

In the popular literature, racial and ethnic discrimination are thought to make the target groups (ethnic minorities, immigrants, and the physically disabled) more likely to form their own businesses under either category A or category B. Clark and Drinkwater (2000) study self-employment among immigrants in England and Wales. From their perspective, self-employment is affected by ethnic characteristics as well as sectoral earnings differentials. They find that differences in an individual's predicted earnings in paid and self-employment are strongly correlated with self-employment. Individuals with low English fluency, and recent immigrants are less likely than other members of ethnic minorities to be self-employed.

Fundamental to the landscape itself is the ethnic entrepreneurship that is often key to many immigrants' economic experiences, either as a means for small-business ownership or employment opportunities or as a way to adapt to the host country and maintain connections with ones' home nation or region. In chapter 10, Oberle studies the morphology and character of the Latino commercial landscape and tells us about the local Hispanic immigrant community and its role in the metropolitan area. The size of the district is directly proportional to the size of the population. The retail landscape yields additional insight about immigrant experience. Accessibility usually implies an orientation towards pedestrians and local traffic, whereas extensive parking lots suggests shoppers arriving from elsewhere. The degree of business specialization and counts of business types (supermarket, bookstore, hardware store, money-wiring, restaurant, or established commercial centre) indicate whether the neighbourhood has a small or emerging Latino population. Shop names, colours, and symbols signify different affiliations, whether it has a pan-Hispanic focus or is an establishment that attract customers from a one particular region. Similarly, the proportion of Spanish to English signage indicates whether the establishment serves non-Hispanic customers, Mexican-Americans, or other native-born Hispanics.

In chapter 11, Chacko and Price study ethnic enterprise by Bolivian and Ethiopian immigrants in the Washington metropolitan area. Bolivian entrepreneurs focus on domestic services (housekeeping and childcare) and construction. Ethiopian entrepreneurs focus on the taxi and limousine, restaurant, real estate, and travel agency businesses. Some of these businesses serve a wider Washington population: for instance, food, taxi and limousine, domestic work, and construction services. Others locate their enterprises near corresponding ethnic enclaves. The co-location of business and residences for each ethnic group means that business owners can more easily tap into the local labour pool as well as access a ready market for their goods and services. Chacko and Price argue that immigrant entrepreneurs help each other. Social networks make it possible to draw on the knowledge, expertise, and experience of others within the community. Non-governmental organizations (NGOs) that cater to the immigrant community provide information, advice, loans, and assistance with bureaucratic paperwork. Chacko and Price find evidence here of integration as time passes: immigrant entrepreneurs develop new ventures in locations that draw on a larger clientele that includes mainstream Americans and immigrants from other ethnic groups. Their evidence suggests that the rise of businesses owned or co-owned by immigrants can help revive local economies.

In chapter 12, Li and Lo examine the roles of financial institutions in immigrant integration. Their exploratory work suggests that mainstream financial institutions do not address the needs of new immigrant businesses. Bank credits are not readily available at the start-up stage. The financial sector's unfamiliarity with (and perhaps scepticism towards) the business practices of immigrant groups is key. Li and Lo offer suggestions to banks that want to expand their assets in ethnic enterprise. First, more training, educational, or bridging programs are needed to give prospective immigrant entrepreneurs knowledge about doing business in Canada and the United States: for example, exposure to regulations and procedures, sources of start-up funding, business proposal writing, and mentoring possibilities. Second, financial institutions need to reinvent themselves as mentors and partners for new immigrant businesses. Finally, banks need to encourage initiatives to help create inter-ethnic networks to help immigrant entrepreneurs learn from one another.

## Conclusion

The purpose of this introduction has been twofold. First, it positions each of the chapters that follow. Second, it provides an agency-based perspective that helps identify important questions as the reader thinks about those papers. Let me summarize the questions.

1. What evidence do the authors present about the goals of immigrants? What were people trying to achieve by immigrating; how do we know this?
2. How did the author assess or incorporate the risks faced by immigrants in the labour market and in entrepreneurship? With respect to earnings and income, remittances, human capital, and financial capital, what strategies did immigrants use to cope with these risks? How did immigrants incorporate the choice between rental housing and homeownership in addressing these risks?
3. What evidence does the author present of the costs borne by immigrants and the impact on their economic experience?
4. In addition to what the immigrants were typically observed to do, what other choices were open to them: as producers and as consumers? To what extent, for example, did immigrants make use of household arrangements, such as lodging or doubling up, to better cope with the economic environment in which they found themselves? How does the author assess the importance of social networks and other mechanisms for enabling the settlement of immigrants?
5. In each chapter, how did the author envisage "structure"? Was structure treated exogenously: as a set of rules? Was structure also treated endogenously: as a set of resources?

## NOTES

1 Emirbayer and Mische (1998, 962) extend the notion of agency to include "selfhood, motivation, will, purposiveness, intentionality, choice, initiative, freedom, and creativity."

2 I am mindful here that this does not allow us to understand interpretive processes in unfolding situations where the outcomes of choices get imagined, evaluated, and reconstructed by the person under study.

3 See Sewell 1992 and Emirbayer and Goodwin 1994.

4 See DiMaggio 1997.

5 See Munshi 2003.

6 Too often forgotten in our understandable enthusiasm about such people is that the gain of the destination country may well be at the expense of the originating country.

7 I leave aside here the issues of congestion and related externalities such as air pollution that might lead some to discourage population growth and immigration.

8 See Greenwood 1969.

9 See Vanderkamp 1971.

10 See Boeri and Brücker 2005.

11 See Stewart 1941, Zipf 1946, Carroll 1955, and Carrothers 1956.

12 See Stouffer 1940.

13 See Harris and Todaro 1970.

14 See Todaro 1969 and Harris and Todaro 1970. See also the update in de Haas 2010.

15 See Stark and Levhari 1982.

16 See Winters et al. 2001 and Munshi 2003.

## REFERENCES

Amuedo-Dorantes, C., and S. Pozo. 2006. "Remittances as Insurance: Evidence from Mexican Immigrants." *Journal of Population Economics* 19 (2): 227–54. http://dx.doi.org/10.1007/s00148-006-0079-6.

Becker, G. 1981. *A Treatise on the Family*. Cambridge, MA: Harvard University Press.

Boeri, T., and H. Brücker. 2005. "Why Are Europeans So Tough on Migrants?" *Economic Policy* 20 (44): 629–703. http://dx.doi.org/10.1111/j.1468-0327.2005.00148.x.

Borjas, G.J. 1987. "Self-selection and the Earnings of Immigrants." *American Economic Review* 77 (4): 531–53.

Carroll, J.D. 1955. "Spatial Interaction and the Urban-Metropolitan Description." *Papers and Proceedings of the Regional Science Association* 1 (1). http://dx.doi.org/10.1111/j.1435-5597.1955.tb01418.x.

Carrothers, G.A.P. 1956. "An Historical Review of the Gravity and Potential Concepts of Human Interaction." *Journal of the American Planning Association* 22 (2): 94–102.

Chiquiar, D., and G.H. Hanson. 2005. "International Migration, Self-Selection, and the Distribution of Wages: Evidence from Mexico and the United States." *Journal of Political Economy* 113 (2): 239–81. http://dx.doi.org/10.1086/427464.

Clark, K., and S. Drinkwater. 2000. "Pushed Out or Pulled In? Self-employment among Ethnic Minorities in England and Wales." *Labour Economics* 7 (5): 603–28. http://dx.doi.org/10.1016/S0927-5371(00)00015-4.

David, P.A. 1974. "Fortune, Risk, and the Microeconomics of Migration." In *Nations and Households in Economic Growth*, ed. P.A. David and M.W. Reder, 21–88. New York: Academic Press. http://dx.doi.org/10.1016/B978-0-12-205050-3.50007-5.

de Haas, H. 2010. "Migration and Development: A Theoretical Perspective." *International Migration Review* 44 (1): 227–64. http://dx.doi.org/10.1111/j.1747-7379.2009.00804.x.

de la Brière, B., E. Sadoulet, A. de Janvry, and S. Lambert. 2002. "The Roles of Destination, Gender, and Household Composition in Explaining Remittances: An Analysis for the Dominican Sierra." *Journal of Development Economics* 68 (2): 309–28. http://dx.doi.org/10.1016/S0304-3878(02)00015-9.

DiMaggio, P. 1997. "Culture and Cognition." *Annual Review of Sociology* 23 (1): 263–87. http://dx.doi.org/10.1146/annurev.soc.23.1.263.

Dustmann, C. 1997. "Return Migration, Uncertainty, and Precautionary Savings." *Journal of Development Economics* 52 (2): 295–316. http://dx.doi.org/10.1016/S0304-3878(96)00450-6.

Emirbayer, M., and J. Goodwin. 1994. "Network Analysis, Culture, and the Problem of Agency." *American Journal of Sociology* 99 (6): 1411–54. http://dx.doi.org/10.1086/230450.

Emirbayer, M., and A. Mische. 1998. "What Is Agency?" *American Journal of Sociology* 103 (4): 962–1023. http://dx.doi.org/10.1086/231294.

Fafchamps, M., and F. Gubert. 2007. "The Formation of Risk Sharing Networks." *Journal of Development Economics* 83 (2): 326–50. http://dx.doi.org/10.1016/j.jdeveco.2006.05.005.

Greenwood, M.J. 1969. "Analysis of Determinants of Geographic Labor Mobility in the United States." *Review of Economics and Statistics* 51 (2): 189–94. http://dx.doi.org/10.2307/1926728.

Harris, J.R., and M.P. Todaro. 1970. "Migration, Unemployment and Development: A Two-Sector Analysis." *American Economic Review* 60 (1): 126–42.

Henley, A. 1998. "Residential Mobility, Housing Equity, and the Labour Market." *Economic Journal* 108 (447): 414–27. http://dx.doi.org/10.1111/1468-0297.00295.

Hoddinott, J. 1994. "A Model of Migration and Remittances Applied to Western Kenya." *Oxford Economic Papers* 46: 459–76.

Katz, E., and O. Stark. 1986. "Labor Migration and Risk Aversion in Less Developed Countries." *Journal of Labor Economics* 4 (1): 134–49. http://dx.doi.org/10.1086/298097.

Lowry, I.S. 1966. *Migration and Metropolitan Growth: Two Analytical Models.* San Francisco: Chandler Publishing Co.

Massey, D.S., J. Arango, G. Hugo, A. Kouaouci, A. Pellegrino, and J.E. Taylor. 1993. "Theories of International Migration – A Review and Appraisal." *Population and Development Review* 19 (3): 431–66. http://dx.doi.org/10.2307/2938462.

Massey, D.S., and K.E. Espinosa. 1997. "What's Driving Mexico-U.S. Migration? A Theoretical, Empirical, and Policy Analysis." *American Journal of Sociology* 102 (4): 939–99. http://dx.doi.org/10.1086/231037.

Miron, J.R. 1978. "Job-search Perspectives on Migration Behaviour." *Environment and Planning A* 10 (5): 519–35. http://dx.doi.org/10.1068/a100519.

Miron, J.R. 2004. "Housing Demand, Coping Strategy, and Selection Bias." *Growth and Change* 35 (2): 220–61. http://dx.doi.org/10.1111/j.0017-4815.2004.00246.x.

Munshi, K. 2003. "Networks in the Modern Economy: Mexican Migrants in the U.S. Labor Force." *Quarterly Journal of Economics* 118 (2): 549–99. http://dx.doi.org/10.1162/003355303321675455.

Ravenstein, E.G. 1885. "On the Laws of Migration." *Journal of the Statistical Society of London* 48 (2): 167–235. http://dx.doi.org/10.2307/2979181.

Ravenstein, E.G. 1889. "The Laws of Migration: Second Paper." *Journal of the Royal Statistical Society* 52 (2): 241–305. http://dx.doi.org/10.2307/2979333.

Sewell, W.H., Jr. 1992. "A Theory of Structure: Duality, Agency, and Transformation." *American Journal of Sociology* 98 (1): 1–29. http://dx.doi.org/10.1086/229967.

Sjaastad, L.A. 1962. "The Costs and Returns of Human Migration." *Journal of Political Economy* 70 (5, part 2): 80–93. http://dx.doi.org/10.1086/258726.

Stark, O., and D. Levhari. 1982. "On Migration and Risk in LDCs." *Economic Development and Cultural Change* 31 (1): 191–6. http://dx.doi.org/10.1086/451312.

Stewart, J.Q. 1941. "An Inverse Distance Variation for Certain Social Influences." *Science* 93 (2404): 89–90. http://dx.doi.org/10.1126/science.93.2404.89.

Stouffer, S.A. 1940. "Intervening Opportunities: A Theory Relating Mobility and Distance." *American Sociological Review* 5 (6): 845–67. http://dx.doi.org/10.2307/2084520.

Todaro, M.P. 1969. "A Model of Labor Migration and Urban Unemployment in Less Developed Countries." *American Economic Review* 59 (1): 138–48.

Vanderkamp, J. 1971. "Migration Flows, Their Determinants and Effects of Return Migration." *Journal of Political Economy* 79 (5): 1012–31. http://dx.doi.org/10.1086/259812.

Wallerstein, J. 1984. "Household Structures and Labor Force Formation in the Capitalist World Economy." In *Households in the World Economy*, ed. J. Smith et al., 17–22. Beverley Hills, CA: Sage.

Winters, P., A. de Janvry, and E. Sadoulet. 2001. "Family and Community Networks in Mexico–U.S. Migration." *Journal of Human Resources* 36 (1): 159–84. http://dx.doi.org/10.2307/3069674.

Zipf, G.K. 1946. "The $P_1P_2/D$ Hypothesis: On the Intercity Movement of Persons." *American Sociological Review* 11 (6): 677–86. http://dx.doi.org/10.2307/2087063.

# 8 The Colour of Money Redux: Immigrant/Ethnic Earnings Disparity in Canada, 1991–2006

KRISHNA PENDAKUR AND
RAVI PENDAKUR

A large body of Canadian research shows that immigrants can face substantial labour market disparity, which may be worsening since 1990 (see, for example, Akbari 1992; Howland and Sakellariou 1993; Stelcner and Kyriazis 1995; Christofides and Swidinsky 1994; Baker and Benjamin 1995; Hum and Simpson 1999; Pendakur and Pendakur 1998; Lian and Matthews 1998). A smaller literature has established that Canadian-born visible minorities also face labour market disparity (see Stelcner 2000; Pendakur and Pendakur 1998, 2002, 2011). These papers have shown that both male and female visible minorities face disadvantage, and that certain visible minority ethnic groups drive this disparity, especially those of South-Asian and Black/Caribbean/African origins (see de Silva and Dougherty 1996; Baker and Benjamin 1995; Hum and Simpson 1999; Pendakur and Pendakur 2002). Further, Skuterud (2010) finds that earnings gaps can remain even after three generations. Similarly, Palameta (2007) and Pendakur and Pendakur (2011) find that gaps do not disappear for Canadian-born visible minorities. Conversely, Reitz, Zhang, and Hawkins (2009) find that some Canadian-born racial minorities, particularly men of Chinese origin, perform better than white workers.

We investigate here how minority and immigrant earning gaps in Canada have evolved from 1991 to 2006. We use four microdata sets containing all the "long form" records collected by Statistics Canada for the 1991, 1996, 2001, and 2006 censuses of Canada. These data sets are very large and allow consistent definitions of variables over the period. It is thus possible to assess the degree to which non-Aboriginal Canadian-born and immigrant minorities face earnings differentials, as well as the degree to which those differentials have changed over time. In this way, we are able to assess how the findings of previous research – which examine a variety of time periods and use different methods – compare to a method that uses consistent data and methodology over time. We are also able to assess how the findings of previous research – which examine a variety of time periods and use different methods – compare to a method that uses consistent data and methodology over time.

We find that visible-minority- and immigrant-based earnings disparity has increased substantially over the fifteen-year period. In contrast, European-origin minorities born in Canada do not face earnings gaps.

This pattern is observed broadly for both men and women, in Canada as a whole and in each of its three largest Census Metropolitan Areas (CMAs), for most white and visible minority immigrant groups, and for most Canadian-born visible minority ethnic groups. The decline in relative earnings is large: it is on the order of twenty percentage points for both white and visible minority immigrants and on the order of ten percentage points for Canadian-born visible minority workers.

**Previous Research**

As noted above, a fairly rich body of research exists that assesses earnings differentials faced by minorities in Canada's labour force. Howland and Sakellariou (1993), using regression techniques to examine the Individual File of Public Use Sample Tape of the 1986 Canadian census, found that the earnings gap ranged from 2 per cent for South Asian men to 21 per cent for Black men, as compared with non-visible minority men. Using the 1989 Labour Market Activity Survey, Christofides and Swidinsky (1994) found that, while British or French immigrant males were not generally disadvantaged in the Canadian labour market, minority immigrant males earned 18 per cent less than non-minority males on average. Using 1991 census Public Use Microdata File (PUMF) data, Pendakur and Pendakur (1998) found the earnings gap to be 2 per cent for immigrant white men and 16 per cent for immigrant visible minority men. Lian and Matthews (1998) similarly used the 1991 census PUMF to conclude that immigrants of Western European ethnic backgrounds actually displayed higher incomes than the average for Canadian-born workers; however, this advantage was not observed for workers of Eastern and Southern European backgrounds, and statistically significant disadvantage was observed for visible minorities ranging from 5 per cent for those of Arab ethnicity to 33 per cent for those of Filipino background (ibid.).

Such earning differentials were confirmed in the later 1990s by Hum and Simpson (1999), who, using data from the 1993 Survey of Labour and Income Dynamics, found entry earnings for visible minority immigrant men to be 37 per cent lower than for Canadian-born males, compared to 9 per cent lower for immigrant males from Europe. Using the Longitudinal Immigrant Data Base (IMDB) to examine immigrant entry earnings and catch-up rates between 1980 and 1996, Li (2003) found that immigrants who arrived in the 1990s did indeed have lower entry earnings than those who arrived in the 1980s. However, his findings indicate that those who arrived more recently took less time to catch up with the average earnings of native-born Canadians. Similarly, Hum and Simpson (1999) estimated that the earnings of immigrant men had the potential to converge within ten years, with the time increasing to thirty years for immigrant women. However, both studies continued to recognize that such optimism was unlikely to be extended to immigrants from non-European countries. Li (2003) found that immigrants from Asia and Africa took longer to catch up, and Hum and Simpson (1999) estimated that convergence was unlikely to occur within the lifetime of visible minorities.

Entering the 2000s, a longitudinal study by Galarneau and Morissette (2004) compared census data from 1991, 1996, and 2001 to reaffirm that no narrowing of the earnings gap had occurred for the immigrant population at large. Where the earnings of recent immigrants were found to be 20 per cent lower than for Canadian-born workers in 1991, the same gap was observed in 2001. As concluded in a literature review conducted by Hum and Simpson (2004), evidence from cross-sections and panel data of studies has contributed to a common acceptance of the notion that immigrant earnings are unlikely to converge with those of native-born Canadians.

Kogan (2006), using the European Labour Force Survey data from 1992 to 2000, revealed that "third-country" immigrants (those from outside the European Union) also experienced greater labour market integration in countries with a higher demand for low-skilled labour, including the United Kingdom and Southern Europe. Differences in immigrants' countries of origin revealed that those from sub-Saharan African countries experienced a substantial disadvantage compared to those from Asian countries (Kogan 2006). Conversely, a US study by Hall and Farkas (2008) applied a random coefficient regression model to the Survey of Income and Program Participation data from 1996 to1999, and 2001 to 2003, to find that many US immigrants are employed in similar occupations to native-born Americans, and experience substantial wage gains over time. However, similarly to the findings of Galarneau and Morissette (2004), Hall and Farkas (2008) found that immigrants earn 24 per cent less than native-born workers, with Latino immigrants suffering from barriers to mobility (moving from one job to another) and more severe wage discrimination. Lysenko and Wang (next chapter) point out that one of the reasons is that immigrants are often underemployed.

A smaller body of literature has focused on patterns across Canadian cities. In particular, Pendakur and Pendakur (1998, 2011) focus attention on the high degree of disparity observed in Montreal relative to Toronto and Vancouver, which cannot be explained by compositional differences. Pendakur and Pendakur (2002, 2011) find that these patterns have existed since at least the 1980s.

## Methodology

The objective of this chapter is to provide a picture of the overall sixteen-year history of minority earnings disparity in Canada, using consistent data definitions and data sources for the period 1991 to 2006 (with incomes reported for 1990 to 2005).

Specifically, we estimate earnings differentials between white and visible minority workers for Canada as a whole and in three large Canadian cities across four census years. In addition, we investigate earnings differentials between British-origin workers and thirty-four ethnic groups (both white and non-white) in Canada as a whole, and in Montreal, Toronto, and Vancouver separately. We choose these groups for two reasons. They are large enough to provide statistically significant results at the Canada-wide level over the sixteen-year period, but not necessarily at the CMA level. Further, this categorization is also compatible with Pendakur and Pendakur (2011), but includes immigrant groups.

Our sample is restricted to workers earning more than $100 per year, whose major source of income is wages and salaries, who are aged between twenty-five and sixty-five, who are not temporary residents of Canada, and who are not Aboriginal persons. In all regressions, the dependent variable is the natural logarithm of annual earnings from wages and salaries in the previous year.[1]

We estimate log-earnings equations controlling only for ethnic origin, personal characteristics, and immigrant characteristics. Ethnic origin, as noted above, is *either* visible minority status *or* a more detailed list of thirty-four ethnic origins. Personal characteristics are: age, education, marital status, official language knowledge, and CMA of residence.

For immigrant workers, we add additional variables: immigrant status, the number of years since migration, and both of these variables interacted with visible minority status. In regressions using the more detailed list of ethnic origins, we interact immigrant status with ethnic origin, rather than with visible minority status, and interact years since migration with visible minority status. We include regressors to control for the amount of time an immigrant has resided in Canada. These controls have a value of 0 for an immigrant whose year of arrival in Canada is five years before the income year reported in that census. Thus, the coefficient on the immigrant variable may be interpreted as the log-earnings disparity faced by an immigrant who has been in Canada for five years. For example, if the immigrant dummy has a value of –0.05 in 1991, this means that a white immigrant who arrived in 1985 faces earnings disparity of about 5 per cent in 1990 (reported in the 1991 census).[2]

In controlling only for the personal characteristics of workers, we intentionally leave out the characteristics of the jobs those workers have. We believe that the work characteristics of workers – such as occupation and industry sector – are at least as susceptible to ethnic discrimination as the wages paid to workers. The case is made by Becker (1996) and others that in competitive labour markets discrimination by employers, workers, or customers results in segregation of workers into different jobs by ethnicity, but does not result in wage differentials for workers in identical jobs. With competitive product and labour markets, this segregation results in a "separate but equal" type of world where ethnic discrimination results in dividing the economy into sub-economies composed of single ethnic groups with identical wage and earnings outcomes across sub-economies.

But, if either of these competitive labour market assumptions is relaxed, the "separate but equal" conclusions do not follow. For example, if product markets are not competitive, so that some firms make excess profits that are partially shared with (possibly unionized) workers, then workers in those firms make more money than seemingly identical workers in other firms with less excess profits. Pendakur and Woodcock (2010) find evidence that immigrants end up in low-wage firms. Here, segregation results in unequal outcomes.

Alternatively, if labour markets are not competitive, a similar "separate but *un*equal" conclusion can follow. For example, consider the occupation of investment banker, a well-remunerated job in large part because investment bankers must have something to lose if their investors are to trust them. If white workers have a better chance of getting these jobs than non-white workers, then occupation segregation results in earnings differentials between white and non-white workers.[3] Our econometric methodology is most suited to this latter interpretation of how labour markets work. We control for the characteristics of workers in assessing earnings disparity. However, we do not control for the characteristics of their jobs because we believe that these characteristics are themselves subject to disparate access on the part of some workers relative to others.

Measured disparity responds to measurement decisions like the exclusion of work characteristics from the model. For example, Pendakur and Pendakur (1998) find that earnings gaps for visible minority immigrants are only about half as large when work characteristics like occupation, industry, and number of weeks worked are included as regressors in the model. However, since we are concerned mainly with the over-time pattern of disparity, the level of measured earnings gaps is not as important as how they evolve over the years. A concern would then be that the deterioration that we observe in a model that does not control for work characteristics might not be evident in a model that does control for them. Unfortunately, because the census classification of occupation and industry changed dramatically between 1991/6 and 2001/6, one cannot create a consistent coding structure that covers all workers over all years. Consequently we do not assess this aspect of measured immigrant disparity.

The results of our analysis are drawn from sixty-four regressions (four time periods * four regions * two ethnic categorizations * two genders), which assess the impacts of immigrant/ethnic earnings differentials after controlling for age, marital status, official language knowledge, household size, CMA of residence, and level of schooling. We run separate regressions for males and females because the census does not have a good indicator of labour force participation. So, our results allow us to compare within – but not across – genders.[4]

We note that in cases where the coefficient has a value of 0.20 or less, differences in the log of wages can be interpreted as per cent differences in earnings. Thus, for example, a coefficient of 0.05 for an ethnic group would suggest that, on average, after controlling for other variables, that ethnic group could expect to earn about 5 per cent more than the comparison ethnic group. However, for coefficients that are greater than either −0.25 or +0.25, this interpretation is not correct. For this reason, in the graphs we have converted all the coefficient values into per cent differences from the comparison group.

## Results

### *Canada as a Whole*

Table 8.1 shows coefficients from estimated log-earnings regressions that control for personal characteristics and visible minority status as described above. We present results for Canada as a whole, with CMA dummies controlling for area of residence, and for Montreal, Toronto, and Vancouver separately.[5]

For Canadian-born persons, the coefficient on visible minority status gives the estimated difference in log-earnings for a visible minority in comparison with a white person. As stated above, for small values (less than 0.25), these may be interpreted as percentage differences in earnings. The coefficient on visible minority for males in the 1991 census data is −0.10, so we conclude that Canadian-born visible minority men earned about 10 per cent less than their white Canadian-born counterparts.

For immigrants, the coefficient on the immigrant dummy gives the estimated difference in log-earnings between a white immigrant who has been in Canada for five years and their white Canadian-born counterparts. For example, the coefficient for immigrant for men in the 1991

Census data is −0.27, indicating that white immigrants who arrived in Canada in 1985 (five years before the 1990 earnings year) earned about 24 per cent less than their white Canadian-born counterparts.[6] The estimated difference in log earnings for a visible minority immigrant is the sum of three effects: the immigrant effect, the visible minority effect, and the visible minority immigrant effect. The estimated difference in log earnings for a visible minority immigrant is the sum of three effects: the immigrant effect, the visible minority effect, and the visible minority immigrant effect. For example, the coefficients on these three variables

Table 8.1 Earnings differentials (coefficients) between white Canadian-born and immigrant and visible minorities by sex, Canada, Montreal, Toronto, and Vancouver, 1991–2006

| | | | 1991 | | 1996 | | 2001 | | 2006 | |
|---|---|---|---|---|---|---|---|---|---|---|
| | | | coef | sig. | coef | sig. | coef | sig. | coef | sig. |
| Canada | Female | Observations | 950,391 | | 980,755 | | 1,045,740 | | 1,136,259 | |
| | | Prob > F | 0.00 | | 0.00 | | 0.00 | | 0.00 | |
| | | R2 | 0.13 | | 0.13 | | 0.13 | | 0.15 | |
| | | Immigrant | –0.26 | *** | –0.43 | *** | –0.43 | *** | –0.49 | *** |
| | | Visible minority | 0.02 | | –0.05 | *** | –0.05 | *** | –0.04 | *** |
| | | Immig * vm | –0.01 | | 0.02 | | –0.05 | *** | –0.11 | *** |
| | | Years since immigrating | 0.02 | *** | 0.03 | *** | 0.03 | *** | 0.03 | *** |
| | | Yrs since immig for vm | 0.01 | *** | 0.01 | *** | 0.01 | *** | 0.02 | *** |
| | Male | Observations | 1,073,026 | | 1,068,370 | | 1,099,470 | | 1,148,612 | |
| | | Prob > F | 0.00 | | 0.00 | | 0.00 | | 0.00 | |
| | | R2 | 0.17 | | 0.18 | | 0.16 | | 0.17 | |
| | | Immigrant | –0.27 | *** | –0.38 | *** | –0.37 | *** | –0.43 | *** |
| | | Visible minority | –0.10 | *** | –0.15 | *** | –0.18 | *** | –0.20 | *** |
| | | Immig * vm | –0.18 | *** | –0.16 | *** | –0.15 | *** | –0.15 | *** |
| | | Years since immigrating | 0.02 | *** | 0.03 | *** | 0.03 | *** | 0.03 | *** |
| | | Yrs since immig for vm | 0.02 | *** | 0.02 | *** | 0.01 | *** | 0.01 | *** |
| Montreal | Female | Observations | 111,831 | | 115,855 | | 123,155 | | 135,106 | |
| | | Prob > F | 0.00 | | 0.00 | | 0.00 | | 0.00 | |
| | | R2 | 0.12 | | 0.13 | | 0.14 | | 0.17 | |
| | | Immigrant | –0.28 | *** | –0.50 | *** | –0.44 | *** | –0.60 | *** |
| | | Visible minority | –0.01 | | –0.15 | *** | –0.15 | *** | –0.10 | *** |

|  |  |  |  |  |  |  |  |  |  |  |
|---|---|---|---|---|---|---|---|---|---|---|
|  |  | Immig * vm | −0.12 | ** | 0.11 | ** | 0.02 |  | −0.01 |  |
|  |  | Years since immigrating | 0.02 | *** | 0.03 | *** | 0.03 | *** | 0.04 | *** |
|  |  | Yrs since immig for vm | 0.02 | *** | 0.00 |  | 0.01 | ** | 0.01 | * |
|  | Male | Observations | 123,763 |  | 123,665 |  | 126,980 |  | 134,668 |  |
|  |  | Prob > F | 0.00 |  | 0.00 |  | 0.00 |  | 0.00 |  |
|  |  | R2 | 0.19 |  | 0.19 |  | 0.17 |  | 0.19 |  |
|  |  | Immigrant | −0.40 | *** | −0.49 | *** | −0.44 | *** | −0.61 | *** |
|  |  | Visible minority | −0.25 | *** | −0.23 | *** | −0.32 | *** | −0.33 | *** |
|  |  | Immig * vm | −0.04 |  | −0.06 |  | −0.05 |  | 0.06 | * |
|  |  | Years since immigrating | 0.01 | *** | 0.03 | *** | 0.02 | *** | 0.03 | *** |
|  |  | Yrs since immig for vm | 0.03 | *** | 0.02 | *** | 0.02 | *** | 0.01 | *** |
| Toronto | Female | Observations | 146,978 |  | 155,910 |  | 174,635 |  | 191,366 |  |
|  |  | Prob > F | 0.00 |  | 0.00 |  | 0.00 |  | 0.00 |  |
|  |  | R2 | 0.10 |  | 0.13 |  | 0.12 |  | 0.15 |  |
|  |  | Immigrant | −0.26 | *** | −0.48 | *** | −0.49 | *** | −0.52 | *** |
|  |  | Visible minority | −0.01 |  | −0.14 | *** | −0.09 | *** | −0.08 | *** |
|  |  | Immig * vm | 0.05 | ** | 0.14 | *** | 0.03 |  | −0.05 | *** |
|  |  | Years since immigrating | 0.02 | *** | 0.03 | *** | 0.03 | *** | 0.04 | *** |
|  |  | Yrs since immig for vm | 0.00 |  | 0.01 | *** | 0.01 | *** | 0.02 | *** |
|  | Male | Observations | 154,781 |  | 159,135 |  | 174,745 |  | 184,843 |  |
|  |  | Prob > F | 0.00 |  | 0.00 |  | 0.00 |  | 0.00 |  |
|  |  | R2 | 0.18 |  | 0.20 |  | 0.18 |  | 0.20 |  |
|  |  | Immigrant | −0.28 | *** | −0.43 | *** | −0.43 | *** | −0.49 | *** |
|  |  | Visible minority | −0.15 | *** | −0.22 | *** | −0.23 | *** | −0.21 | *** |

Table 8.1 (*continued*)

| | | | 1991 | | 1996 | | 2001 | | 2006 | |
|---|---|---|---|---|---|---|---|---|---|---|
| | | | coef | sig. | coef | sig. | coef | sig. | coef | sig. |
| | | Immig * vm | –0.06 | *** | 0.00 | | –0.02 | | –0.07 | *** |
| | | Years since immigrating | 0.02 | *** | 0.03 | *** | 0.03 | *** | 0.03 | *** |
| | | Yrs since immig for vm | 0.01 | *** | 0.01 | *** | 0.01 | *** | 0.01 | *** |
| Vancouver | Female | Observations | 57,734 | | 63,695 | | 69,280 | | 76,699 | |
| | | Prob > F | 0.00 | | 0.00 | | 0.00 | | 0.00 | |
| | | R2 | 0.08 | | 0.12 | | 0.11 | | 0.13 | |
| | | Immigrant | –0.20 | *** | –0.37 | *** | –0.32 | *** | –0.40 | *** |
| | | Visible minority | 0.12 | *** | 0.10 | *** | 0.04 | ** | 0.06 | *** |
| | | Immig * vm | –0.15 | *** | –0.16 | *** | –0.25 | *** | –0.27 | *** |
| | | Years since immigrating | 0.02 | *** | 0.03 | *** | 0.02 | *** | 0.03 | *** |
| | | Yrs since immig for vm | 0.02 | *** | 0.02 | *** | 0.03 | *** | 0.02 | *** |
| | Male | Observations | 63,126 | | 66,730 | | 69,925 | | 74,408 | |
| | | Prob > F | 0.00 | | 0.00 | | 0.00 | | 0.00 | |
| | | R2 | 0.18 | | 0.20 | | 0.18 | | 0.18 | |
| | | Immigrant | –0.24 | *** | –0.35 | *** | –0.33 | *** | –0.29 | *** |
| | | Visible minority | –0.04 | * | –0.07 | *** | –0.14 | *** | –0.14 | *** |
| | | Immig * vm | –0.34 | *** | –0.32 | *** | –0.30 | *** | –0.37 | *** |
| | | Years since immigrating | 0.02 | *** | 0.03 | *** | 0.02 | *** | 0.02 | *** |
| | | Yrs since immig for vm | 0.02 | *** | 0.02 | *** | 0.01 | *** | 0.02 | *** |

Note: Other controls include CMA, age, level of schooling, marital status, official language knowledge, and household size.
Source: 1991, 1996, 2001, and 2006 Census of Canada.

for men in the 1991 census year are −0.27, −0.10, and −0.18, respectively, indicating that visible minority immigrants who had arrived in Canada in 1985 had earnings about 42 per cent lower (coefficient of −0.55) than their Canadian-born white counterparts.

As noted above, for visible minority immigrants, three estimated coefficients are relevant to their earnings disparity. Thus, reading data results off of the table can be cumbersome. In addition, for large negative values, proportionate differences in earnings are smaller in magnitude than the coefficients, and for large positive values, proportionate differences are larger in magnitude than the coefficients. In order to simplify the tracking of disparity over time, we present the information in the table graphically, and transformed into exact proportionate differences in figures 8.1–8.4.

Figure 8.1 presents the information from the top panel of table 8.1 graphically. It gives the proportionate differences in earnings for male and female minorities across four income years (1990, 1995, 2000, and 2005). The proportionate difference in earnings for Canadian-born visible minority males was −9 per cent in 1990 in comparison with Canadian-born white workers with otherwise similar personal characteristics. For visible minority immigrant males, the proportionate difference was 42 per cent (which, as noted above, is smaller in magnitude than −0.55, the sum of the three relevant coefficients).

Past research suggests that labour force prospects for immigrants and visible minorities have deteriorated over time (see for example, Akbari 1992; Howland and Sakellariou 1993; Stelcner and Kyriazis 1995; Christofides and Swidinsky 1994; Baker and Benjamin 1995; Hum and Simpson 1999; Pendakur and Pendakur 1998; Lian and Matthews 1998). Our findings reiterate this. Among both men and women, disparity increased from 1991 to 1996 and stayed high for all immigrants and Canadian-born visible minorities. These results are easily seen in figure 8.1, which provides results for Canada-wide regressions. First, we will consider visible minority Canadian-born women. As did Pendakur and Pendakur (2011), we see a deterioration in relative earnings over time, from a (statistically insignificant) 2 per cent earnings advantage in 1990 to a (statistically significant) 4 per cent earnings *gap* in 2005. This is not a dramatic change, but it is definitely a downward trend.

Turning to the results for white immigrant women, it is apparent that immigrants saw their earnings prospects fall dramatically, from

Figure 8.1 Earnings differentials between selected groups and Canadian-born white men and women, 1990–2005, Canada

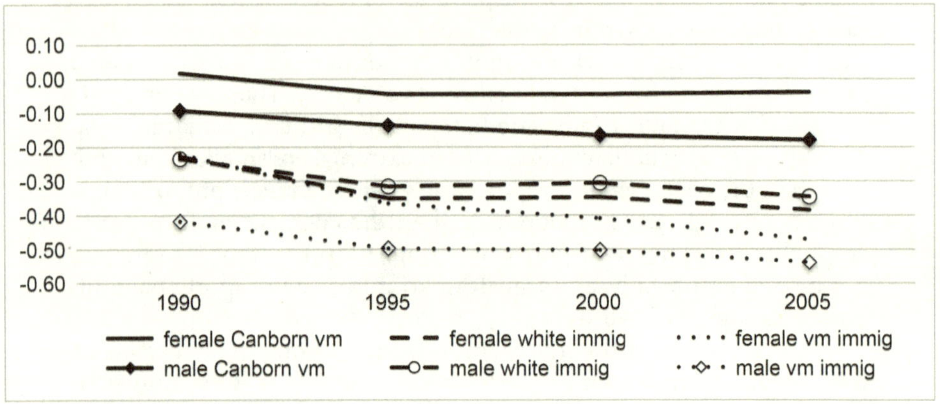


Source: 1991, 1996, 2001, and 2006 Census of Canada.

Figure 8.2 Earnings differentials between selected groups and Canadian-born white men and women, 1990–2005, Montreal

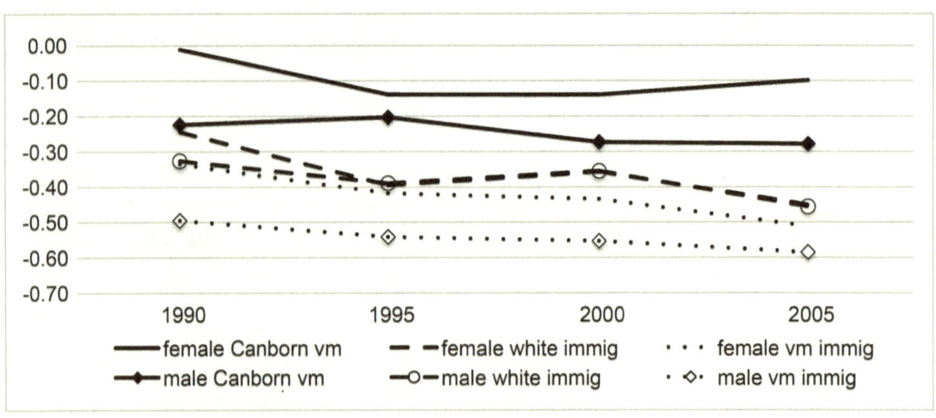


Source: 1991, 1996, 2001, and 2006 Census of Canada.

Figure 8.3 Earnings differentials between selected groups and Canadian-born white men and women, 1990–2005, Toronto

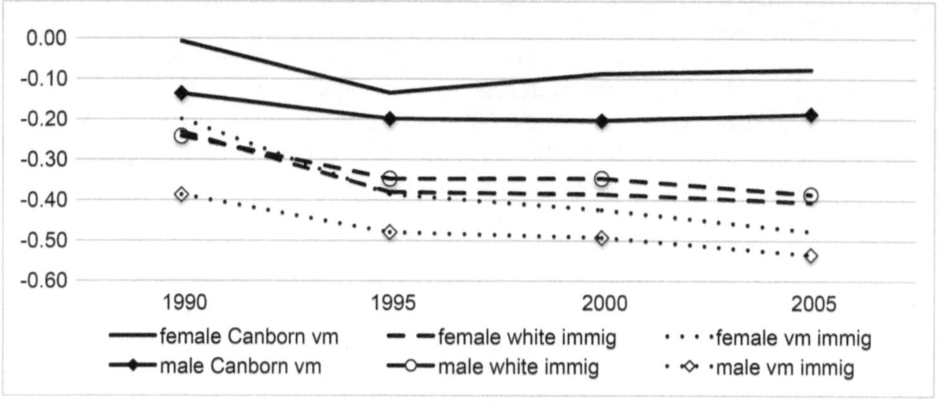


Source: 1991, 1996, 2001, 2006 Census of Canada.

Figure 8.4 Earnings differentials between selected groups and Canadian-born white men and women, 1990–2005, Vancouver

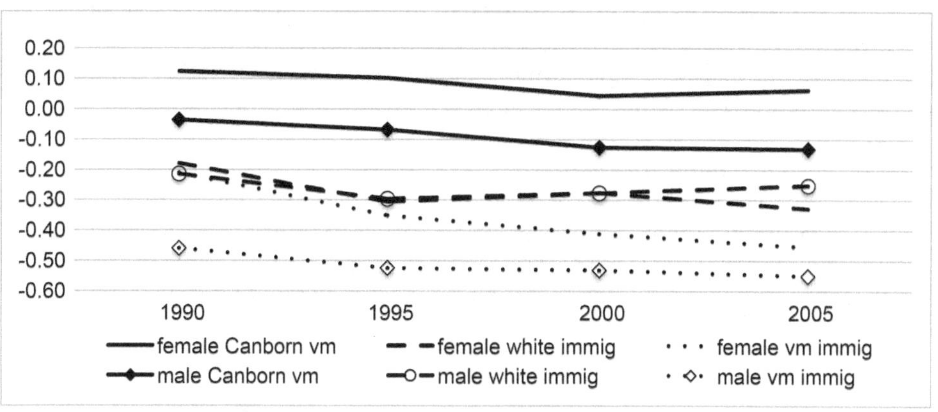


Source: 1991, 1996, 2001, and 2006 Census of Canada.

an earnings gap of 22 per cent in 1990 to 35 per cent in 1995 to 39 per cent in 2005. This is a very large change; among white immigrants with five years of residence in Canada, the relative earnings gap almost doubled.

Now moving to visible minority immigrant women, we see an even more dramatic deterioration in relative earnings. In 1990 their earnings gap was 22 per cent, the same as that of white immigrant women. By 1995 it had increased to 37 per cent, and then it continued to increase unabated over the next two periods, to reach 47 per cent by 2005. For visible minority immigrant women, relative earnings disparity more than doubled from 1990 to 2005.

Canadian-born visible minority men saw their earnings gap grow from 9 per cent in 1990 to 18 per cent in 2005. This increase in disparity was also reported in Pendakur and Pendakur (2011). However, that paper did not consider immigrants. White immigrant men also saw an increase in earnings disparity; their earnings gap was 24 per cent in 1990, but grew to 35 per cent by 2005, with the bulk of that increase occurring between 1990 and 1995.

Considering now the results for visible minority immigrant men, the earnings gap in 1990 was 42 per cent. In 1995 it had grown to 50 per cent, and by 2005 it had reached 54 per cent.

All in all, we see a very strong pattern of increasing earnings disparity for all minority groups: for both males and females, Canadian-born visible minorities, immigrant whites, and immigrant visible minorities all saw a big decline in their relative earnings over the four census periods. For most groups, the bulk of this deterioration occurred in the early 1990s.

### *Montreal, Toronto, and Vancouver*

Given the dynamics of the Canadian labour market across regions, it is possible that the Canada-wide picture might average out important variation across cities. Running regressions at the CMA level suggests that the overall patterns in the evolution of disparity identified above are evident in each of the three cities we study: Montreal, Toronto, and Vancouver.

Figure 8.2 shows analogous results to those of figure 8.1 for the Montreal CMA. The over-time pattern we observe in figure 8.1 is seen also for Montreal: all groups saw a decline in their relative earnings from 1990 to 2005. The deterioration is large for some groups. For female and male visible minority immigrants, the earnings gap grew by seventeen and nine percentage points, respectively. White immigrants also saw their relative earnings decline, though not by as much as visible minority immigrants did. Finally, we see a drop of about six percentage points in the relative earnings of Canadian-born visible minority men.

An important difference between the Canada-wide results shown in figure 8.1 and the Montreal results shown in figure 8.2 is that disparity tends to be worse in Montreal. For most minority groups, the earnings gap in Montreal is about ten percentage points larger than the analogous national average earnings gap. That is, minority earnings disparity is larger in Montreal than in Canada taken as a whole, a result that echoes the findings of numerous previous papers (Pendakur and Pendakur 1998, 2002, 2007, 2011).

In figure 8.3, we present results for the Toronto CMA, the largest urban agglomeration in Canada. Since the Toronto CMA accounts for nearly one-fifth of the data, and for nearly one-third of visible minorities, it should not be surprising that the results for Toronto closely track the national averages presented in figure 8.1. All the patterns observed for Canada as a whole are evident for the Toronto CMA: visible minorities and immigrants saw substantial declines in their relative earnings from 1990 to 2005, and much of this deterioration occurred in the early part of that period.

Figure 8.4 presents results for the Vancouver CMA. Here, we see much the same pattern as in figure 8.1 for visible minority immigrants. Earnings gaps widened from about 20 per cent to more than 45 per cent for women from 1990 to 2005. For men, they widened from about 45 per cent to about 55 per cent over the same period. The bulk of this decline occurred in the early 1990s, but decline continued throughout the rest of the period.

In contrast, for white immigrants, the decline was much smaller, and changed direction over the period. Earnings gaps for men and women were about 20 per cent in 1990. By 2005 these had grown to about 32 per cent and 25 per cent for women and men, respectively. Interestingly, although both groups showed a marked deterioration in relative earnings over the early 1990s, male white immigrants showed some convergence in earnings between 1995 and 2005, with their earnings gap narrowing from 30 per cent to 25 per cent.

Canadian-born visible minorities tended to have higher relative earnings in Vancouver than in other CMAs, or in Canada as a whole.

Indeed, in the early part of the period of 1990 to 2005, Canadian-born visible minority women earned much *more* than their Canadian-born white counterparts. By 2005, this earnings advantage had contracted, but remained statistically significant. Canadian-born visible minority men earned about the same as Canadian-born white men in 1990 (a fact noted in Pendakur and Pendakur 1998), but by 2005 their earnings were statistically significantly lower than those of their Canadian-born white counterparts. In 2005 Canadian-born visible minority men faced an earnings gap of 13 per cent in Vancouver, which, although lower than the 20 per cent gap faced at the Canada-wide level, represents a substantial labour market disadvantage.

*Canada as a Whole: Disparity across Fine-Grained Ethnic Groups*

Results shown in table 8.2 are drawn from regressions that break out the immigrant and visible minority categories from table 8.1. This set of analyses provides detailed earnings differentials for immigrant and Canadian-born workers across twenty-three European origins and ten non-European origins for each of the four census periods. For each census period, we show two columns of coefficients. The first shows coefficients for the Canadian-born population, and the second shows the equivalent immigrant ethnic group. We are therefore able to compare, for example, the 1990 earnings differential for Canadian-born Greek women (−0.06) to immigrant Greek women. Greek women born in Canada get the coefficient for Greek women (−0.06 or about −6%). Greek immigrant women get both the coefficient for Greek women (−0.06) as well as the coefficient for being an immigrant (−0.23) as well as −0.03 for being a female Greek immigrant. This means the earnings differential for Greek immigrant women is large, exceeding 30 per cent (−0.06 + −0.23 + −0.03 = −0.32). This translates to a difference of −27 per cent. Figures 8.5 through 8.10 show results for selected ethnic groups at the Canada-wide level. As above, these figures show differences expressed as per cent differences; however, in this case, the comparison group is individuals who report only British origin on the ethnic origin question. We note that, in 2006, there is no significant difference between the earnings of people reporting British, French, or Canadian as ethnic origin, so we interpret our results as the difference in earnings between minority and majority ethnic origin workers (British, French, Canadian, or any combination).

Table 8.2 Earnings differentials between selected ethnic groups and British-origin, Canadian-born workers by immigrant status and sex, Canada and MTV, 1991–2006

| | | | 1991 | | | | 1996 | | | | 2001 | | | | 2006 | | | |
|---|---|---|---|---|---|---|---|---|---|---|---|---|---|---|---|---|---|---|
| | | | Canadian-born | | Immigrant | | Canadian-born | | Immigrant | | Canadian-born | | Immigrant | | Canadian-born | | Immigrant | |
| | | | coef | sig. | coef | sig. | coef | sig. | coef | sig. | coef | sig. | coef | sig. | coef | sig. | coef | sig. |
| Canada | female | Observations | 950,391 | | | | 980,755 | | | | 1,045,740 | | | | 1,136,259 | | | |
| | | Prob > F | 0.00 | | | | 0.00 | | | | 0.00 | | | | 0.00 | | | |
| | | R2 | 0.13 | | | | 0.13 | | | | 0.13 | | | | 0.15 | | | |
| | | Immigrant | | | –0.23 | *** | | | –0.38 | *** | | | –0.38 | *** | | | –0.43 | *** |
| | | French | 0.03 | *** | –0.01 | | 0.02 | *** | 0.00 | | 0.00 | | 0.05 | ** | 0.00 | | 0.07 | *** |
| | | Canadian | 0.05 | *** | 0.00 | | 0.00 | | –0.02 | | –0.02 | *** | –0.06 | ** | 0.00 | | –0.09 | ** |
| | | Baltic | 0.02 | | –0.10 | * | 0.05 | | –0.13 | ** | –0.03 | | –0.04 | | 0.00 | | –0.06 | |
| | | Austrian German | 0.00 | | –0.10 | *** | 0.01 | | –0.10 | *** | 0.00 | | –0.10 | *** | 0.01 | * | –0.12 | *** |
| | | Czech Slovak | 0.07 | ** | –0.06 | | 0.07 | ** | –0.15 | *** | –0.01 | | –0.08 | * | 0.03 | | –0.11 | ** |
| | | Scandinavian | 0.02 | | –0.09 | *** | 0.02 | | –0.02 | | 0.01 | | –0.07 | ** | 0.03 | * | –0.04 | |
| | | Dutch | –0.01 | | –0.12 | *** | 0.02 | | –0.10 | *** | –0.01 | | –0.12 | *** | –0.02 | | –0.06 | ** |
| | | Polish | 0.05 | *** | –0.17 | *** | 0.06 | *** | –0.20 | *** | 0.05 | *** | –0.14 | *** | 0.03 | ** | –0.09 | *** |
| | | Jewish | 0.00 | | –0.02 | | –0.05 | *** | –0.01 | | –0.07 | *** | –0.01 | | –0.11 | *** | –0.02 | |
| | | Hungarian | –0.01 | | –0.06 | * | 0.05 | ** | –0.15 | *** | 0.00 | | –0.09 | ** | –0.01 | | –0.02 | |
| | | Russian | –0.03 | | –0.19 | ** | 0.02 | | –0.43 | *** | –0.04 | | –0.37 | *** | –0.02 | | –0.30 | *** |
| | | Portuguese | 0.09 | *** | 0.03 | | 0.11 | *** | 0.02 | | 0.06 | *** | –0.04 | | 0.07 | *** | –0.01 | |
| | | Italian | 0.10 | *** | –0.11 | *** | 0.08 | *** | –0.15 | *** | 0.03 | *** | –0.09 | *** | 0.04 | *** | –0.07 | *** |

Table 8.2 (*continued*)

| | 1991 | | | | 1996 | | | | 2001 | | | | 2006 | | | |
|---|---|---|---|---|---|---|---|---|---|---|---|---|---|---|---|---|
| | Canadian-born | | Immigrant | | Canadian-born | | Immigrant | | Canadian-born | | Immigrant | | Canadian-born | | Immigrant | |
| | coef | sig. | coef | sig. | coef | sig. | coef | sig. | coef | sig. | coef | sig. | coef | sig. | coef | sig. |
| Greek | –0.06 | ** | –0.03 | | –0.02 | | –0.17 | *** | –0.05 | *** | –0.19 | *** | –0.06 | *** | –0.11 | *** |
| Balkan | 0.08 | *** | –0.03 | | 0.05 | ** | –0.12 | *** | 0.09 | *** | –0.17 | *** | 0.07 | *** | –0.11 | *** |
| Spanish | –0.07 | | –0.09 | | 0.11 | * | –0.25 | *** | –0.07 | | –0.13 | * | –0.06 | | –0.18 | *** |
| Ukrainian | 0.06 | *** | –0.02 | | 0.06 | *** | –0.12 | *** | 0.03 | *** | –0.30 | *** | 0.05 | *** | –0.23 | *** |
| Belgian | 0.04 | | 0.00 | | 0.06 | * | –0.06 | | 0.09 | *** | –0.06 | | 0.11 | *** | –0.18 | ** |
| Other European | 0.06 | * | –0.07 | * | 0.02 | | –0.10 | *** | 0.01 | | –0.02 | | –0.03 | | –0.01 | |
| Arab W Asian | 0.06 | | –0.27 | *** | 0.02 | | –0.28 | *** | –0.04 | | –0.26 | *** | –0.04 | | –0.28 | *** |
| S. Asian | 0.02 | | –0.05 | | –0.04 | | –0.04 | | –0.06 | *** | –0.07 | ** | –0.01 | | –0.20 | *** |
| Chinese | 0.14 | *** | –0.04 | * | 0.10 | *** | –0.07 | *** | 0.05 | *** | –0.13 | *** | 0.06 | *** | –0.24 | *** |
| SE Asian | 0.01 | | 0.06 | | –0.23 | *** | 0.24 | *** | –0.09 | ** | 0.03 | | 0.07 | *** | –0.16 | *** |
| Other Asian | 0.16 | *** | –0.34 | *** | 0.14 | *** | –0.39 | *** | 0.08 | *** | –0.42 | *** | 0.06 | ** | –0.45 | *** |
| African Black | –0.17 | | 0.03 | | –0.11 | | –0.10 | | –0.15 | ** | –0.15 | ** | –0.24 | *** | 0.00 | |
| Black | –0.12 | *** | 0.05 | | –0.18 | ** | 0.06 | | –0.18 | *** | –0.03 | | –0.20 | *** | 0.00 | |
| Caribbean | –0.09 | | 0.10 | | –0.19 | *** | 0.12 | *** | –0.14 | *** | –0.01 | | –0.09 | *** | –0.04 | * |
| Spanish Latin American | –0.17 | | –0.04 | | –0.21 | | –0.03 | | –0.15 | * | –0.09 | | –0.05 | | –0.19 | *** |
| American Australian NZ | –0.16 | * | 0.16 | | –0.04 | | 0.05 | | 0.01 | | –0.07 | | 0.00 | | 0.00 | |

| | | | | | | | | | | | | | | | | | | |
|---|---|---|---|---|---|---|---|---|---|---|---|---|---|---|---|---|---|---|
| | | Br Fr Can multiple | –0.01 | *** | –0.01 | | 0.00 | | –0.03 | ** | –0.02 | *** | –0.04 | ** | 0.00 | | –0.05 | *** |
| | | Vismin w white | –0.05 | ** | –0.01 | | –0.11 | *** | 0.01 | | –0.07 | *** | –0.09 | *** | –0.08 | *** | –0.05 | *** |
| | | White multiple | –0.01 | ** | –0.06 | *** | 0.01 | *** | –0.12 | *** | –0.02 | *** | –0.06 | *** | 0.00 | | –0.06 | *** |
| Canada | male | Observations | 1,073,026 | | | | 1,068,370 | | | | 1,099,470 | | | | 1,148,612 | | | |
| | | Prob > F | 0.00 | | | | 0.00 | | | | 0.00 | | | | 0.00 | | | |
| | | R2 | 0.18 | | | | 0.18 | | | | 0.16 | | | | 0.17 | | | |
| | | Immigrant | | | –0.15 | *** | | | –0.25 | *** | | | –0.27 | *** | | | –0.31 | *** |
| | | French | 0.02 | *** | –0.11 | *** | 0.03 | *** | –0.10 | *** | 0.00 | | –0.03 | | 0.00 | | –0.07 | *** |
| | | Canadian | 0.04 | *** | –0.16 | *** | 0.01 | *** | –0.11 | *** | –0.01 | | –0.17 | *** | 0.01 | ** | –0.20 | *** |
| | | Baltic | 0.01 | | –0.07 | * | 0.06 | * | –0.15 | *** | 0.04 | | –0.18 | *** | 0.04 | | –0.29 | *** |
| | | Austrian German | 0.03 | *** | –0.13 | *** | 0.08 | *** | –0.17 | *** | 0.04 | *** | –0.08 | *** | 0.06 | *** | –0.09 | *** |
| | | Czech Slovak | 0.11 | *** | –0.27 | *** | 0.10 | *** | –0.20 | *** | –0.01 | | –0.11 | ** | –0.01 | | –0.07 | |
| | | Scandinavian | 0.05 | *** | –0.15 | *** | 0.06 | *** | –0.09 | *** | 0.06 | *** | –0.11 | *** | 0.05 | *** | –0.06 | * |
| | | Dutch | 0.06 | *** | –0.16 | *** | 0.11 | *** | –0.18 | *** | 0.07 | *** | –0.11 | *** | 0.08 | *** | –0.13 | *** |
| | | Polish | 0.05 | *** | –0.32 | *** | 0.07 | *** | –0.29 | *** | 0.03 | ** | –0.18 | *** | 0.04 | *** | –0.16 | *** |
| | | Jewish | 0.04 | *** | –0.18 | *** | 0.01 | | –0.21 | *** | 0.05 | *** | –0.21 | *** | 0.02 | | –0.12 | *** |
| | | Hungarian | 0.03 | | –0.23 | *** | 0.05 | ** | –0.19 | *** | 0.00 | | –0.13 | *** | 0.02 | | –0.16 | *** |
| | | Russian | 0.02 | | –0.21 | *** | 0.04 | | –0.52 | *** | 0.08 | *** | –0.44 | *** | 0.03 | | –0.26 | *** |
| | | Portuguese | –0.03 | | –0.07 | ** | 0.04 | | –0.13 | *** | 0.02 | | –0.13 | *** | 0.02 | | –0.09 | *** |
| | | Italian | 0.01 | | –0.18 | *** | 0.04 | *** | –0.21 | *** | –0.03 | *** | –0.14 | *** | 0.00 | | –0.11 | *** |
| | | Greek | –0.21 | *** | –0.23 | *** | –0.19 | *** | –0.27 | *** | –0.22 | *** | –0.21 | *** | –0.17 | *** | –0.24 | *** |
| | | Balkan | 0.02 | | –0.19 | *** | 0.06 | *** | –0.26 | *** | 0.00 | | –0.18 | *** | 0.01 | | –0.21 | *** |

Table 8.2 (*continued*)

| | 1991 | | | | 1996 | | | | 2001 | | | | 2006 | | | |
|---|---|---|---|---|---|---|---|---|---|---|---|---|---|---|---|---|
| | Canadian-born | | Immigrant | | Canadian-born | | Immigrant | | Canadian-born | | Immigrant | | Canadian-born | | Immigrant | |
| | coef | sig. | coef | sig. | coef | sig. | coef | sig. | coef | sig. | coef | sig. | coef | sig. | coef | sig. |
| Spanish | –0.09 | * | –0.24 | *** | –0.13 | ** | –0.20 | *** | –0.12 | ** | –0.22 | *** | –0.07 | * | –0.26 | *** |
| Ukrainian | 0.05 | *** | –0.22 | *** | 0.05 | *** | –0.21 | *** | 0.01 | | –0.29 | *** | 0.06 | *** | –0.33 | *** |
| Belgian | 0.09 | *** | –0.19 | *** | 0.07 | ** | –0.14 | *** | 0.02 | | –0.04 | | 0.05 | * | –0.07 | |
| Other European | 0.04 | | –0.19 | *** | 0.04 | * | –0.21 | *** | 0.02 | | –0.13 | *** | 0.00 | | –0.20 | *** |
| Arab W Asian | –0.03 | | –0.38 | *** | –0.03 | | –0.41 | *** | –0.10 | *** | –0.34 | *** | –0.13 | *** | –0.36 | *** |
| S. Asian | –0.16 | *** | –0.18 | *** | –0.21 | *** | –0.15 | *** | –0.22 | *** | –0.15 | *** | –0.20 | *** | –0.20 | *** |
| Chinese | –0.07 | *** | –0.24 | *** | 0.00 | | –0.35 | *** | –0.10 | *** | –0.31 | *** | –0.11 | *** | –0.37 | *** |
| SE Asian | –0.27 | ** | –0.07 | | –0.28 | *** | –0.06 | | –0.12 | *** | –0.24 | *** | –0.22 | *** | –0.15 | *** |
| Other Asian | 0.07 | *** | –0.48 | *** | 0.08 | *** | –0.47 | *** | –0.02 | | –0.48 | *** | –0.01 | | –0.54 | *** |
| African Black | –0.18 | ** | –0.31 | *** | –0.29 | *** | –0.28 | *** | –0.27 | *** | –0.27 | *** | –0.28 | *** | –0.17 | *** |
| Black | –0.25 | *** | –0.17 | *** | –0.24 | *** | –0.21 | ** | –0.24 | *** | –0.25 | *** | –0.40 | *** | 0.06 | |
| Caribbean | –0.15 | ** | –0.23 | *** | –0.31 | *** | –0.10 | *** | –0.24 | *** | –0.14 | *** | –0.26 | *** | –0.12 | *** |
| Spanish Latin American | –0.16 | | –0.32 | ** | –0.37 | *** | –0.08 | | –0.22 | *** | –0.14 | ** | –0.15 | *** | –0.21 | *** |
| American Australian NZ | 0.02 | | 0.00 | | 0.02 | | –0.10 | * | 0.03 | | –0.06 | | 0.01 | | –0.08 | |
| Br Fr Can multiple | 0.01 | | –0.14 | *** | 0.03 | *** | –0.09 | *** | 0.00 | | –0.05 | *** | 0.00 | | –0.05 | *** |
| Vismin w white | –0.08 | *** | –0.28 | *** | –0.09 | *** | –0.23 | *** | –0.14 | *** | –0.18 | *** | –0.11 | *** | –0.20 | *** |
| White multiple | 0.03 | *** | –0.16 | *** | 0.06 | *** | –0.20 | *** | 0.03 | *** | –0.17 | *** | 0.04 | *** | –0.14 | *** |

Figure 8.5 Earnings differentials among women, selected European origins versus Canadian-born British origin, 1990–2005, Canada

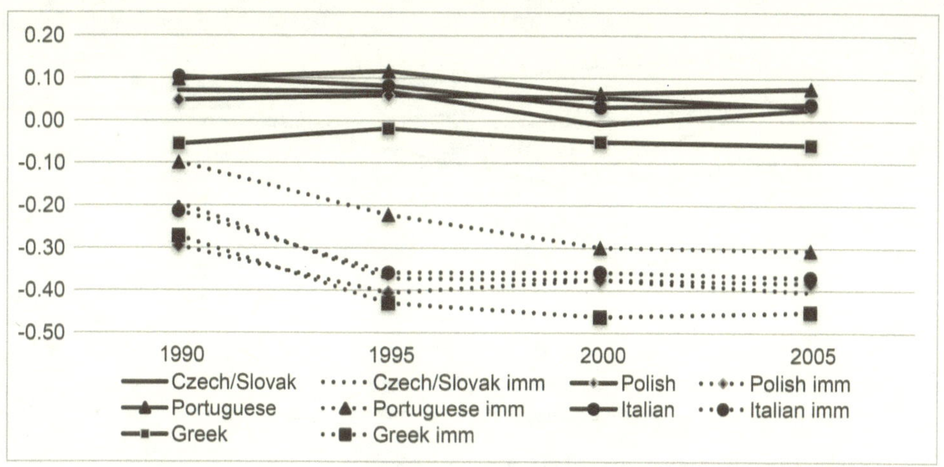


Source: 1991, 1996, 2001, and 2006 Census of Canada.

Looking first at the results for women of European origin born in Canada, we do not see any significant changes over the fifteen-year period (see figure 8.5). Women reporting Northern European ancestry did not face any earnings disadvantage controlling for personal characteristics. Women of Portuguese and Balkan origin born in Canada, however, enjoy a bonus of 7 per cent. Immigrant women from Europe do not fare as well. For example, Czech/Slovak immigrant women saw their earnings prospects deteriorate from a gap of –20 per cent in 1990 to –40 per cent in 2005. The earnings gap faced by Portuguese immigrant women also increased from –10 per cent to –30 per cent.

The results for non-European origin women are quite uneven. Women of Chinese origin born in Canada enjoyed an earnings bonus of about 6 per cent in 2005 (14% in 1990), suggesting that they are paid more than similarly qualified women of British origin (see figure 8.6). Women of South Asian and Arab and West Asian origin earn about the same as women of British origin after controlling for personal characteristics; however, Black women (reporting African, Black, or Caribbean origins) face earnings gaps of between 9 per cent to 24 per cent in 2005. Over all, this represents a deterioration in the wage prospects of visible minority

Figure 8.6 Earnings differentials among women, selected non-European origins (1) versus Canadian-born British origins, 1990–2005, Canada

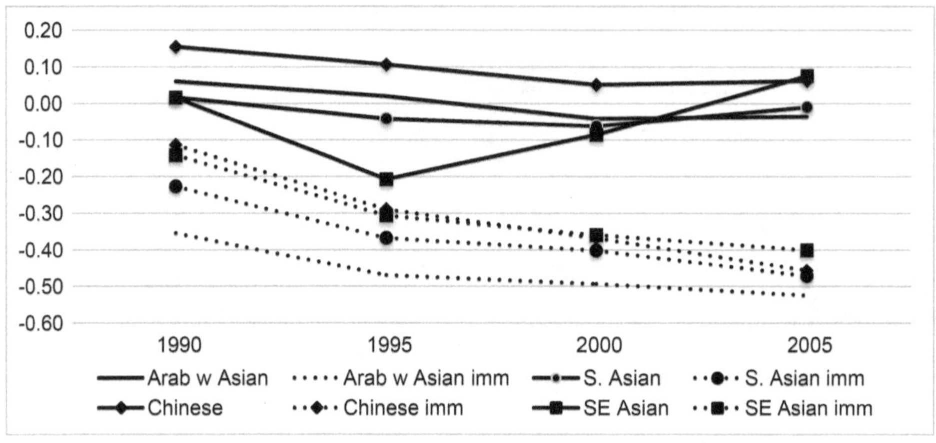


Source: 1991, 1996, 2001, and 2006 Census of Canada.

Figure 8.7 Earnings differentials among women, selected non-European origins (2) versus Canadian-born British origin, 1990–2005 Canada

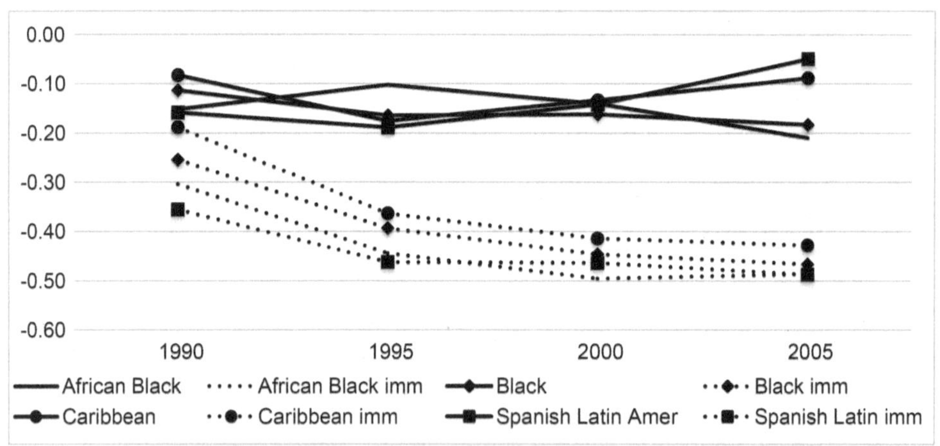


Source: 1991, 1996, 2001, and 2006 Census of Canada.

Figure 8.8 Earnings differentials among men, selected European origins versus Canadian-born British origin, 1990–2005, Canada

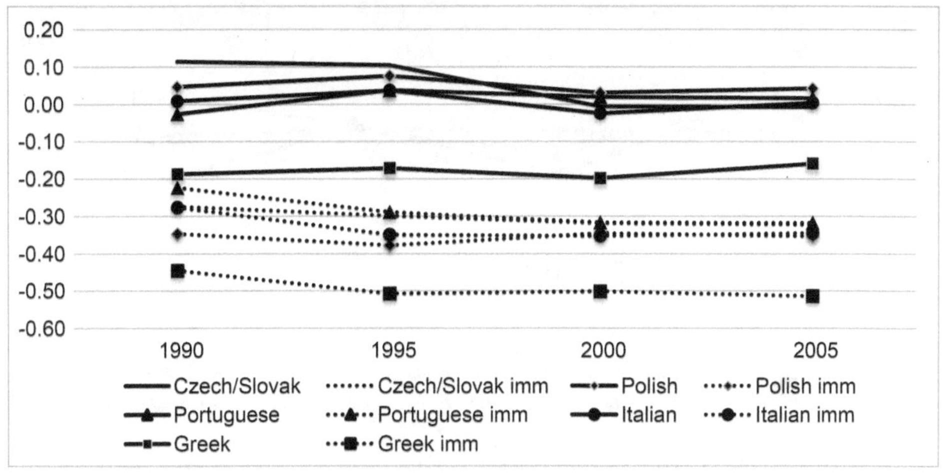


Source: 1991, 1996, 2001, and 2006 Census of Canada.

Figure 8.9 Earnings differentials among men, selected non-European origins (1) versus Canadian-born British origin, 1990–2005, Canada

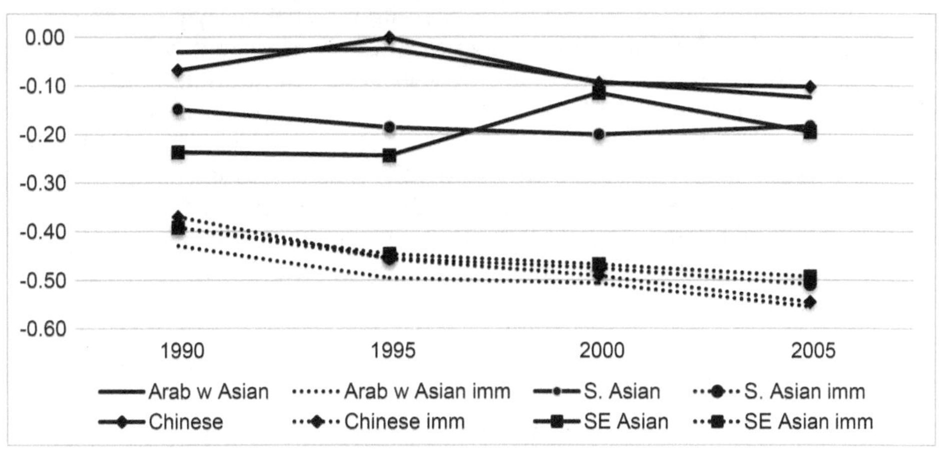


Source: 1991, 1996, 2001, and 2006 Census of Canada.

Figure 8.10 Earnings differentials among men, selected non-European origins (2) versus Canadian-born British origin, 1990–2005, Canada

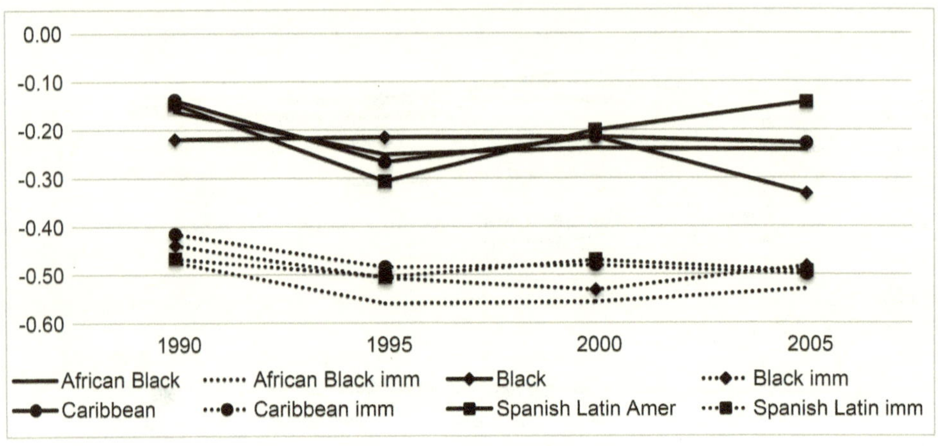


Source: 1991, 1996, 2001, and 2006 Census of Canada.

women since 1990, when the gaps ranged from −9 per cent to −17 per cent. However, women with South East Asian origins have seen substantial improvement in their prospects, with the earnings gap decreasing from −20 per cent in 1995 to +7 per cent in 2005. The situation for visible minority immigrant women is worse. Visible minority immigrant women have seen a steady decline in their earnings prospects decline over the entire fifteen-year period of this study. In 1995, Chinese immigrant women, for example, earned about 29 per cent less than women of British origin born in Canada. By 2005, they earned about half of what similarly qualified women of British origin earned. Immigrant women from Africa and the Caribbean also saw sharp declines in their earnings relative to women of British origin born in Canada. Caribbean immigrant women, for example, earned about 20 per cent less in 1990. In 2005, earnings had declined to about 43 per cent of similarly qualified women of British origin born in Canada.

Looking at the data for men, we can see sharp differences in results by origin and immigrant status. Men of European origin often enjoyed a small bonus over the entire fifteen-year period. Men of German, Dutch, and Polish origin born in Canada earned about 4 per cent to 8 per cent more than similarly qualified men of British origin born in Canada over the entire fifteen-year period. Immigrant men from the same origins faced earnings disparity ranging from –9 per cent to –16 per cent in 2005. With the exception of immigrant men from Scandinavia, Czechoslovakia, and Belgium, all other European immigrant groups faced earnings gaps between –9 per cent and –33 per cent in comparison to men of British origin born in Canada over the entire fifteen-year period.

The situation for visible minority men, however, was quite different. Visible minority men born in Canada faced earnings gaps between –11 per cent and –40 per cent in 2005. For some groups, this represents a decline in prospects, and for others, an improvement. South East Asian men born in Canada, for example, have seen the earnings gap improve, from –27 per cent in 1995 to –22 per cent in 2005. Men of Chinese origin, however, have seen a decline in their prospects, with the earnings gap increasing from –7 per cent to –11 per cent.

Visible minority immigrant men faced large earnings gaps over the entire time, which, depending on the group, either remained steady or got worse. Men of Arab and West Asian, Black, Caribbean, or African Black origin saw their prospects deteriorate by about 10 per cent from 1990 to 2005. Men of South Asian origin saw a minor decline in their prospects, from –16 per cent in 1990 to –20 per cent in 2005.

## Conclusions

The findings of this study may be summarized simply: minorities faced increasing earnings disadvantage over 1990 to 2005, with the bulk of the decline in their prospects occurring over the early 1990s. This pattern is observed broadly for both men and women, in Canada as a whole and in each of its three largest CMAs, for most Canadian-born visible minority ethnic groups, and for most white and visible minority immigrant groups. The decline in relative earnings is large: it is on the order of ten percentage points for Canadian-born visible minorities, and on the order of twenty percentage points for both white and visible minority immigrants.

The fact that immigrants face earnings disparity is widely known (see, for example, Stelcner 2000; Lian and Matthews 1998; and Hum and Simpson 2004). However, the magnitude of the decline in prospects is not as well studied. Our results suggest that the relative labour market outcomes of recent immigrants are substantially worse today as compared to two decades ago.

In the same way, the fact that some minorities born in Canada face gaps in their labour market attainment is fairly well documented (although somewhat more contentious). Christofides and Swidinsky (1994) and Hum and Simpson (1999) argue that minorities born in Canada do not face earnings gaps at all. However, their data sets are relatively small, implying relatively imprecise estimated coefficients. Our findings, using large and consistent data sets, suggest that not only are these earnings gaps concentrated in non-European (visible minority) ethnic groups, but also they are remarkably consistent over time and in some cases getting larger. These results are consistent with those of Skuterud (2010) and Palameta (2007).

That immigrant earnings disparity has increased over time is troubling in the context of Canada's steadily large intake of immigrants and steady increase in ethnic diversity. Canada's cities are crucibles of superdiversity, and are seen worldwide as roadmaps to a future of cohesive diversity. Growing immigrant and visible minority labour market disparity threatens this. The social and economic capital of immigrants may be the culprit here, as also demonstrated in chapter 9 of this volume in the case of United States. But the welcoming – or excluding – nature of Canada's urban communities may also be at play. Unravelling these sources of disparity is an important goal of future research.

Increasing disparity faced by minorities born in Canada is troubling because these individuals are educated and socialized in Canada. In comparison with immigrants, these individuals do not face the same barriers related to language knowledge, recognition of credentials, accent penalties, or lack of networks.

Understanding the evolving barriers facing immigrants and ethnic minorities in Canada will require new empirical strategies. The mechanisms outlined above are not amenable to study using traditional data sources like the Censuses of Canada. Field experimental work, such as that of Oreopoulous (2011), has illuminated name discrimination as an important player in immigrant and visible minority disparity. New lab experimental research may also shed light on processes of exclusion in our superdiverse urban environments. Data sources targeting particular mechanisms may also have an important role to play in uncovering the magnitudes of particular channels of immigrant disparity, as in the work of Ferrer et al. (2006) on literacy skills. Finally, panel data sources must be brought to bear on these questions to reasonably assess a dynamic phenomenon like the integration of immigrants into economic life.

**Appendix A**

The personal characteristics in our regressions are coded as follows:

| | |
|---|---|
| Age: | Eight age cohorts as dummy variables (age 25 to 29, 30 to 34, 35 to 39, 40 to 44, 45 to 49, 50 to 54, 55 to 59, and 60 to 64). Age 25 to 29 is the left-out dummy variable. |
| Schooling: | We control for 12 levels[7] of certificates as dummy variables (none, high school, trades certificate, college certificate less than 1 year, college certificate less than 3 years, college certificate 3 or more years, university certificate less than bachelor's, bachelor's degree, BA+, medical degree, master's degree, and doctorate). No certificate is the left-out dummy variable. We note that although schooling information is available by place of schooling (foreign or domestic) in the 2006 census, it is not available in other years. Thus, we do not control for variation in the location of schooling. |
| Marital status: | Five dummy variables indicating marital status (single – never married, married, separated, divorced, widowed). Single is the left-out dummy variable. |
| Household size: | A dummy variable indicating a single-person household and a continuous variable indicating the number of family members for other households. |
| Official language: | Four dummy variables (English, French, both official languages – English and French, and neither official language). English is the left-out dummy variable. Note that although nearly all Canadian-born persons (minority or not) speak either or both official language(s), many immigrants (especially recent ones) speak neither. |
| CMA: | In regressions that pool all the cities together, we use 12 dummy variables indicating the Census Metropolitan Area / Region (Halifax, |

| | |
|---|---|
| | Montreal, Ottawa, Toronto, Hamilton, Winnipeg, Calgary, Edmonton, Vancouver, Victoria, or not in one of the 10 listed CMAs). Toronto is the left-out dummy variable. |
| Ethnic origin: | Two dummy variables indicating employment equity group status (white, visible minority). White is the left-out dummy variable. Alternatively, 34 dummy variables indicating ethnic origin (with separate dummies for various multiple-origin groups), with British-only as the left-out ethnic origin. For these, we do not report on the eight Aboriginal-origin groups. Note that any regression that includes the employment equity group status dummy does not include the 34 ethnic origin dummies, and vice versa. |
| Immigrant status: | Immigrant dummy equal to 1 if the person is not a citizen by birth. |
| Years since migrating: | Continuous variable equal to 0 for persons with Immigrant Status equal to 0, and equal to the number of years in Canada, less 6 years. Noting that census earnings relate to the previous year, an immigrant who arrived in Canada in 1995 in the 2001 census database would have Years Since Migration equal to 0 and would have reported their earnings for the 2000 calendar year. Thus, the coefficient on the immigrant dummy may be interpreted as the log-earnings disparity faced by an immigrant who has been in Canada for 5 years.[8] We note that citizenship status is highly correlated with years since migration for permanent residents, since a very large proportion of permanent residents take up citizenship soon after they are eligible. Consequently, to avoid collinearity problems, we do not include citizenship status as a regressor. |

Coding of ethnic origin is based on census variables eth1 and eth2 (Ethnic Origin Component, Question 17: First ethnic origin and Second ethnic origin).

Note: if the first ethnic origin is Canadian, we use the second ethnic origin.

*Single origins*

| | |
|---|---|
| British | English, Irish Manx, Scottish, Welsh, British Isles |
| French | Acadian French, Quebecois |
| Canadian | Canadian, Newfoundlander, Nova Scotian, Ontarian, Other provincial or regional groups |
| Baltic | Estonian, Latvian, Lithuanian, Byelorussian |
| German Austrian | German, Austrian |
| Czech Slovak | Czech, Slovak, Czechoslovakian |
| Scandinavian | Finnish, Danish, Icelandic, Norwegian, Swedish, Scandinavian |
| Dutch | Dutch (Netherlands), Flemish, Frisian |
| Polish | Polish |
| Jewish | Jewish |
| Hungarian | Hungarian |
| Russian | Russian |
| Portuguese | Portuguese |
| Italian | Italian, Sicilian |
| Greek | Cypriot, Greek |
| Balkan | Romanian, Albanian, Bosnian, Bulgarian, Croatian, Macedonian, Maltese, Montenegrin, Serbian, Slovenian, Yugoslav |
| Spanish | Spanish, Basque |
| Ukrainian | Ukrainian |
| Belgian | Belgian |
| Other European | Luxembourger, Swiss, Kosovar, Gypsy, Slav, Afrikaner |
| Arab W Asian | Egyptian, Iraqi, Jordanian, Kuwaiti, Lebanese, Libyan, Algerian, Berber, Moroccan, Tunisian, Maghrebi, Palestinian, Saudi, Syrian, Yemeni, Arab, Afghan, Armenian, Assyrian, Azerbaijani, Georgian, Iranian, Israeli, Kurd, Pashtun, Tatar, Turk, Arab |
| S Asian | Bangladeshi, Bengali, East Indian, Goan, Gujarati, Kashmiri, Nepali, Pakistani, Punjabi, Sinhalese, Sri Lankan, Tamil, South Asian |
| Chinese | Chinese, Taiwanese |

| | |
|---|---|
| SE Asian | Burmese, Cambodian, Hmong, Khmer, Laotian, Malaysian, Singaporean, Thai, Vietnamese, East or SE Asian |
| Other Asian | Filipino, Indonesian, Japanese, Korean, Mongolian, Tibetan, Asian n.o.s. Fijian, Polynesian, Samoan, Pacific Islander |
| African Black | Akan, Amhara, Angolan, Ashanti, Bantu, Burundian, Cameroonian, Chadian, Congolese, Dinka, East African, Eritrean, Ethiopian, Gabonese, Gambian, Ghanaian, Guinean, Harari, Ibo, Ivorian, Kenyan, Malagasy, Malian, Mauritian, Nigerian, Oromo, Peulh, Rwandan, Senegalese, Seychellois, Sierra Leonean, Somali, South African, Sudanese, Tanzanian, Tigrian, Togolese, Ugandan, Yoruba, Zambian, Zimbabwean, Zulu, African |
| Black | Black |
| Caribbean | Antiguan, Bahamian, Barbadian, Bermudan, Carib, Cuban, Dominican, Grenadian, Guyanese, Haitian, Jamaican, Kittitian/Nevisian, Martinican, Montserratan, Puerto Rican, St Lucian, Trinidadian/Tobagonian, Vincentian/Grenadinian, West Indian Caribbean |
| Latin American | Indigenous Latin American, Argentinian, Belizean, Bolivian, Brazilian, Chilean, Colombian, Costa Rican, Ecuadorian, Guatemalan, Hispanic, Honduran, Maya, Mexican, Nicaraguan, Panamanian, Paraguayan, Peruvian, Salvadorean, Uruguayan, Venezuelan, Latin Central or S. America |
| American, Australia or New Zealander | American, Australian, New Zealander, Hawaiian, Maori |

*Multiple origins*

| | |
|---|---|
| British/French/Canadian Combinations | Any combination of British, French, and/or Canadian |

| | |
|---|---|
| Visible minority with White | any combination of a visible minority origin with European and/or majority origins |
| White multiple | Any combination or European or majority origins |

NOTES

1 Coefficients from log-earnings can be interpreted as approximately equal to percentage disparities between the group of interest and the specified reference (i.e., "left out") category, holding constant all the personal characteristics in the regression. In our regressions, the left-out category is "white" when considering visible minorities as a whole, and is "British" when considering the collection of 34 ethnic groupings. Thus, if a reported coefficient on visible minority is –0.05, then one could say that visible minorities earn 5% less than whites with similar personal characteristics.

2 Details on the full list of variables included in the regression can be found in appendix A.

3 It should also be noted that changes in industry and occupation coding make comparison of work-related characteristics impossible over the four census periods.

4 The census does not include a variable that describes the amount of time workers have been active in the labour force. Generally for men, there is an assumption that this is equal to years of age minus years of schooling. However women are more likely to be in and out of the labour force. This means that it is difficult to directly compare the earnings of men and women. For this reason we limit our analysis to comparisons within rather than across genders.

5 We do not report data means because the data are drawn from the confidential files of the censuses of Canada. However, the mean values of our data are very similar to analogous data means drawn from the public-use data files of each year's census.

6 The coefficients on immigrant dummies give the estimated disparity faced by an immigrant who has been in Canada five years. Immigrants with other durations in Canada are, of course, in the empirical model. To compute their estimated earnings disparity, one would add the estimated coefficients on years in Canada and its square (plus its interactions with the visible minority indicator). We do not discuss the estimated disparities

for immigrants with other durations in Canada in the text, as they make summarizing the results more confusing.

7 Empty cells are not a problem in regression analysis: empty cells are dropped; small cells are retained, but their parameters are estimated with high variance.

8 We note that both immigrants who have been in Canada for 5 years and less and Canadian-born persons have a value of 0 for the years-since-migration regressor. However, immigrants have the immigrant status dummy equal to 1, while Canadian-born persons have it equal to 0. Thus, the immigrant status dummy carries the immigrant effect for an immigrant who has been in Canada for 5 years and less.

## REFERENCES

Akbari, A. 1992. *Economics of Immigration and Racial Discrimination: A Literature Survey (1970–1989)*. Ottawa: Multiculturalism and Citizenship Canada.

Baker, M., and D. Benjamin. 1995. "Ethnicity, Foreign Birth and Earnings: A Canada/US Comparison." In *Transition and Structural Change in the North American Labour Market*, ed. M. Abbott, C. Beach, and R. Chaykowski, 281–313. Kingston, ON: IRC Press, Queen's University.

Becker, G.S. 1996. *Accounting for Tastes: Cambridge and London*. Cambridge, MA: Harvard University Press.

Christofides, L.N., and R. Swidinsky. 1994. "Wage Determination by Gender and Visible Minority Status: Evidence from the 1989 LMAS." *Canadian Public Policy* 20 (1): 34–51. http://dx.doi.org/10.2307/3551834.

de Silva, A., and C. Dougherty. 1996. *Discrimination against Visible Minority Men*. Ottawa: HRDC Applied Research Branch, Strategic Policy Document, 96-6E.

Ferrer, A., D. Green, and W.C. Riddell. 2006. "The Effect of Literacy on Immigrant Earnings." *Journal of Human Resources* 41 (2): 380–410.

Galarneau, D., and R. Morissette. 2004. "Immigrants: Settling for Less?" *Perspectives on Labour and Income* 5 (6): 5–16.

Hall, M., and G. Farkas. 2008. "Does Human Capital Raise Earnings for Immigrants in the Low-Skill Labour Market?" *Demography* 45 (3): 619–39. http://dx.doi.org/10.1353/dem.0.0018.

Howland, J., and C. Sakellariou. 1993. "Wage Discrimination, Occupational Segregation and Visible Minorities in Canada." *Applied Economics* 25 (11): 1413–22. http://dx.doi.org/10.1080/00036849300000146.

Hum, D., and W. Simpson. 1999. "Wage Opportunities for Visible Minorities in Canada." Income and Labour Dynamics Working Paper Series. Ottawa: Statistics Canada. http://dx.doi.org/10.2307/3551526.

Hum, D., and W. Simpson. 2004. "Economic Integration of Immigrants to Canada: A Short Survey." *Canadian Journal of Urban Research* 13 (1): 46–62.

Kogan, I. 2006. "Labour Markets and Economic Incorporation among Recent Immigrants in Europe." *Social Forces* 85 (2): 697–721. http://dx.doi.org/10.1353/sof.2007.0014.

Li, P.S. 2003. "Initial Earnings and Catch-up Capacity of Immigrants." *Canadian Public Policy* 29 (3): 319–27. http://dx.doi.org/10.2307/3552289.

Lian, J., and D. Matthews. 1998. "Does the Vertical Mosaic Still Exist? Ethnicity and Income in Canada, 1991." *Canadian Review of Sociology and Anthropology / La Revue Canadienne de Sociologie et d'Anthropologie* 35 (4): 461–81. http://dx.doi.org/10.1111/j.1755-618X.1998.tb00732.x.

Oreopoulous, P. 2011. "Why Do Immigrants Struggle in the Labour Market?" *American Economic Journal: Economic Policy* 3 (4): 148–71.

Palameta, B. 2007. *Economic Integration of Immigrants' Children*. Perspectives on Labour and Income. Ottawa: Statistics Canada. www.statcan.gc.ca/pub/75-001-x/2007110/article/10372-eng.htm.

Pendakur, K., and R. Pendakur. 1998. "The Colour of Money: Earnings Differentials among Ethnic Groups in Canada." *Canadian Journal of Economics / Revue Canadienne d'Économique* 31 (3): 518–48. http://dx.doi.org/10.2307/136201.

Pendakur, K., and R. Pendakur. 2002. "Colour My World: Have Earnings Gaps for Canadian-Born Ethnic Minorities Changed Over Time?" *Canadian Public Policy* 28 (4): 489–511. http://dx.doi.org/10.2307/3552211.

Pendakur, K., and R. Pendakur. 2007. "Minority Earnings Disparity across the Distribution." *Canadian Public Policy* 33 (1): 41–62. http://dx.doi.org/10.3138/cpp.v33.1.041.

Pendakur, K., and R. Pendakur. 2011. "Colour by Numbers: Minority Earnings in Canada 1996–2006." *Journal of International Migration and Integration* 12 (3): 305–29.

Pendakur, K., and S. Woodcock. 2010. "Glass Ceilings or Glass Doors? Wage Disparity within and between Firms." *Journal of Business & Economic Statistics* 28 (1): 181–9. http://dx.doi.org/10.1198/jbes.2009.08124.

Reitz, J., H. Zhang, and A. Hawkins. 2009. "Comparisons of the Success of Racial Minority Immigrant Offspring in the United States, Canada and Australia." Unpublished working paper.

Skuterud, M. 2010. "Visible Minority Earnings Gap across Generations of Canadians." *Canadian Journal of Economics / Revue Canadienne d'Économique* 43 (3): 860–81.

Stelcner, M. 2000. "Earnings Differentials among Ethnic Groups in Canada: A Review of the Research." *Review of Social Economy* 58 (3): 295–317. http://dx.doi.org/10.1080/00346760050132346.

Stelcner, M., and N. Kyriazis. 1995. "Empirical Analysis of Earnings among Ethic Groups in Canada." *International Journal of Contemporary Sociology* 32 (1): 41–79.

# 9 Immigrant Underemployment in the US Urban Labour Markets

TETIANA LYSENKO AND QINGFANG WANG

Compared to US-born workers, immigrants are more likely to be found working in jobs that are below employees' full working capacity (Jensen and Slack 2003; Rebhun 2008; Batalova et al. 2008; Chiswick and Miller 2009). The phenomenon is often referred as "brain waste," "overeducation," or "underemployment." According to a recent study, more than 1.3 million college-educated immigrants are unemployed or working in unskilled jobs such as dishwashers, security guards, and taxi drivers in the United States (Batalova et al. 2008). As in the case of Canada, such a situation contributes to lower income among immigrants, with persistent income disparity even after a long settlement process (chapter 8 of this volume).

The most documented outcome of underemployment is job dissatisfaction. Research conducted with diverse samples has consistently found a robust, negative relationship between underemployment and job satisfaction (e.g., Feldman and Turnley 1995; Johnson and Johnson 2000). A large number of studies have found that underemployed workers are less likely to be committed to their jobs than adequately employed individuals (Leana and Feldman 1995; Maynard, Joseph, and Maynard 2006). Therefore, underemployed workers are likely to perceive their current job as temporary and continue job-hunting activities; or they may choose an alternative route of self-employment instead (chapters 10–12, this volume). In addition, the negative consequences of underemployment extend beyond the working realm. Researchers have argued that underemployment may be associated with decreased quality of marital, family, and social relationships (Feldman 1996; Jones-Johnson 1989). Furthermore, within the global context, underemployment of highly educated nationals could make brain drain more severe for the sending countries by undercutting remittances or the circulation of knowledge or expertise (Batalova et al. 2008; Batalova and Lowell 2007).

Therefore, underemployment could significantly impact immigrants and their families' socio-economic well-being and their experiences of moving upward in the receiving society (Li and Teixeira, chapter 1). The objective of this chapter is to investigate the patterns of job mismatch among immigrants in the United States and the factors associated with such discrepancies. This chapter is focused on the over-education aspect of job mismatch, which is defined as having more formal education than is required for the job (McGoldrick and Robst 1996). Immigrants are defined as those who were born outside of the United States, and interchangeably used with "the foreign born" in this chapter. The next section reviews the existing literature on underemployment. Following that, the data and methods used in this study are explained. Then, the results describe the rates of underemployment for all immigrants and the differences by country of birth. Further, personal and household socio-economic characteristics associated with underemployment are examined. Finally, this chapter concludes with some remarks about policy implications.

## Literature Review on Underemployment

There are different perspectives in explaining job mismatch among immigrants. For example, human capital accumulated in home countries often loses value as individuals cross national borders, and thus a degree obtained in a home country may not be recognized in the receiving countries (Carneiro, Fortuna, and Varejão 2012; Chiswick 1979; Friedberg 2000). The depreciation of the human capital acquired in the home country is particularly significant for non-English speakers (Chiswick and Miller 2009; Piracha and Vadean 2012). Batalova et al. (2008) show that high-skilled immigrants with limited English proficiency are twice as likely to work in unskilled jobs than those who speak English fluently. Chiswick and Taengnoi (2007) argue that highly educated immigrants without fluent English also tend to concentrate in social service occupations that do not require a lot of English knowledge. Other factors, such as job-searching strategies, immigrants' unfamiliarity with the receiving country's labour regulations, visa status, and other economic and cultural differences between the source and destination countries, have all impacted the transferability of human capital across countries (Chiswick and Miller 2009; Chiswick 1979).

Accordingly, a number of studies suggest that underemployment is a temporary phenomenon and is caused by a transitional period of immigrants' adjustment to the new labour market. This perspective implies that newcomers have limited information about the labour market and job opportunities in the receiving country. Thus, they are more likely to take jobs for which they have more education than needed. Eventually, however, after individuals acquire more labour market experience, they are able to get better jobs that match their education attainment level. At the same time, employers may not understand how to evaluate education obtained abroad, and thus immigrants could experience underemployment. Again, over time, the negative "screening effect" on the employer side is expected to diminish. In sum, these studies suggest that a change in underemployment depends on immigrants' integration into the labour market in the receiving country (Hartog 2000; Chiswick 2005; Piracha et al. 2012).

The rate of underemployment also differs significantly depending onthe country of origin. Chiswick, Lee, and Miller (2003) and Batalova et al. (2008) report that immigrants from countries that are similar in language, occupational requirements, and labour market structure are less likely to be overeducated in the receiving country. Since European immigrants are the only group that closely resembles the employment patterns of the US-born, most other immigrants tend to have much higher rates of underemployment than the US-born. This gap is especially wide among doctoral degree holders born in Latin America and Africa compared to US-born populations, with the underemployment rate among PhD holders born in Asia and Europe being the lowest compared to other education attainment groups. However, there is a significant variation within Asian and European groups as well. For instance, immigrants from China and India are less likely to be underemployed than Filipinos or those from the rest of Asia. The rates of overeducation also vary greatly within the European region, with Eastern Europeans being the most likely to be underemployed and Western Europeans the least (similarly to a persistent earnings gap between these two groups described in Pendakur and Pendakur, chapter 8 above).

In addition to differences of language, human capital, and familiarity with cultures between different labour markets, discrimination could play a role in immigrants' job searching-matching process. Employers may judge the capacities of immigrants (who in most cases are ethnic minorities as well) based on group characteristics rather than individual merit, which could translate into a devaluation of their education

(Chiswick and Miller 2008; Pichler 2011; Subirós 2011). For instance, De Jong and Madamba (2001) argue that hiring discrimination against immigrants is another influential factor that significantly increases chances of job mismatch for the immigrant population. A European study also finds that migrants and ethnic minorities are particularly susceptible to unemployment, inactivity, and job mismatch, and some of these phenomena are associated with perceived discrimination and negative views on migration (CEDEFOP, 2011; Pendkur and Pendkur, chapter 8).

Finally, social capital and social networks are particularly important for immigrant labour market experiences. Quite often, information about employment in certain sectors is disseminated by workers in that sector to co-ethnic job-seekers, causing a particular occupation to become an ethnic niche (Loury 1977; Light and Bonacich 1988; Waldinger 1996; Nakhaie and Kazemipur 2013; Paul 2013; Oberle, chapter 10; Chacko and Price, chapter 11; Li and Lo, chapter 12). Some employers also strategically recruit new workers through networking of current employees, thus homogenizing the racial or ethnic diversity in job sectors (Rosenfeld and Tienda 1999; Waldinger and Der-Martirosian 2001; Johnson-Webb 2003; Ellis et al. 2004; Fernandes 2011). In these cases, the occupations in ethnic niche sectors may not be commensurate with the workers' educational attainment.

Overall, these studies suggest that many factors are significantly associated with immigrants' underemployment experiences, such as individual immigrant's human capital, social capital, length of stay in the receiving society, and country of origin. At the same time, how their human capital is perceived and evaluated and how they themselves have used their social networking and social capital are unavoidably subject to the power relationship between the sending and receiving countries, globalized economic restructuring, and localized social, political, and economic dynamics.

## Data and Methodology

The data used in this study are extracted from the 2006–10 American Community Survey (ACS) five-year Public Usable Microdata Sample (PUMS) (Ruggles et al. 2010). We restricted our data samples to those in the urban area who are at least sixteen years old, employed in the civilian labour force, and work at least forty hours a week.

The underemployment variable is defined dichotomously (underemployed = 1, not underemployed = 0). The person is considered underemployed/overeducated when he or she possesses more years of education than the job requires. We will use these two terms interchangeably for the rest of the paper. In order to determine whether a worker is overeducated, we used the method of "realized matches" that is the most commonly used with US census data. According to this method, we first calculate the average educational attainment (measured by the years in school) for each occupation; then employees are considered as overeducated when they have more education than one standard deviation above the mean for their occupation (Chiswick and Miller 2009; Hartog 2000).

Descriptive statistics are first used to compare labour force characteristics between the US-born and the foreign-born, and among different country of origins. Then the following binary logistic model is used to estimate the characteristics associated with the likelihood of underemployment:

$$Y = \beta X+\varepsilon,\ Y_i \in \{0,1\},\ \varepsilon \sim N(0,\sigma^2);$$

where dependent variable Y is the (log)odds to being underemployed versus not. With the associated coefficient β, the vector X is the set of independent variables that includes personal demographic and household characteristics: age, gender, marital status, family size, hours worked per week, English proficiency, years in the United States; and industrial sectors, including agriculture, mining, construction, wholesale trade, retail trade, transportation and warehousing, information and communications, finance, insurance, and real estate (FIRE), service, public administration, manufacturing, and others (see table 9.3 for the details). The choice of these variables is based on the discussion of existing literature in this chapter. For instance, age, gender, and marital status were shown to have a different influence on underemployment depending on race or nationality, while the level of English proficiency and hours worked per week are predominantly associated with higher rates of underemployment for all groups. Under economic restructuring and the decline of manufacturing industries in the United States, underemployment also significantly varies by industrial sector.

The logistic regression modelling is conducted separately for the following groups: (1) the entire foreign-born as one group; (2) US-born non-Hispanic whites as the reference group; (3) the top ten largest immigrant groups including those from Mexico, India, El Salvador, China, the Philippines, Vietnam, Korea, Cuba, Germany, and Guatemala; and

(4) countries whose immigrants have the highest underemployment rate and a large presence in the United States that are not included in the above groups, namely, Poland and Russia. The US-born non-Hispanic white labour force is included as a reference group. Selecting only one group can avoid noise caused by diversity among the entire US-born group. At the same time, as a racial and ethnic majority group, we assume non-Hispanic whites have fewer social and structural barriers in entering labour markets when compared to other minority groups; therefore, we choose non-Hispanic whites as a "benchmark." At the same time, due to the fact that the time period of this study is covers the economic recession in the United States, we will compare the rate of underemployment with the data in 2000 for a more comprehensive picture.

## Results and Discussion

### *Overview: General Labour Force Characteristics of the Groups in Study*

The immigrant labour force has demographic characteristics distinct from those of US-born non-Hispanic whites and blacks, and the characteristics vary significantly by country of origin (table 9.1). Specifically, the foreign-born as one group are younger (average years old = 41.6) than the US-born (mean = 42.3). However, within foreign groups, the average age ranges from 35.4 for Mexicans to 44.4 for Cubans. Generally speaking, the immigrant labour force from Asia (except for India) and Eastern Europe are older and those from Latin America are younger. The sex component follows a similar pattern. The percentage of the female labour force ranges from 25.2 per cent for Guatemala to 54.1 per cent for the Philippines. The average number of years living in the United States also differs from country to country. Germans, Cubans, Koreans, and Vietnamese on average have stayed longer in the United States than immigrants from Eastern Europe and the Caribbean.

There are significant differences among immigrant groups in their English proficiency and educational attainment. The majority of immigrants (presented in table 9.1) have high levels of English proficiency. The exceptions are those from Mexico, El Salvador, and Guatemala, with around 50 per cent of the labour force who can speak English well or very well, compared to more than 90 per cent for immigrants from India, the Philippines, Germany, and Russia. Thirty-nine per cent of the

Table 9.1 Demographic characteristics of immigrants in the US urban areas

| | Mean age | Percentage of females | Percentage with bachelor's degree | Mean family size | Mean years in the United States | Percentage speaking English well or very well |
|---|---|---|---|---|---|---|
| Foreign-born | 41.6 | 39.6 | 39.0 | 3.1 | 19.1 | 77.1 |
| US-born | 42.3 | 40.3 | 41.6 | 2.6 | 0.0 | 99.4 |
| Mexico | 37.0 | 26.8 | 6.1 | 3.9 | 16.9 | 50.5 |
| India | 38.8 | 31.3 | 82.8 | 3.1 | 13.3 | 96.2 |
| Philippines | 43.7 | 54.2 | 54.0 | 3.5 | 19.8 | 96.9 |
| El Salvador | 37.9 | 35.2 | 7.4 | 3.6 | 16.7 | 56.1 |
| China | 42.4 | 45.3 | 61.7 | 3.1 | 17.7 | 77.3 |
| Vietnam | 41.9 | 44.1 | 30.9 | 3.7 | 20.4 | 72.5 |
| Korea | 42.4 | 45.7 | 57.9 | 3.0 | 20.8 | 77.9 |
| Cuba | 44.4 | 39.9 | 27.4 | 3.1 | 22.3 | 67.5 |
| Germany | 42.4 | 44.2 | 44.5 | 2.6 | 32.6 | 98.8 |
| Guatemala | 35.4 | 25.2 | 8.1 | 3.2 | 14.0 | 48.7 |
| Russia | 41.2 | 46.5 | 67.2 | 2.7 | 14.2 | 94.2 |
| Poland | 43.0 | 40.8 | 34.4 | 2.8 | 18.0 | 80.9 |

Source: Calculated by authors.

entire foreign-born labour force and 41.6 of the US-born have a college degree. Among them, only 6.1 per cent of the Mexican labour force has a college degree, followed by El Salvador (7.4%) and Guatemala (8.1%). In contrast, 82.8 per cent of the foreign-born Indian labour force has a college degree, followed by Russia (67.2%) and China (61.7%). These socio-economic characteristics are expected to be closely related to the rate of underemployment for each group.

*Rate of Underemployment across Groups*

The overall pattern is depicted in figure 9.1. During 2006–10, the foreign-born group has a slightly higher rate (14.0%) of underemployment than the US-born white labour force (13.5%). However, immigrant groups from different countries present tremendous differences in the rate. While most groups from Latin America, such as Mexico (4.4%),

Figure 9.1 Underemployment rate of immigrants in US urban areas by country of origin

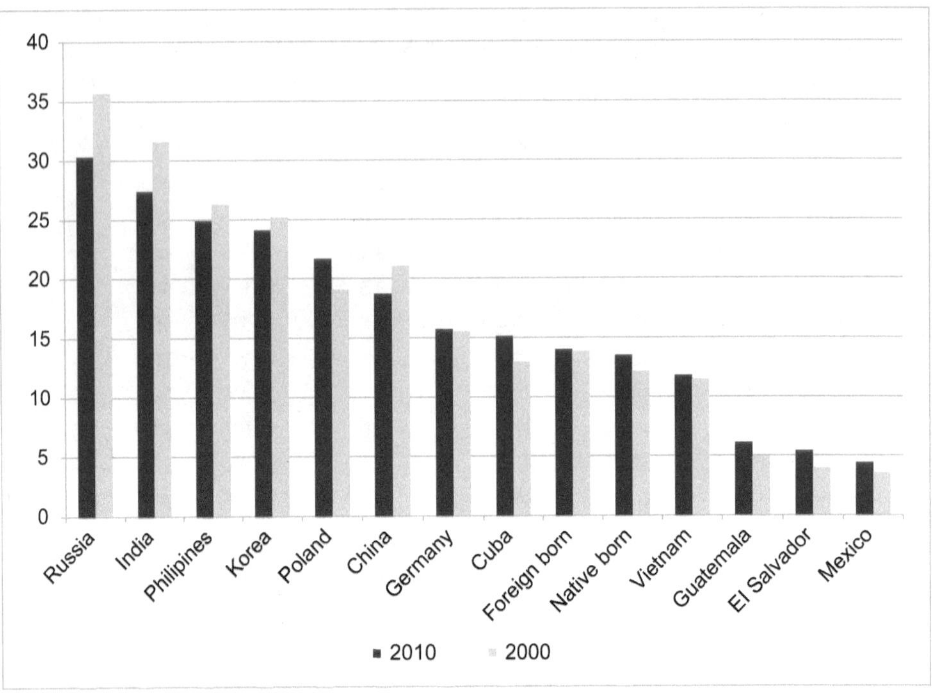


Source: Calculated by the authors.

El Salvador (5.4%), and Guatemala (6.1%), have much lower rates than the national average, groups from Asia, such as India (27.4%), the Philippines (24.9%), and Korea (24.1%), have considerably higher rates.

This pattern is consistent with the level of human capital for each group. As we discussed in the preceding section, English proficiency and educational attainment are both lower for the foreign-born Mexican, Salvadoran, and Guatemalan. It is reasonable that the rate of over-education (the measure of underemployment in this study) will be lower for these groups than for those who have higher level of human capital endowment, such as Asians. This pattern is also consistent with earlier studies that have observed a higher underemployment rate among Asian immigrants (e.g., Madamba 1998).

Not shown in the graph, some foreign-born groups have lower total-number-of-employed labour force, but have a very high rate of underemployment, such as those from Egypt (31%), Ukraine (31.5%), Bangladesh (32.4%), Kazakhstan (37.6%), and Georgia (41.1%). One possibility is the bias caused by their small sample size. At the same time, these groups could possibly suffer from a shorter length of stay in the United States and lack of supports. These groups deserve further study.

The rate of underemployent differs between 2000 and 2006–10 for each group. For both the entire foreign-born group and US-born whites, the rate of underemployment was lower in 2000. This change can be attributed to the recent economic recession that negatively affected the labour market as a whole. A similar pattern holds true for the Latin American groups such as the Mexican, El Salvadoran, and Guatemalan. However, the trend is opposite for most immigrants from Asia. For instance, the underemployment rate was 31.5 per cent among Indians, 25.2 per cent among Koreans, and 26.3 per cent among Philippinos in 2000, higher than the five-year average between 2006 and 2010. The decrease of the rate between the two years is most obvious for Russians.

### *Industrial Distribution of the Underemployed*

Although the overall rate of underemployment for the foreign-born is higher than the rate for the US-born labour force, it is not the case if we look at individual industrial sectors. As shown in table 9.2, both the foreign-born and the US-born labour force have a higher rate in retail trade (around 20%). For the immigrant group, they have much higher rates than US-born non-Hispanic whites in information and technology, FIRE, education, transportation and warehousing, and the health and social services sectors. At the same time, US-born non-Hispanic whites have much higher rates than the foreign-born in the agriculture and mining, construction, and art, entertainment, and recreation sectors. Such distinction suggests that overeducation is an industry-specific phenomenon for both the US-born and foreign-born populations.

### *Regression Results: Characteristics Associated with Underemployment*

In understanding the personal characteristics associated with the foreign-born labour force's underemployment, we conducted an individual-level logistic regression with the (log)odds of underemployment as the dependent variables and other individual-level characteristics as independent variables. For the entire foreign-born group, we particularly

Table 9.2 Underemployment rate of immigrants in US urban areas by industry

| | All | | Foreign-born | | Native NH white | |
|---|---|---|---|---|---|---|
| | No. | % | No. | % | No. | % |
| Agriculture and mining | 85,224 | 10.78 | 14,241 | 4.26 | 70,983 | 15.56 |
| Construction | 647,537 | 12.04 | 162,468 | 8.87 | 485,069 | 13.68 |
| Manufacturing | 1,113,409 | 13.25 | 331,053 | 13.85 | 782,356 | 13.01 |
| Wholesale trade | 267,232 | 10.61 | 75,400 | 12.42 | 191,832 | 10.03 |
| Retail trade | 1,142,727 | 19.49 | 284,280 | 19.47 | 858,447 | 19.50 |
| Transportation and warehousing | 409,213 | 14.74 | 142,375 | 17.70 | 266,838 | 13.53 |
| Information and communications | 259,262 | 14.51 | 64,342 | 18.97 | 194,920 | 13.46 |
| FIRE | 768,131 | 14.33 | 195,588 | 18.46 | 572,543 | 13.31 |
| Professional services | 1,003,677 | 12.58 | 296,363 | 13.99 | 707,314 | 12.07 |
| Education, health, and social services | 1,189,433 | 10.26 | 322,922 | 12.56 | 866,511 | 9.58 |
| Art, entertainment, and recreation | 613,487 | 16.12 | 205,740 | 12.47 | 407,747 | 18.92 |
| Other services | 323,107 | 12.08 | 109,897 | 11.87 | 213,210 | 12.20 |

Source: Calculated by authors.

include the ethnic/racial categories (i.e., whites, blacks, Hispanics, and Asian). For comparison purposes, regression is also conducted separately for US-born non-Hispanic whites, the groups with the largest foreign-born labour force, and the groups with a high rate of underemployment. The results are provided in table 9.3.

### PERSONAL AND HOUSEHOLD CHARACTERISTICS FOR THE FOREIGN-BORN AND US-BORN

For all the foreign-born, those who are married, have a larger family size, and stay longer in the United States are less likely to be underemployed. Those who are older, work longer hours, and speak more fluent English are more likely to be underemployed. These findings indicate that, on the one hand, longer stay in the receiving society could improve the possibility to match educational attainment with occupational requirements. Specifically, a ten years longer stay in the United States will decrease the likelihood of underemployment by 25 per cent for

Table 9.3 Regression results: Characteristics associated with underemployment for immigrants in US urban areas

| Variable | Foreign- born | US-born whites | Mexico | India | Philippines | El Salvador | China | Vietnam | Korea | Cuba | Germany | Guatemala | Russia | Poland |
|---|---|---|---|---|---|---|---|---|---|---|---|---|---|---|
| Age | 1.217*** | −0.004*** | 1.522*** | 1.204*** | 1.934*** | 1.290*** | 0.759*** | −0.342*** | 0.815*** | 1.775*** | 1.011*** | 1.011*** | 2.079*** | 0.813*** |
| Gender | 0.013 | −0.012*** | 0.051 | 0.207*** | 0.006 | 0.045 | −0.018 | −0.172*** | −0.168*** | 0.039 | −0.227*** | 0.013 | 0.295*** | 0.126* |
| Marital status | −0.089*** | −0.008*** | −0.118* | −0.153* | −0.014 | −0.032 | −0.194** | −0.009 | −0.133 | −0.178 | −0.003 | −0.21 | 0.111 | 0.125 |
| Family size | −0.145*** | −0.018*** | −0.235*** | −0.002 | 0.001 | −0.186*** | −0.157*** | −0.134*** | −0.023 | −0.047 | −0.144*** | −0.196*** | −0.129* | −0.218*** |
| Hours worked per week | 0.286*** | 0.077*** | 0.818*** | 0.024 | −0.09 | 0.291 | −0.161 | 0.146 | 0.034 | 0.807*** | 0.588*** | −0.003 | −0.027 | −0.109 |
| English | 0.882*** | | 0.939*** | 1.202*** | 0.631*** | 1.294*** | 1.047*** | 0.230*** | 0.534*** | 0.076*** | 0.387 | 0.631*** | 0.138 | 0.694*** |
| Years in the US | −0.029*** | | −0.014*** | −0.041*** | −0.037*** | −0.014*** | 0.000 | −0.026*** | −0.046*** | −0.006*** | −0.021*** | −0.015** | −0.042*** | −0.034*** |
| Construction | −0.108*** | 0.007*** | −0.244*** | 0.379*** | 0.214* | −0.122 | −0.067 | 0.382** | 0.412*** | 0.12 | 0.164 | −0.177 | 0.738*** | −0.074 |
| Trade | 0.206*** | 0.037*** | 0.021 | 0.453*** | 0.194*** | −0.123 | 0.127* | 0.312*** | 0.423*** | 0.226** | 0.196** | 0.007 | 0.378*** | 0.273** |
| Service | −0.311*** | −0.014*** | −0.118*** | −0.516*** | −0.245*** | −0.119 | −0.689*** | −0.08 | −0.104* | −0.107 | −0.440*** | −0.058 | −0.196** | −0.290*** |
| Public sector | 0.414*** | 0.070*** | 0.921*** | −0.085 | 0.109 | 0.707*** | 0.300*** | 0.450*** | 0.385*** | 0.595*** | 0.447*** | 1.079*** | 0.38 | 0.288 |
| Other | −0.680*** | 0.019*** | −1.020*** | 0.185 | −0.500* | −1.033* | −0.155 | −0.05 | 0.324 | −0.493 | 0.343 | −0.063 | −1.124 | 0.772 |
| Manufacturing | −0.044*** | 0.000 | −0.083* | 0.189*** | 0.106* | −0.167 | −0.057 | 0.1 | −0.078 | 0.1 | 0.069 | −0.258* | 0.117 | 0.028 |
| Hispanic | −.713*** | | | | | | | | | | | | | |
| Black | −0.324*** | | | | | | | | | | | | | |
| Asian | 0.191*** | | | | | | | | | | | | | |
| Intercept | −7.101*** | −0.128*** | −11.038*** | −6.224*** | −8.373*** | −8.095*** | −4.069*** | −1.973*** | −3.962*** | −10.865*** | −7.173*** | −6.149*** | −8.033*** | −3.720*** |
| chi2 | 36134.04 | 15916.64 | 2791.874 | 960.843 | 1690.159 | 170.68 | 1183.558 | 487.926 | 402.859 | 610.58 | 417.726 | 150.634 | 375.142 | 180.37 |
| R2 | 0.07 | 0.008 | 0.046 | 0.025 | 0.047 | 0.021 | 0.058 | 0.032 | 0.022 | 0.047 | 0.029 | 0.025 | 0.061 | 0.023 |
| *N* | 677742 | 2293030 | 174338 | 36394 | 37701 | 20536 | 25772 | 24649 | 17552 | 16072 | 17929 | 12157 | 5575 | 8178 |

the foreign-born. On the other hand, human capital is not consistently being transferred to a better-matched job, especially for those who are highly educated (in most cases, higher education and higher English proficiency also go hand in hand). For instance, compared to those who do not speak fluent English, speaking English well or very well will increase the odds of underemployment by 2.4 times. The overall pattern supports some existing studies (De Jong and Madamba 2001; Madamba 1998; McKee-Ryan and Harvey 2011) and holds true in general for the US-born. However, older workers are less likely to be underemployed. This means that for the US-born, more experience in the job market can be transferred more effectively for a better-matched job.

While gender effect is not significant for the foreign-born group overall, women in the US-born labour force are less likely to be underemployed. Specifically, compared to males, the female labour force is less likely to be underemployed by about 10 per cent if one holds other conditions the same. This finding is surprising because a large literature in social sciences has demonstrated that women are more likely to be segregated in semi- or low-skilled job sectors ("pink-collar" jobs) which quite probably do not match their level of education (Hudson 2002). In fact, overall, the US-born female labour force has higher percentage (43.5%) of college degree holders than men (40.2%) do. However, US-born female college holders (average = 41 years old) are younger than women who do not have a college degree (average = 43 years old). In other words, among the older age cohorts, women tend to have a lower level of education; at the same time, as shown earlier, older US-born Americans tend to be less likely to be overeducated. This could possibly explain the gender difference for US-born labour force. It would be very interesting to view the gender difference in a few years, when more and more female college graduates enter the labour force.

Among all the foreign-born, compared to non-Hispanic whites, both Hispanic and blacks have a significantly lower probability of underemployment; however, the possibility for the Asian group is much higher. Specifically, among the entire foreign-born labour force, when all other conditions are the same, Hispanics are 50 per cent less likely to be underemployed and blacks are 30 per cent less likely to be underemployed, when compared with non-Hispanic whites. In contrast, Asians are 20 per cent more likely to be underemployed than whites. On the one hand, this pattern is consistent with the fact that the Asian immigrant group has a much higher percentage of college graduates than other groups; on the other hand, it indicates that other factors related to race and ethnicity, in addition to personal characteristics, may play a role in the job searching/matching and pay process.

### INDUSTRIAL DIFFERENCES FOR THE FOREIGN-BORN AND THE US-BORN

As mentioned earlier, underemployment is industrial-specific. Compared to working in professional, business-services, and FIRE industries (which we grouped into "service sectors with high-status"), the foreign-born who work in the construction, social and personal services, and manufacturing industries are less likely to be overeducated; but, those who work in trade (wholesale and retail) and public sector are more likely to be overeducated. Specifically, when other conditions are the same, working in social and personal services will decrease the odds of underemployment by almost 30 per cent if compared with the high-status services. In contrast, working in trade will increase the odds by 1.2 times when compared to those who work in the high-status service sectors. The increase is 1.5 times for the public sector.

The overall pattern indicates that immigrants are more likely to be overeducated in the professional, trade, and public sectors. In many cases, the labour force in these sectors has a higher English proficiency. Therefore, it is consistent with the earlier section on personal characteristics, which shows a positive relationship between English proficiency and underemployment. As suggested by the literature, the reasons could be very complex, such as a devaluation of education credentials from the home country, failure to adjust to the differences in the receiving countries' labour markets, lack of appropriate social networking, or discriminatory practices in the labour markets. Quite interestingly, the findings are also consistent with a large number of existing studies for the highly educated labour force and those in high-skilled job sectors, where a "glass ceiling" is commonly reported for immigrants and ethnic minorities (Woo 2000; Varma 2004).

The industrial effects are different for the US-born. For instance, unlike immigrants in the construction sector, who are less likely to be self-employed when compared to those in high-status services, the US-born workers in this industry are more likely to be underemployed. While manufacturing jobs are better matched with the foreign-born's level of education, for the US-born, working in this sector does not make much difference from working in the high-status service sectors. The difference between the foreign-born and the US-born indicates that immigrants may do better in job-education matching in the traditional niche sectors such as manufacturing and construction. In this sense, in the low-skilled or semi-skilled immigrant-concentrated industrial sectors, social networking, in many cases bounded by ethnicity and nationality as well, may help the foreign-born not only to find a job, but also to find a job that could best match their educational attainment. However, as we will discuss later, it may not be true for other industrial or occupational sectors across ethnic and national groups, especially the highly skilled sectors.

### DIFFERENCES ACROSS GROUPS

Demographic characteristics associated with the propensity for underemployment or overeducation vary significantly by country of origin. Age is positively associated with underemployment for most foreign-born groups except for the Vietnamese. Like the US-born white group, older Vietnamese-born workers are less likely to be underemployed. While gender is not a significant variable for immigrant workers from Mexico, the Philippines, El Salvador, China, Cuba, and Guatemala, women are more likely to be underemployed than men for those from India, Russia, and Poland, but less likely to be underemployed for those from Vietnam, Korean, and Germany. Indian- and Russian-born immigrants have a very high percentage of college degree holders (80% for Indian women, 84% for Indian men, 70.8% for Russian women, and 64.1% for Russian men). For Russia and Poland, women include a significantly higher percentage of college degree holders than co-nationality men do. For Korea and Germany, the situation is totally opposite.

Across all groups, English proficiency is associated with a higher likelihood of underemployment, and longer stay in the United States decreases the odds of mismatch, while holding other conditions constant. This reinforces the earlier discussion that more experience in the receiving country will help the foreign-born to find jobs that better match their educational attainment. Obviously, familiarity with the culture and environment, broadened social networking, and improved "soft skills" could all play a positive role for immigrants' labour market experiences in the United States.

Similarly to both the entire foreign-born group and US-born whites, foreign-born Mexican, Indian, and Chinese who are married are less likely to be underemployed. But, marital status is not a significant characteristic for other groups. Likewise, for most groups, a larger family size is positively associated with a higher rate of underemployment. Across all groups, the more hours worked per week, the more likely that the worker is underemployed.

Compared to the high-status service sectors (which is the reference sector including professional business services and FIRE industries), Mexicans are less likely to be underemployed in construction; however, all other groups, just like US-born whites, are more likely to be underemployed in this sector. Most groups are more likely to be underemployed in trade except for those from Mexico, El Salvador, and Guatemala. With the same reference sector, those who work in social and personal services are less likely to be underemployed except for those from El Salvador, Vietnam, Cuba, and Guatemala. Most workers in public sectors are more likely to be overeducated except for those from India, the Philippines, Russia, and Poland. Association between manufacturing jobs and underemployment is the most diverse across the groups. Compared to the reference sector, those from Mexico and Guatemala who work in manufacturing sectors are more likely to match their education with the jobs. The same association holds true for the entire foreign-born group, probably due to Mexicans' large share of the total foreign-born. However, Indians and Philippinos working in manufacturing sectors are more likely to be overeducated. For other groups, there is no significant difference between manufacturing sectors and the high-status service sectors. We believe the group's differences in the probability of underemployment across industrial sectors have to do with labour market concentration and segmentation along the lines of ethnicity. A large number of studies have shown that immigrants and ethnic minorities in the United States are highly segmented in particular occupations and industrial sectors (Borjas 2006; Ellis and Wright 1999; Ettlinger and Kwon 1994; Logan, Alba, and McNulty 1994; Wang 2006; Wilson 2003). While ethnic groups are niched into particular job sectors for very different reasons, the educational level required by the job may not necessarily be commensurate with the human capital they endow.

## Conclusions

We analysed the phenomenon of underemployment among the foreign-born, with US-born non-Hispanic whites as a reference group. This research shows that immigrants have higher rate of underemployment than US-born whites. When comparing the average rate for the foreign-born to the US-born, the difference is insignificant. However, the difference becomes much more significant if we look into the subgroups among the foreign-born. Generally speaking, immigrants with a lower level of educational attainment have a lower rate of underemployment or being overeducated. This group includes immigrants from Mexico, Guatemala, and

El Salvador in the study. Chinese, Filipinos, and Koreans, by contrast, are 1.5 times more likely to be underemployed than US-born whites.

Personal and household characteristics are significantly associated with the rate of underemployment. While those people who speak better English and have higher level of educational attainment are more likely to be overeducated or underemployed, a longer stay in the United States could significantly increase the chance to better match their job with their education. As previous studies have suggested, the foreign-born could have difficulty in transferring their educational credentials and human capital obtained from their home country to their labour market experiences in the United States. However, with further adjustment in the receiving society, the situation can be significantly improved. This calls for better public policy programs such as validating foreign academic credentials, helping newcomers with language training, and providing employers with more knowledge and cultural competence in evaluating and hiring internationally trained labour force.

After controlling for other individual characteristics, there is a significant difference across ethnic and racial groups among the foreign-born in the rate of underemployment which is consistent with the differences across countries of origin. Unfortunately, this current study does not provide a causal relationship between country of origin, or race and ethnicity, and underemployment, nor does it provide the explanations of the reasons why there are gender differences across country-of-origin groups. Obviously, further research in this area is much needed. That said, based on existing literature (as we discussed in the literature review), we can expect that discriminative practices by employers, ethnic- or nationality-based social networking, and social capital accumulation, in addition to individual characteristics, could all play a role in the job searching and job matching process. Similarly, the gender effect still differs between the foreign-born and US-born whites, and among subgroups of the foreign-born after considering human capital and other personal and household characteristics. We can expect that gender stereotyping, gender-based social networking, and especially the interaction between gender and cultural-social practices based on country of origin could help provide significantly different paths for immigrant women in the labour market. While further research in these areas is called for, public policies should be aimed at creating more tolerant and open local labour markets to make progress in reducing discrimination against visible minorities and women (see more in chapter 14).

A very important aspect of underemployment is that its rates among the foreign-born across all the industrial sectors are significantly different from those among the US-born. Each subgroup of the foreign-born also demonstrates its unique presence in industrial sectors. There has been a large literature in documenting ethnic concentration and segmentation in the labour markets over time and space. We believe this pattern is highly related to the fact that ethnic groups (in many cases the boundaries of ethnicity are consistent with the boundaries of nationality) are highly segmented from each other across industrial sectors. With underemployment as an industry-specific phenomenon, further research in this area should be conducted to understand how local economic structure plays a role in producing the underemployment patterns. Further, while each individual labour force is embedded in a local labour market, the conditions of each labour market are very important to consider. In this sense, integrating immigration policies and labour market equity policies targeting individual labour forces (or groups) into regional economic development agenda could provide the best mix of both people- and place-based policy responses.

## REFERENCES


Batalova, J., M. Fix, and P.A. Creticos. 2008. *Uneven Progress: The Employment Pathways of Skilled Immigrants in the United States*. Washington, DC: Migration Policy Institute.

Batalova, J., and B.L. Lowell. 2007. "Immigrant Professionals in the United States." *Society* 44 (2): 26–31. http://dx.doi.org/10.1007/BF02819923.

Borjas, G.J. 2006. "Native Internal Migration and the Labor Market Impact of Immigration." *Journal of Human Resources* 41 (2): 221–58.

Carneiro, A., N. Fortuna, and J. Varejão. 2012. "Immigrants at New Destinations: How They Fare and Why." *Journal of Population Economics* 25 (3): 1165–85. http://dx.doi.org/10.1007/s00148-011-0387-3.

CEDEFOP (European Centre for the Development of Vocational Training). 2011. *Migrants, Minorities, Skill Mismatch among Migrants and Ethnic Minorities in Europe*. Luxembourg: Publications Office of the European Union.

Chiswick, B.R. 1979. "The Economic Progress of Immigrants: Some Apparently Universal Patterns." In *Contemporary Economic Problems*, ed. W. Feller, 357–99. Washington, DC: American Enterprise Institute.

Chiswick, B.R. 2005. *The Economics of Immigration: Selected Papers of Barry R. Chiswick*. Edward Elgar Publishing. Google e-book.

Chiswick, B.R., Y.L. Lee, and P.W. Miller. 2003. "Patterns of Immigrant Occupational Attainment in a Longitudinal Survey." *International Migration* (Geneva, Switzerland) 41 (4): 47–69. http://dx.doi.org/10.1111/1468-2435.00252.

Chiswick, B.R., and P.W. Miller. 2008. "Occupational Attainment and Immigrant Economic Progress in Australia." *Economic Record* 84 (9): S45–56. http://dx.doi.org/10.1111/j.1475-4932.2008.00482.x.

Chiswick, B.R., and P.W. Miller. 2009. "The International Transferability of Immigrants' Human Capital." *Economics of Education Review* 28 (2): 162–9. http://dx.doi.org/10.1016/j.econedurev.2008.07.002.

Chiswick, B.R., and S. Taengnoi. 2007. "Occupational Choice of High Skilled Immigrants in the United States." *International Migration* (Geneva, Switzerland) 45 (5): 3–34. http://dx.doi.org/10.1111/j.1468-2435.2007.00425.x.

De Jong, G.F., and A.B. Madamba. 2001. "A Double Disadvantage? Minority Group, Immigrant Status, and Underemployment in the United States." *Social Science Quarterly* 82 (1): 117–30. http://dx.doi.org/10.1111/0038-4941.00011.

Ellis, M., and R. Wright. 1999. "The Industrial Division of Labor among Immigrants and Internal Migrants to the Los Angeles Economy." *International Migration Review* 33 (1): 26–54. http://dx.doi.org/10.2307/2547321.

Ellis, M., R. Wright, and V. Parks. 2004. "Work Together, Live Apart? Geographies of Racial and Ethnic Segregation at Home and at Work." *Annals of the Association of American Geographers* 94 (3): 620–37. http://dx.doi.org/10.1111/j.1467-8306.2004.00417.x.

Ettlinger, N., and S. Kwon. 1994. "Comparative Analysis of U.S. Urban Labor Markets: Asian Immigrant Groups in New York and Los Angeles." *Tijdschrift voor Economische en Sociale Geografie* 85 (5): 417–33. http://dx.doi.org/10.1111/j.1467-9663.1994.tb00701.x.

Feldman, D.C. 1996. "The Nature, Antecedents and Consequences of Underemployment." *Journal of Management* 22 (3): 385–407. http://dx.doi.org/10.1177/014920639602200302.

Feldman, D.C., and W.H. Turnley. 1995. "Underemployment among Recent Business College Graduates." *Journal of Organizational Behavior* 16 (S1): 691–706. http://dx.doi.org/10.1002/job.4030160708.

Fernandes, D. 2011. "Social Networks of Migrant Construction Workers in Goa." *Indian Journal of Industrial Relations* 47 (1): 65.

Friedberg, R. 2000. "You Can't Take It with You? Immigrant Assimilation and the Portability of Human Capital." *Journal of Labor Economics* 18 (2): 221–51. http://dx.doi.org/10.1086/209957.

Hartog, J. 2000. "Over-Education and Earnings: Where Are We, Where Should We Go?" *Economics of Education Review* 19 (2): 131–47. http://dx.doi.org/10.1016/S0272-7757(99)00050-3.

Hudson, M. 2002. "Modeling the Probability of Niche Employment: Exploring Workforce Segmentation in Metropolitan Atlanta." *Urban Geography* 23 (6): 528–59. http://dx.doi.org/10.2747/0272-3638.23.6.528.

Jensen, L., and T. Slack. 2003. "Underemployment in America: Measurement and Evidence." *American Journal of Community Psychology* 32 (1-2): 21–31. http://dx.doi.org/10.1023/A:1025686621578.

Johnson, G., and R. Johnson. 2000. "Perceived Overqualification and Dimensions of Job Satisfaction: A Longitudinal Analysis." *Journal of Psychology* 134 (5): 537–55. http://dx.doi.org/10.1080/00223980009598235.

Johnson-Webb, K.D. 2003. *Recruiting Hispanic Labor: Immigrants in Non-traditional Areas.* New York: LFB Scholarly Pub.

Jones-Johnson, G. 1989. "Underemployment, Underpayment, and Psychosocial Stress among Working Black Men." *Western Journal of Black Studies* 13 (2): 57–65.

Leana, C., and D. Feldman. 1995. "Finding New Jobs after a Plant Closing: Antecedents and Outcomes of the Occurrence and Quality of Reemployment." *Human Relations* 48 (12): 1381–1401. http://dx.doi.org/10.1177/001872679504801201.

Light, I.H., and E. Bonacich. 1988. *Immigrant Entrepreneurs: Koreans in Los Angeles, 1965–1982.* Oakland: University of California Press.

Logan, J.R., R.D. Alba, and T.L. McNulty. 1994. "Ethnic Economies in Metropolitan Regions: Miami and Beyond." *Social Forces* 72 (3): 691–724. http://dx.doi.org/10.1093/sf/72.3.691.

Loury, G. 1977. "A Dynamic Theory of Racial Income Differences." In *Women, Minorities, and Employment Discrimination*, ed. P.A. Wallace and A. Le Mund, 153–86. Lexington, MA: Lexington Books.

Madamba, A.B. 1998. *Underemployment among Asians in the United States: Asian Indian, Filipino, and Vietnamese Workers.* New York: Routledge.

Maynard, D.C., T.A. Joseph, and A.M. Maynard. 2006. "Underemployment, Job Attitudes, and Turnover Intentions." *Journal of Organizational Behavior* 27 (4): 509–36. http://dx.doi.org/10.1002/job.389.

McGoldrick, K., and J. Robst. 1996. "Gender Differences in Overeducation: A Test of the Theory of Differential Overqualification." *American Economic Review* 86 (2): 280–4.

McKee-Ryan, F.M., and J. Harvey. 2011. "'I Have a Job, But ...': A Review of Underemployment." *Journal of Management* 37 (4): 962–96. http://dx.doi.org/10.1177/0149206311398134.

Nakhaie, M.R., and A. Kazemipur. 2013. "Social Capital, Employment and Occupational Status of the New Immigrants in Canada." *Journal of International Migration and Integration* 14 (3): 419–37. http://link.springer.com/article/10.1007/s12134-012-0248-2.

Paul, A.M. 2013. "Good Help Is Hard to Find: The Differentiated Mobilisation of Migrant Social Capital among Filipino Domestic Workers." *Journal of Ethnic and Migration Studies* 39 (5): 719–39. http://dx.doi.org/10.1080/1369183X.2013.756660.

Pichler, F. 2011. "Success on European Labor Markets: A Cross-National Comparison of Attainment between Immigrant and Majority Populations." *International Migration Review* 45 (4): 938–78. http://dx.doi.org/10.1111/j.1747-7379.2011.00873.x.

Piracha, M., M. Tani, and F. Vadean. 2012. "Immigrant Over- and Under-education: The Role of Home Country Labour Market Experience." *IZA Journal of Migration* 1 (1): 1–21.

Piracha, M., and F. Vadean. 2012. "Migrant Educational Mismatch and the Labour Market." IZA Discussion Paper no. 6414.

Rebhun, U. 2008. "A Double Disadvantage? Immigration, Gender, and Employment Status in Israel." *European Journal of Population / Revue Européenne de Démographie* 24 (1): 87–113. http://dx.doi.org/10.1007/s10680-007-9137-3.

Rosenfeld, M., and M. Tienda. 1999. "Mexican Immigration, Occupational Niches, and Labor-Market Competition: Evidence from Los Angeles, Chicago, and Atlanta, 1970–1990." In *Immigration and Opportunity*, ed. F.D. Bean and S. Bell-Rose, 64–105. New York: Russell Sage Foundation.

Ruggles, S., J.T. Alexander, K. Genadek, R. Goeken, M.B. Schroeder, and M. Sobek. 2010. *Integrated Public Use Microdata Series: Version 5.0*. [Machine-readable database.] Minneapolis: University of Minnesota.

Subirós, P. 2011. "'Don't Ask Me Where I'm From': Thoughts of Immigrants to Catalonia on Social Integration and Cultural Capital." *International Journal of Urban and Regional Research* 35 (March): 437–44.

Varma, R. 2004. "Asian Americans: Achievements Mask Challenges." *Asian Journal of Social Science* 32 (2): 290–307. http://dx.doi.org/10.1163/1568531041705103.

Waldinger, R.D. 1996. *Still the Promised Land*. Cambridge, MA: Harvard University Press.

Waldinger, R.D., and C. Der-Martirosian. 2001. "The Immigrant Niches: Pervasive, Persistent, Diverse." In *Strangers at the Gate: New Immigrants in Urban America*, ed. R.D. Waldinger, 228–71. Berkeley: University of California Press.

Wang, Q. 2006. "Linking Home to Work: Ethnic Labor Market Concentration in the San Francisco Consolidated Metropolitan Area." *Urban Geography* 27 (1): 72–92. http://dx.doi.org/10.2747/0272-3638.27.1.72.

Wilson, F. 2003. "Ethnic Niching and Metropolitan Labor Markets." *Social Science Research* 32 (3): 429–66. http://dx.doi.org/10.1016/S0049-089X(03)00015-2.

Woo, D. 2000. *The Glass Ceiling and Asian Americans*. Walnut Creek, CA: AltaMira Press.

# 10 The Latino Commercial Landscape and Evolving Hispanic Immigrant Population in Two Midwestern Metropolitan Areas

ALEX OBERLE

In recent years, much of the focus on Latino immigration has been centred on new "gateway cities," particularly in the American South and West, as well as small, rural communities that have been transformed by immigration from Mexico and elsewhere in Latin America (Singer, Hardwick, and Brettell 2008; Odem and Lacy 2009; Millard and Chapa 2004). Yet, in the US Midwest there has also been extensive immigration and a key, visible part of the changing population patterns consists of businesses that specifically cater to the Latino immigrant population and serve as a concrete and functional representation of immigrants' economic experiences. In the mid-sized metropolitan area of Des Moines, Iowa, this consists of a compact business district that extends along a major thoroughfare. In greater Chicago the Pilsen neighbourhood in Chicago proper is a long-standing Mexican-American and Mexican immigrant community and the Latino commercial landscape here is both extensive and specialized, drawing clientele from the local neighbourhood and elsewhere in the metropolitan area. On the far suburban fringe is the exurban community of Carpentersville, a now majority Latino community whose spread-out Latino commercial district is a magnet for both local Latino residents and Hispanic immigrants in the northwestern Chicago suburbs and exurban communities that are functionally linked to the larger metropolitan area.

This chapter explores immigrants' economic experiences by evaluating these Latino commercial landscapes in greater Chicago and in Des Moines, Iowa, thus focusing on example business districts that are representative of others elsewhere in the US Midwest. Assessing the morphology and character of such business landscapes provides a means for comparing and contrasting different types of immigrant communities and the relationship between those communities and the larger metropolitan area, particularly their varying function as commercial centres. Before discussing the research and case study locations, the chapter begins by providing a broad overview of Latino immigration in the Midwest. Next, the chapter discusses the current state of immigrant entrepreneurship and Hispanic-owned businesses, as well as presenting key literature relating to immigrant and ethnic entrepreneurship. Because of the landscape focus of this research, the next section presents existing research in the area of immigrant and ethnic commercial landscapes. The chapter then outlines the methods used in this research. Differently from chapters 8 and 9, which utilize large national census data sets for comprehensive analysis, this chapter takes a case study approach consisting of a field-based landscape analysis supplemented by an analysis of census data. Discussing the rationale for the selection of the case study locations, the chapter provides a context for selecting the Des Moines and Chicago metropolitan areas. Following this is a brief overview of Des Moines and the case study neighbourhood, then a detailed analysis of the Latino commercial landscape. The next section focuses on two case study areas in greater Chicago, beginning with an overview of the larger metropolitan area, followed by an analysis of the Latino business landscape in both case study areas. The following section presents an analysis of key demographic data for the case study areas and compares and contrasts the three locations. The chapter ends with a conclusion about directions for future research.

## Hispanic Settlement in the United States and Latino Immigration in the Midwest

Used interchangeably, the terms "Hispanic" and "Latino" refer to ethnic and nationality groups from Spain or Latin America. In the United States, as of 2010, 16.4 per cent of the population is Latino; in Canada, in 2006, Hispanics represented only 1 per cent of the population (Motel 2010; Statistics Canada 2012), which marks a key difference in where immigrants to the two countries come from and how the respective countries admit immigrants. More than one in three Latinos in the United States are foreign-born, compared to 12.9 per cent of all US residents (Grieco et al. 2012). Among Latinos in the United States, the Mexican ancestry population is the largest group, representing approximately 65 per cent of all Latinos (Motel and Patten 2012b). All other Hispanic groups, including Puerto Rican, Cuban, Salvadoran, and Colombian groups each account for less than 10 per cent of the total Latino population (ibid.). Currently, almost half of the Latino population lives in just ten metropolitan areas, ranging from traditionally Hispanic cities like Los Angeles, San Antonio, and Miami to metropolitan areas such Houston and Phoenix that had only a modest Latino population a few decades ago (ibid.). In addition to these larger urban centres, rapid Hispanic population growth has recently occurred in many other metropolitan areas such as Charlotte, Las Vegas, and Orlando (Suro and Singer 2002).

The Latino population in the US Midwest,[1] like elsewhere in the nation, includes native-born Mexican-Americans, Puerto Ricans, and other Latinos as well as a large percentage of foreign-born Hispanics, especially from Mexico and Central America. Across the region, there is an uneven pattern of Latino immigration. Among some metropolitan areas, such as Chicago and Kansas City, there are both long-standing Hispanic neighbourhoods and a vibrant Latino immigrant community, while in St Louis, Cleveland, and other major cities there is only a small total Hispanic population. This pattern also exists outside of the largest metropolitan areas, with sizeable Latino populations in mid-sized metropolitan areas and smaller cities such as Omaha, Nebraska, Beloit, Wisconsin, Des Moines, Iowa, as well as dozens of rural counties where agribusiness has attracted substantial numbers of foreign-born workers. From 2000 to 2010, the change in Hispanic population in the Midwest approximated that of the national average, but with fast-growing Latino populations in some smaller states like Iowa and South Dakota, and much slower rates in Michigan and Illinois (Pew Research Center 2011).

## Immigrant Entrepreneurship and Latino-Owned Businesses

Immigrant entrepreneurship and Hispanic-owned businesses continue to expand despite the recent economic downturn, including a greater than two-tenths of 1 per cent increase in immigrant entrepreneurship between 2006 and 2010 (Fairlie 2012, 32). Nationwide, there are higher rates of business ownership among immigrants than among the native-born population and business formation rates are substantially higher among immigrant groups (Fairlie 2012). Hispanic-owned business, including establishments owned by both immigrants and non-immigrants, increased by over 43 per cent between 2002 and 2007 and account for more than 8 per cent of non-farm business in the United States (US Bureau of the Census 2007). In the Midwest, the increase was similar to the national average, with the Chicago Census Statistical Area (CSA) having the sixth largest numbers of Hispanic-owned businesses in the United States and by far the largest number in the Midwest (ibid.).

The consistent increase in Hispanic-owned businesses and high rates of immigrant entrepreneurship mirror both demographic trends and increasing levels of income among Latinos. In the United States, the collective purchasing power of Latinos has grown dramatically in recent decades. The Association of Hispanic Marketing Agencies estimated that purchases by US Hispanics increased from $104 billion to $221 billion in just ten years between 1983 and 1993 (Dávila 2001, 68). In 2008, Latino buying power was $951 billion and it was forecast to reach $1.4 trillion in 2013 (Korzenny and Korzenny 2012).

Many, but not all, of these businesses can likely be classified as an ethnic economy, a business where a self-employed "ethnic" hires other co-ethnic employees (Light et al. 1994). Within the larger Hispanic ethnic economy, there is extensive research on Cuban ethnic economies (Portes 1987; Portes and Jensen 1989; Zsembik 2000), but comparatively little research on the ethnic economies of Mexican immigrants. An exception to this is Guarnizo (1998), who developed a typology of Mexican-owned enterprises in Los Angeles and calculated that there are 3.4 self-employed Mexican immigrants per every self-employed Cuban immigrant, dispelling previously held notions that the Mexican ethnic economy was inconsequential (Guarnizo 1998, 2).

## Latino Commercial Landscapes

The Mexican ethnic economy is not only substantial, but the built environment – including Hispanic-owned and immigrant-owned businesses, and other retail establishments catering to Hispanic immigrants – is a fundamentally visible aspect of the urban environment. This Latino commercial landscape is evident in cities across the United States, both in established immigrant neighbourhoods in the central city as well as in emerging suburban locales with growing foreign-born populations (as shown in Washington, DC, area by Chacko and Price in chapter 11 of this volume). Landscapes, because of their inherent visibility, can serve as a key diagnostic for understanding patterns, processes, and change in cities. This is particularly true in regard to ethnic and immigrant communities such as the Vietnamese commercial landscape in northern Virginia (Wood 1997), post–Second World War Italian immigration to a Toronto neighbourhood (Buzzelli 2001), and Ethiopian settlement in Washington, DC (Chacko 2003; Chacko and Price, this volume).

The Latino commercial landscape consists of retail establishments that cater to a primarily Hispanic immigrant clientele. These landscapes are immediately visible not only because of their use of Spanish-language signage, but also as a result of the widespread use of particular colours and symbols, as well as business types that are either a significant variation on mainstream North American counterparts or represent a unique business type. Example businesses that commonly constitute this landscape include butcher shops/small groceries (*carnicerías*), bakeries (*panaderías*), shops that sell traditional herbal medicines and religious items (*yerberías*), and specialized restaurants.

The spatial arrangement of these business types in an urban area can serve as an indicator of the size and character of the Latino immigrant population. For example, in the context of Phoenix, Arizona, and likely other large, fast-growing metropolitan areas with a sizeable population of Hispanic immigrants, *carnicerías* are "pioneer" businesses that are the first to locate in newer Latino neighbourhoods, suburban locales, or other areas where there originally was little or no Hispanic population until perhaps a few years prior (Oberle 2006). In English, the direct translation of *carnicería* is "butcher shop," but these establishments typically sell other items in addition to specialized cuts of meat. Examples may include Latin American and North American food products, freshly cooked, ready-to-eat items, money wiring services, household products, international calling cards, or clothing. Other business types such as *llanteras* (tire shops) and *yerberías* (herbal medicines and religious items) are more indicative of older, more mature Latino immigrant neighbourhoods (Oberle 2004).

The Latino commercial landscape can also signify the character of the local Hispanic immigrant population, such as its nationality or even region of origin. Oberle and Arreola (2008) assert that business names and signage typically broadcast broad pan-Hispanic connections, strong national pride, or regional identity. For example, despite critiques of terms like "Hispanic" or "Latino" as being contrived and overly simplistic, the use of such terms in business names or the incorporation of the names of national colours of multiple Latin American countries signify a Latino community that is diverse, though perhaps quite small. In other cases, *Mexicanidad* or "Mexicanness" prevails, symbolizing Mexican pride and a sizeable Mexican immigrant community. The incorporation of the green, white, and red tricolour flag is a common means for depicting these connections to Mexico. In other communities or neighbourhoods, the use of national colours may likewise indicate a substantial El Salvadoran or Dominican immigrant presence. In areas with a large Mexican foreign-born population, regional identify is often visible in the Latino commercial landscape. Among businesses in Phoenix, the use of state and city names, outlines, and symbols reflect hometowns and state pride. Examples include the states of Chihuahua, Sinaloa, Michoacán and the cities of Culiacán and Hermosillo.

## Methods and Case Study Areas

This research employs field-based landscape analysis and the analysis of census data to evaluate the Latino commercial landscape in the Des Moines, Iowa, and in greater Chicago (figure 10.1). These two metropolitan areas include three distinct types of population patterns and Latino commercial landscapes. In Des Moines, the Hispanic population is relatively small and is dispersed throughout the city, with only two census tracts that have a majority Latino population. The Latino commercial landscape is compact and serves the local Hispanic neighbourhood as well as Latinos throughout the city and metropolitan area. As such, Des Moines is representative of many medium-sized cities throughout the Midwest. In greater Chicago, the Latino population exceeds two million and includes long-standing Mexican-American districts, central city neighbourhoods that are home to a recent influx of Latino immigrants, and a burgeoning – yet dispersed – Hispanic immigrant population in suburban and exurban Chicago. The areas of Chicago included in this study are the long-standing Mexican-American and Mexican immigrant neighbourhood of Pilsen, located in Chicago proper as well as Carpentersville on the periphery of greater Chicago, an area that has received a substantial influx of Latino immigrants in the past two decades. Pilsen is representative of other similar neighbourhoods in Midwestern cities and Carpentersville in exurban Kane County exemplifies Latino settlement on the suburban fringe.

Specific methods employed in this research include a landscape analysis that is based on existing research (Wood 1997; Buzzelli 2001; Walcott 2002; Chacko 2003; Oberle 2004; Oberle and Arreola 2008). In an urban context, such landscape analyses include the entire ethnic landscape in a city or metropolitan area or a transect or sample set. In Des Moines, because of the compact nature of its Latino commercial landscape, the entire area is included in the analysis. In the Pilsen

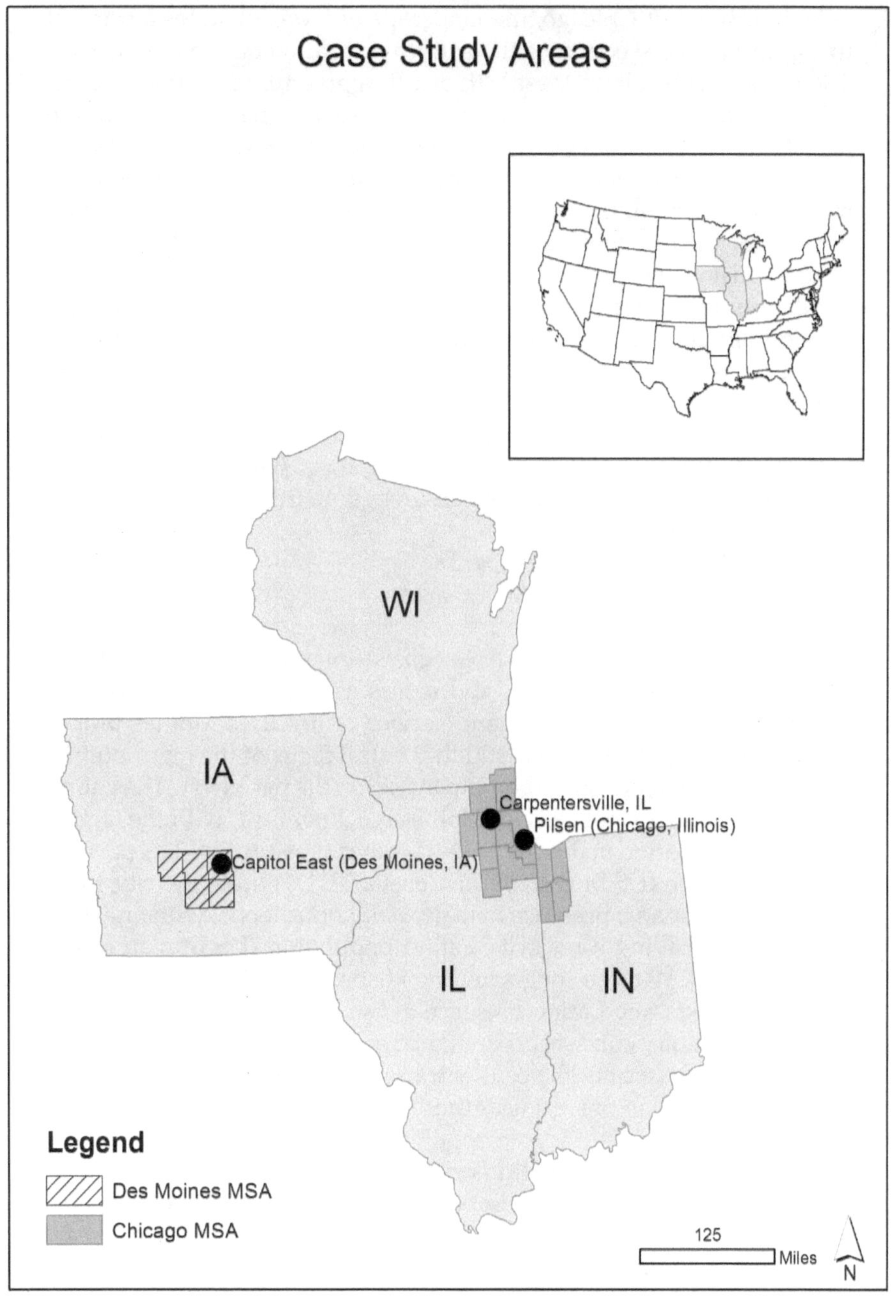


Figure 10.1 Case study area

neighbourhood of Chicago, the landscape analysis includes a transect to capture a cross-section of the extensive Latino commercial district. The transect runs along West 18th Street, representing a sample of the larger commercial district. Because of the dispersed nature of the Latino population and commercial businesses in Kane County, the landscape analysis here focuses on business clusters in the majority Latino community of Carpentersville, representing the total universe of Latino businesses in that community. Landscape analysis includes a count of total businesses, a categorization of their type and function, and a tabulation of key features that may indicate the character of the local Latino immigrant population. These are based on field observation of the businesses and, in some cases, required photographing businesses or visiting inside the establishment to gain additional information that cannot be readily ascertained from field observation. Supporting this landscape assessment is an analysis of local census data, including a GIS analysis of demographic changes over time from 1990 to 2010.

### *The Latino Commercial Landscape: The Capitol East Neighbourhood in Des Moines, Iowa*

The Des Moines Metropolitan Area consists of more than one-half million residents in central Iowa, and serves as the state capitol as well as a major regional economic and service centre. Des Moines proper remains the largest city and is still the central focus of the metropolitan area despite extensive suburban expansion on the periphery. The Latino population in metropolitan Des Moines is 6.2 per cent, with the largest Hispanic population in the city of Des Moines, constituting 11.9 per cent of its population (US Bureau of the Census 2011). Within the city, there is a modest Hispanic presence in many neighbourhoods, with only two census tracts having a majority Latino population (US Bureau of the Census 2010a). With no long-standing Hispanic or Mexican-American neighbourhoods, the Latino presence in the city is relatively new. The Latino community consists of foreign-born Hispanics as well as native-born Latinos. Several of the census tracts with the largest foreign-born populations overlap or exist near those tracts with the highest Hispanic population.

The vast majority of Latino-oriented business establishments in greater Des Moines form a compact Latino commercial landscape in the Capitol East neighbourhood, near the boundary of the two census tracts with majority Hispanic populations and near the intersection of four census tracts which collectively account for over 11 per cent of the city's foreign-born population (US Bureau of the Census 2010b). The Capitol East neighbourhood is part of the Des Moines "East Side," the portion of the city to the east of Des Moines River that historically was somewhat marginalized and detached from the city centre. Capitol East and surrounding neighbourhoods emerged as a working-class community, and the many cottages and bungalows attest to its modest origins, housing workers who were employed in local industries such as a dairy plant, flour mills, factories, and warehouses (Jacobsen 2002). The vibrancy of the early neighbourhood begin to diminish after the Second World War, and the area declined more quickly with the construction of Interstate 235, which split the community and provided easy access to emerging retail establishments in suburban Des Moines (Oltrogge 2009). The closing of small businesses in the neighbourhood occurred concurrently with residential decline, and by the 1970s the area was known for blight and a high crime rate. The city promoted infill, and large tracts of the old neighbourhood were razed. In 1997, the city began a major redevelopment initiative in East Village, just to the west of Capitol East (Jacobsen 2002). These nearby efforts, along with a growing Latino presence in the neighbourhood, resulted in a revitalized commercial landscape consisting of more than two dozen Latino-oriented businesses.

The Latino commercial landscape in Des Moines's Capitol East neighbourhood consists of twenty-six businesses that are located on or near Grand Avenue, a major east to west thoroughfare in the city. Within this landscape, there are three major clusters of businesses, with the remaining establishments scattered across several blocks. The business clusters include five establishments in La Tapatía Plaza, a relatively new shopping centre anchored by a large Hispanic grocery store; three businesses that are part of a small, newly built Latino-themed plaza; and five shops that occupy a cluster of original early-twentieth-century era brick storefronts. Ample parking is available at all establishments, and the business clusters are additionally accessible by pedestrians, particularly the Latino-themed plaza, which has walkways and side entrances, suggesting that it was designed to encourage foot traffic once the customer has parked in the lot behind the stores.

The Latino-oriented businesses include a diverse mix such as grocery stores, restaurants, beauty salons, and *panaderías* (bakeries), as well as establishments that do not fit a category, such as a liquor store, automobile repair shop, laundromat, and ice cream/candy shop. The anchor

Figure 10.2 La Tapatía Tienda Mexicana in Des Moines, Iowa

store in the area is the La Tapatía Tienda Mexicana, a large and busy grocery store that carries both North American and Latin American products (figure 10.2). The business typology developed in Phoenix does not apply in the Des Moines context due to the relatively small Latino population and the compact, centralized nature of the Latino commercial landscape. For example, *carnicerías*, widespread in metropolitan Phoenix and indicative of emerging Latino immigrant neighbourhoods, do not exist in Des Moines where such similar populations are largely confined to the central city and not sizeable enough to attract such businesses. Other businesses that are significant in Phoenix are either absent in Des Moines, such as *llanteras* (tire shops) or *yerberías* (medicinal herbs and religions items), or – like *panaderías* (bakeries) – only occur as part of the Latino commercial landscape in the Capitol East neighbourhood and do not serve as signifiers of nascent Hispanic immigrant neighbourhoods as they do in the Phoenix context. Despite the compressed nature of the Hispanic commercial district, there is still a variety of establishments that are representative of Latino neighbourhoods. These include beauty salons, boutiques and other clothing stores, restaurants and shops that sell candy or ice cream. Here, unlike some of the other establishments that cater specifically to Hispanic immigrant clientele, most Capitol East restaurants serve a cross over population of Latino immigrants, native-born Hispanics, and non-Hispanics alike.

Even with a diverse Latino population, nationality or region of origin are not readily evident in the landscape, with a few exceptions. The largest Hispanic subgroup in Des Moines consists of individuals of Mexican ancestry, and three businesses include variations of Mexico in their name, two variety stores/markets and the large supermarket. There is one Salvadoran restaurant, a *pupusería*, that serves the Salvadoran national dish (*pupusas*) and other related menu items. Regional or state affiliation from Mexico include a taco shop bearing the Mexican state name Jalisco, a common source area for many Mexican immigrants in the United States, and another restaurant that incorporates the Mexican state of Aguascalientes into its name. While an ice cream shop uses the state name of Michoacán, this is in the spirit of the state's Mexico-wide recognition for its ice cream and popsicles rather than any particular connection to that state.

### *The Latino Commercial Landscape: Chicago's Pilsen Neighbourhood and Carpentersville, Illinois*

Greater Chicago includes more than 9.5 million residents in three states, encompassing Chicago proper and its suburbs as well as exurban communities like Joliet and Aurora in Illinois, and roughly extending from Michigan City, Indiana, to Kenosha, Wisconsin. The Latino population in the metropolitan area is over two million, constituting 21.1 per cent of the total population (US Bureau of the Census 2011). Nearly two out of three Chicago-area Hispanics have Mexican ancestry, yet the Latino population is diverse, with a sizeable Puerto Rican, Guatemalan, and South American community. The foreign-born population is 17.8 per cent of the total, with Latinos constituting 48 per cent of the immigrant population (ibid.). Within the metropolitan area the largest Latino population is in the city of Chicago, but some suburban and exurban communities also have sizeable Hispanic populations, including the suburban communities of Cicero and Addison, as well as Waukegan,

Aurora, and Carpentersville on the periphery of the metropolitan area (Hudson 2006).

The Pilsen neighbourhood of Chicago, just to the southwest of downtown, is the heart of Chicago's Latino community, particularly Mexican-Americans and Mexican immigrants. Typical of many Chicago neighbourhoods, Pilsen has undergone substantial changes, beginning as a German and Irish immigrant neighbourhood, changing to majority Czech and Polish just before the turn of the twentieth century, and then emerging as a Mexican immigrant neighbourhood just after the Second World War (Gellman 2005). Throughout its history, the neighbourhood has been a working-class community connected to the nearby rail yards and adjacent factories. Currently, Pilsen is a vibrant commercial and residential neighbourhood that retains its cultural character despite gentrification pressures that are due to its proximity to downtown Chicago and the University of Illinois–Chicago as well as its historical significance and architectural heritage (Price et al. 2011).

The Latino commercial landscape in Pilsen is extensive, running along three major thoroughfares and dozens of smaller streets. A twenty-one-block transect along West 18th Street captures the heart of the commercial district and provides a representative sample of businesses. Here, there are a total of seventy-three Hispanic-oriented businesses, with no real clusters since the district runs largely uninterrupted through Pilsen. In such a well-established neighbourhood that is the heart of Chicago's Mexican immigrant and Mexican-American community, there is a great diversity of business types, ranging from more common establishments like restaurants, supermarkets, and *panaderías* to specialized businesses such as appliance stores, insurance agencies, and law firms as well as a bookstore, hardware store, and tortilla factory. However, since Pilsen has become an attraction due to extensive murals and the National Museum of Mexican Art – as well as gentrification attracting non-Hispanic residents – quite a few of the businesses serve a dual clientele. In some cases these ethnic businesses retain more of their character, while in other examples establishments appear to attract a largely non-Hispanic clientele. Some "boutiqueification" is evident in the neighbourhood, a term for shops that have opened or evolved to sell upscale goods to newer, more affluent residents who are driving the gentrification process (Lees et al. 2007). The business typology in Pilsen differs from that of Phoenix, primarily because as a long-standing central city neighbourhood it lacks the large lot size and open space that characterize equivalent neighbourhoods in Phoenix, where *llanteras*, tire shops that are common in its established Latino neighbourhoods, are prevalent. Shops that sell herbs and religions items are, however, present in Pilsen and here, as in Phoenix, they signify a larger, more established Hispanic community. Pilsen's Latino commercial landscape may in fact be unique, with neighbourhood businesses, specialized establishments that attract both Hispanic immigrants and Mexican-Americans from across the city, and shops and restaurants that cater to non-Hispanic visitors and new residents. Murdie and Teixeira (2011 and chapter 5 of this volume) document a similar ethnic business landscape in Toronto's traditional Portuguese neighbourhood, where some establishments have closed or moved to the suburbs, others still serve a largely Portuguese clientele, and some are considering diversifying to attract gentrifiers and other new residents.

With a dominantly Mexican-ancestry Latino population, local and state pride is evident in several Pilsen businesses, and there is a distinctive absence of establishments that promote a pan-Hispanic identity. The use of the Mexican tricolour and national symbols is common, but among business names direct reference to Mexico is only evident in six establishments. Of the references to Mexican states or towns, the majority connect with west-central and northwest Mexico, all regions with strong migration streams to the United States. Examples of state names include a restaurant and tortilla factory that reference the northeastern state of Nuevo León, as well as Supermercado Monterrey, named after the capital city of this state and also symbolized by the outline of La Silla, the saddle-shaped mountain that dominates that city's skyline. Among names referring to west-central Mexico are Carnitas Uruapán (Uruapán, Michoacán), Birrieria Reyes de Ocotlán (Ocotlán, Jalisco), and Tacos Don Chon, a restaurant that specializes in Capulhuac-style cuisine (Capulhuac, Mexico) (figure 10.3).

The case study area of Carpentersville, Illinois, is a small city of approximately 37,000 in Kane County on the fringe of metropolitan Chicago about thirty-five miles northwest of downtown Chicago. The city began as a crossing on the Fox River and emerged as a working-class settlement connected to local industry and manufacturing. At first lacking direct rail connections, and now offset by several miles from Interstate 90 and other major highways that lead directly to the heart of the metropolitan area, Carpentersville did not grow as large as other towns in the area and it has not experienced the same rapid population growth that has characterized many exurban communities elsewhere in the county or in nearby parts of the metropolitan edge (City of

Figure 10.3 Birrieria Reyes de Octotlán Restaurant in the Pilsen neighbourhood in Chicago, Illinois

Carpentersville 2012). Today, Carpentersville is a largely divided city that is 50 per cent Hispanic, with most Latino residents living east of the Fox River and a small Hispanic presence on the west side (US Bureau of the Census 2010a). This relatively recent surge in Hispanic immigration to a community that previously lacked a Latino population is due to relatively low home prices and monthly rents, a high demand for service-sector employment along the booming suburban fringe, and the existence of some local manufacturing in and around Carpentersville.

The Latino commercial district in Carpentersville consists of eighteen businesses spread throughout the eastern half of the community, chiefly along two state highways. While there are several businesses that stand alone, three clusters of businesses occur in the community, one in an older strip mall, one in a large but declining shopping mall, and one in a

Figure 10.4 Strip mall illustrating part of the Latino commercial landscape in Carpentersville, Illinois

newer strip mall. In all cases easy automobile access is a key feature and there is ample parking. One particularly notable business is distinctive because of its size: Village Fresh Market is an expansive Latino-oriented supermarket that compares in size to one of the larger mainstream supermarkets and includes a separate bakery.

The Latino-oriented businesses in Carpentersville include an array of supermarkets, restaurants, and a few small specialized establishments such as a jewellery store, travel agent, and furniture store (figure 10.4). Even with the giant Village Fresh Market, there is still a high percentage of markets, likely due to Carpentersville serving as a shopping destination for the small population of Latinos in other exurban communities in the northwestern reach of the Chicago metropolitan area. In this way, the Phoenix typology has some transferability, although while the term *carnicería* is not used as a primary business name in Carpentersville, the small markets serve a similar function. Otherwise, because of the relatively small size and more residential nature of the community, there are a limited number of commercial districts of any kind, contrasting with Phoenix's extensive commercial properties. The Latino population in Carpentersville is overwhelmingly of Mexican ancestry, but any signifier of nationality or regional identify is reflected in one store name, where Tamazula Tacos is likely named for a Mexican community in the state of Jalisco.

## Characteristics and Comparisons of the Case Study Areas

Each of the case study areas has a distinctive function based on its morphology and business composition as well as its local demographic characteristics (table 10.1). The Latino business landscape in Des Moines's Capitol East neighbourhood serves smaller but steadily growing Latino neighbourhoods in the adjacent area, including both foreign-born and native-born Hispanics. Encompassing the only notable Latino-oriented businesses in the metropolitan area, the compact district also attracts customers from elsewhere in the central city and suburbs. While there are diverse business types in this neighbourhood, there is a lack of specialized establishments that would typically exist in a larger Latino community. Because the Hispanic community is largely of Mexican ancestry, businesses tend to incorporate names or symbols that connect to Mexico or in some cases regions or states within that country.

The Pilsen neighbourhood in Chicago exists as an extensive commercial centre in a well-established Mexican-American and Mexican immigrant neighbourhood. With such a large and vibrant Latino community in the nearby residential neighbourhoods, Pilsen's Latino commercial landscape could prosper as just a neighbourhood district, but instead also serves as an attraction for Hispanics elsewhere in Chicago. Furthermore, with gentrification and proximity to other amenities, many of the businesses here serve a non-Hispanic population, and three decades of census data indicate a slowly increasing non-Latino population in adjacent census tracts. With such a large base population from Mexico or of Mexican heritage, establishments in Pilsen frequently accentuate Mexican regional and state affiliations, with virtually no evidence of promoting a pan-Hispanic identity.

Clusters of establishments characterize the Latino commercial landscape in Carpentersville, Illinois, where the Latino population has

Table 10.1 Select characteristics of case study areas

| Case study areas* | Commercial district classification | % Foreign-born 2010/2011 | % Latino 1990 | % Latino 2000 | % Latino 2010 |
|---|---|---|---|---|---|
| Capitol East Des Moines, IA | Central city compact | 19 | 5 | 14 | 34 |
| Pilsen Chicago, IL | Central city extensive | 36 | 86 | 76 | 67 |
| Carpentersville, IL | Exurban dispersed | 26 | 13 | 31 | 54 |

* Data were calculated based on census tracts within a ½-mile radius around Latino-oriented businesses. Census tract boundary changes through census years were aggregated/disaggregated to match 2010 boundaries, so percentages are approximate.
Source: US Bureau of the Census 1990, 2000, 2010b, 2011.

increased steadily in adjacent census tracts over the past three decades. The Latino-oriented businesses in this community, located on the far edge of greater Chicago, also attract Latino customers from communities scattered across Kane County and nearby suburban or exurban towns and cities. The prevalence of groceries and restaurants and comparative lack of specialized shops or services is indicative of a smaller, more dispersed Latino community. The typology developed in Phoenix has limited transferability to Midwestern cities, as exemplified with the three case study areas and their differential morphology and function. However, if additional neighbourhoods in metropolitan Chicago had been analysed, there may prove to be greater similarities in emerging Latino immigrant neighbourhoods on the outer edges of Chicago proper or in inner-ring suburbs. Still, while many Midwestern cities have sizeable Latino immigrant populations, the morphology and character of these metropolitan areas differ from those in Phoenix or other Sun Belt or Western cities such as Denver, Dallas, or Las Vegas, which may better fit the business typology.

## Conclusions and Directions for Future Research

The morphology and character of the Latino commercial landscape are diagnostic of the local Hispanic immigrant community and its role within the larger metropolitan area. Clearly the size of the district is directly proportional to the size of the population, but the nature of its landscape yields additional insight about immigrants' economic experiences. Accessibility can include a definitive orientation towards pedestrians and local traffic, whereas extensive parking-lot space suggests that many customers commute from elsewhere in the city. The degree of business specialization and counts of business types indicate whether the neighbourhood has a small or emerging Latino population that needs a modest supermarket or *carnicería*, or whether it is an established commercial centre featuring a specialized bookstore or hardware store. Shop names, colours, and symbols signify different affiliations, whether it is a pan-Hispanic focus for a money-wiring outlet that seeks to attract immigrants from a dozen Latin American nations or a restaurant that seeks to attract customers from a local region. Similarly, the proportion of Spanish to English signage reinforces that many establishments serve some local non-Hispanic residents, non-Latino visitors from outside the neighbourhood, or a Mexican-American or other native-born Hispanic clientele. Together, the Latino commercial landscape is a visible indicator of the dynamism and complexity that exists in Hispanic neighbourhoods and communities, particularly in regard to immigrants' economic experiences, where the majority of the US foreign-born population has arrived within only the past twenty years (Teixeira and Li, introduction to this volume).

Fundamental to the landscape itself is the ethnic entrepreneurship that is often key to many immigrants' economic experiences, either as a means for small-business ownership or employment opportunities or as a way to adapt to the host country and maintain connections with one's home nation or region. As noted by Teixeira and Li, Chacko and Price, and Li and Lo (introduction and chapters 11 and 12 respectively from this volume), entrepreneurship is an avenue for immigrants to offset the many challenges they face in the traditional labour market. Business ownership is, however, fraught with major obstacles, chiefly issues of financing and access to capital, that are particularly acute among small, immigrant-owned establishments (Li and Lo, chapter 12). Still, employment opportunities and challenges for owners and employees are not the only aspect of entrepreneurship, as these commercial establishments create a place for social interaction and a means to reinforce cultural identities, whether Mexican state pride, national identity, or broader pan-Hispanic connections. Such *ethnic sociocommerscapes*, a term developed and further discussed by Chacko and Price in this volume (chapter 11), certainly apply to this context, particularly in areas of dispersed Latino settlement such as Carpentersville, and exist in other metropolitan areas characterized by decentralized, suburban settlement (Chacko and Price, chapter 11; Oberle 2006).

There are two primary avenues for future research in relation to Latino commercial landscapes. First, how do other Latino commercial landscapes compare to these Midwestern examples or the case of Phoenix? With increasing Latino immigrant settlement in metropolitan areas like Washington, DC, Charlotte, and Atlanta, do these cities follow a pattern like Phoenix, like Chicago, or are they perhaps distinctive from both Midwestern and Southwestern cities? Second, additional research is needed to evaluate how these landscapes evolve over time in relation to immigrants' year of entry into the United States. For example, if the influx of new immigrants slows considerably, as has happened in many communities during the recent economic downturn, how do the morphology and the function of the landscape change when the majority of the foreign-born population has been in the host country for twenty or more years? Similarly, how does the landscape evolve as the native-born children of immigrants begin to outnumber those who are foreign-born? Undoubtedly, as immigration status or length of residency progress, so too do immigrants' economic experiences.

## NOTE

1 In the context of this chapter, the Midwest is defined as a twelve-state area that extends from Michigan to Kansas and North Dakota to Ohio.

## REFERENCES

Buzzelli, M. 2001. "From 'Little Britain' to 'Little Italy': An Urban Ethnic Landscape Study in Toronto." *Journal of Historical Geography* 27 (4): 573–87. http://dx.doi.org/10.1006/jhge.2001.0355.

Chacko, E. 2003. "Ethiopian Ethos and the Making of Ethnic Places in the Washington Metropolitan Area." *Journal of Cultural Geography* 20 (2): 21–42. http://dx.doi.org/10.1080/08873630309478274.

City of Carpentersville. 2012. "Community History." http://vil.carpentersville.il.us/Community/History.asp.

Dávila, A. 2001. *Latinos, Inc.: The Marketing and Making of a People*. Berkeley: University of California Press.

Fairlie, R. 2012. *Immigrant Entrepreneurs and Small Business Owners, and Their Access to Financial Capital.* Washington, DC: US Small Business Association, Office of Advocacy.

Gellman, E. 2005. "The Electronic Encyclopedia of Chicago." http://www.encyclopedia.chicagohistory.org/pages/2477.html.

Grieco, E., Y. Acosta, G. de la Cruz, C. Gambino, T. Gryn, L. Larsen, E. Trevelyan, and N. Walters. 2012. *The Foreign-Born Population in the United States.* Washington, DC: US Bureau of the Census.

Guarnizo, L. 1998. *The Mexican Ethnic Economy in Los Angeles: Capitalist Accumulation, Class Restructuring, and the Transnationalism of Migration.* Davis: California Communities Program of the University of California.

Hudson, J. 2006. *Chicago: A Geography of the City and Its Region.* Chicago: University of Chicago Press.

Jacobsen, J. 2002. *In the Shadow of the State Capitol: An Architectural and Historical Survey of the Capitol East Neighborhood, Des Moines, Iowa.* Des Moines: History Pays.

Korzenny, F., and B. Korzenny. 2012. *Hispanic Marketing: Connecting with the Latino Consumer.* New York: Routledge.

Lees, L., T. Slater, and E. Wyly. 2007. *Gentrification.* New York: Routledge.

Light, I., G. Sabagh, M. Bozorgmehr, and C. Der-Martirosian. 1994. "Beyond the Ethnic Enclave Economy." *Social Problems* 41 (1): 65–80. http://dx.doi.org/10.2307/3096842.

Millard, A., and J. Chapa. 2004. *Apple Pie and Enchiladas: Latino Newcomers in the Rural Midwest.* Austin: University of Texas Press.

Motel, S. 2010. *Statistical Portrait of Hispanics in the United States, 2010.* Washington, DC: Pew Hispanic Center.

Motel, S., and E. Patten. 2012a. *Characteristics of the 60 Largest Metropolitan Areas by Hispanic Population.* Washington, DC: Pew Hispanic Center.

Motel, S., and E. Patten. 2012b. *The Ten Largest Hispanic Origin Groups: Characteristics, Rankings, Top Counties.* Washington, DC: Pew Hispanic Center.

Murdie, R., and C. Teixeira. 2011. "The Impact of Gentrification on Ethnic Neighborhoods in Toronto: A Case Study of Little Portugal." *Urban Studies* (Edinburgh, Scotland) 48 (1): 61–83. http://dx.doi.org/10.1177/0042098009360227.

Oberle, A. 2004. "Se Venden Aquí: The Latino Commercial Landscape in Phoenix, Arizona." In *Hispanic Spaces, Latino Places,* ed. D. Arreola, 239–54. Austin: University of Texas Press.

Oberle, A. 2006. "Phoenix's Hispanic Business Landscape: Latino Outposts on the Anglo Suburban Frontier." In *Landscapes of the Ethnic Economy,* ed. D. Kaplan and W. Li, 149–63. Lanham, MD: Rowman and Littlefield.

Oberle, A., and D. Arreola. 2008. "Resurgent Mexican Phoenix." *Geographical Review* 98 (2): 171–96.

Odem, M., and E. Lacy. 2009. *Latino Immigrants and the Transformation of the U.S. South.* Athens: University of Georgia Press.

Oltrogge, S. 2009. *East Village.* Chicago: Arcadia Publications.

Pew Research Center. 2011. *Census 2010: 50 million Latinos.* Washington, DC: Pew Research Center.

Portes, A. 1987. "The Social Origins of the Cuban Enclave Economy of Miami." *Sociological Perspectives* 30 (4): 340–72. http://dx.doi.org/10.2307/1389209.

Portes, A., and L. Jensen. 1989. "The Enclave and the Entrants: Patterns of Ethnic Enterprise in Miami before and after Mariel." *American Sociological Review* 54 (6): 929–49. http://dx.doi.org/10.2307/2095716.

Price, P., C. Lukinbeal, R. Gioiso, D. Arreola, D. Fernández, T. Ready, and M. de los Angeles Torres. 2011. "Placing Latino Civic Engagement." *Urban Geography* 32 (2): 179–207. http://dx.doi.org/10.2747/0272-3638.32.2.179.

Singer, A., S. Hardwick, and C. Brettell. 2008. *Twenty-first Century Gateways: Immigrant Incorporation in Suburban America.* Washington, DC: Brookings Institution Press.

Statistics Canada. 2012. *Canada at a Glance.* Ottawa: Statistics Canada.

Suro, R., and A. Singer. 2002. *Latino Growth in Metropolitan America: Changing Patterns, New Locations.* Washington, DC: The Brookings Institution.

US Bureau of the Census. 1990. 1990 Census of Population.

US Bureau of the Census. 2000. 2000 Census of Population.

US Bureau of the Census. 2007. "Survey of Business Owners – Hispanic owned firms, 2007." http://www.census.gov/econ/sbo/getsof.html?07hispanic.

US Bureau of the Census. 2010a. 2010 Census of Population.

US Bureau of the Census. 2010b. American Community Survey 2010, 5-year estimates.

US Bureau of the Census. 2011. American Community Survey 2011, 1-year estimate.

Walcott, S. 2002. "Overlapping Ethnicities and Negotiated Space: Atlanta's Buford Highway." *Journal of Cultural Geography* 20 (1): 51–75. http://dx.doi.org/10.1080/08873630209478281.

Wood, J. 1997. "Vietnamese-American Place-Making in Northern Virginia." *Geographical Review* 87 (1): 58–72. http://dx.doi.org/10.2307/215658.

Zsembik, B. 2000. "The Cuban Ethnic Economy and Labor Market Outcomes of Latinos in Metropolitan Florida." *Hispanic Journal of Behavioral Sciences* 30 (2): 230–50.

# 11 Immigrant Entrepreneurship in the Washington Metropolitan Area: Opportunities and Challenges Facing Ethnic Minorities

ELIZABETH CHACKO AND MARIE PRICE

During the 1990s metropolitan Washington, DC, emerged as a relatively new but increasingly important immigrant gateway city (Chacko 2008; Price and Singer 2008; Price et al. 2005). According to the most recent data from the US Census Bureau, the metropolitan region has 1.2 million foreign-born residents out of a total population of 5.6 million, and so 21.4 per cent of the population are immigrants (US Bureau of the Census 2009–11). In many ways the immigrant experience in Washington is representative of other twenty-first-century immigrant destinations in the United States and Canada, where immigrant flows are highly diverse in terms of countries of origin, skill levels, and the mix of economic migrants, refugees, family members, and the undocumented (Singer, Hardwick, and Brettell 2008; Teixeira, Li, and Kobayashi 2012). Likewise, the pattern of immigrant settlement tends to be more suburban and dispersed, thus provoking a range of local government policies and responses to newcomers.

This chapter focuses on two immigrant populations in the Washington area, one African (Ethiopian) and one Hispanic (Bolivian), and their engagement in entrepreneurship. As of 2011, Ethiopians are the tenth largest foreign-born group in the metropolitan area and Bolivians are the eleventh. Although not hailing from the largest source countries, both Ethiopians and Bolivians had made Washington their primary area of settlement in the United States, and so it is an important hub for these immigrant communities. In addition, these immigrant groups are comparable in their absolute population size. As a whole, Ethiopians and Bolivians are not ethnic groups known for their entrepreneurial tendencies, but entrepreneurship has emerged as a critical adaptive strategy for these populations as they cope with structural barriers in the urban labour market and discrimination that many immigrants and visible minorities face, a point that Lysenko and Wang make in chapter 9 when documenting immigrant underemployment. Structural and linguistic barriers in the labour market help to explain higher-than-average rates of self-employment for both Ethiopians and Bolivians. These factors make them ideal populations in which to study the emergence and evolution of immigrant entrepreneurship in an urban context.

This study is anchored in the literature on immigrant entrepreneurship, especially the scholarship that focuses upon the importance of immigrant social networks, institutional contexts of reception at the local level, and the creation of new multi-ethnic suburban landscapes. In the Washington context, where there are few single-ethnic population clusters, a heterolocal model of immigrant settlement prevails in which a sense of community is created more through immigrant institutions and gathering places than dense residential clusters (Zelinsky and Lee 1998). We argue that immigrant entrepreneurs have created *ethnic sociocommerscapes* mostly in the suburbs, with the dual purpose of creating business opportunities and sites of social interaction between co-ethnics (Chacko 2003), a notion similar to what Oberle discussed in the previous chapter on Latino business landscapes in the Midwest. Others have developed businesses that cater to the general population such as construction, childcare, or high-tech companies. The localities in which these entrepreneurs conduct their businesses and the spatial concentrations of ethnic groups and ethnic businesses within an urban area all contribute to the success or failure of these enterprises. In addition to the spatial context of immigrant businesses, this analysis highlights the role of local governance in supporting or suppressing immigrant integration and entrepreneurship. This is especially significant for metropolitan Washington, as it is a region that exhibits a mix of inclusionary and exclusionary policies that vary sharply by jurisdiction (Walker and Leitner 2011).

## Immigrants, Entrepreneurship, and the Spatial Context

There is an established literature on how and why certain groups of immigrants become entrepreneurs and the impact that their entrepreneurial ventures have on their communities as well as the social, cultural, and economic landscapes of the places in which they settle (Aldrich and Waldinger 1990; Light and Bonacich 1991; Li 2001; Zhou 2004; Kaplan and Li 2006; Oberle 2006; Wang and Li 2007). In general, immigrants in North America tend to have higher rates of self-employment than the native-born (Borjas 1986; Kaplan and Li 2006; Hou and Wang 2011). Several hypotheses have been proposed for the greater entrepreneurship among immigrants. The Middleman Minority thesis proposes that immigrant sojourners often turn to small businesses and occupations that are not directly in competition with those of the native majority group. This strategy is partly to combat their marginal status as foreigners in the receiving society, but is also facilitated by ethnic solidarity, which leads to cooperation within the group, even to the extent of lending each other money to start a business or raising the needed capital as a group through rotating credit associations (Bonacich 1973).

Ethnic market niches have also been offered as a reason for immigrant entrepreneurial activity. According to Aldrich and Waldinger (1990), three basic elements help us understand the role of ethnic market niches in immigrant businesses: opportunity structures, group characteristics, and ethnic strategies. Opportunity structures are the various types of opportunities that are presented to immigrants by the economy in the receiving country; group characteristics include forms of capital that the immigrants have, including financial, human, and social capital, while ethnic strategies refer to tactics that emanate from combinations of opportunity structure and group characteristics that benefit ethnic immigrants.

Other theoretical frameworks that help understand the prevalence of self-employment among immigrants include the "enclave thesis" and "blocked mobility thesis." The enclave thesis stresses that immigrant entrepreneurship often arises from the demand for goods and services within the immigrant community itself, especially when it is spatially concentrated. In cities as diverse as New Orleans (Airriess 2006), Los Angeles (Li et al. 2006), and Toronto (Lo 2006; Teixeira 2006), scholars have demonstrated that immigrants turn to co-ethnics in forming ethnic economies. The "blocked mobility thesis," by contrast, posits that immigrants may have to create their own jobs, as they have limited options due to the receiving society's employment structure or prejudices. Thus, if the immigrant's academic and professional credentials are not accepted, if their receiving-country language skills are limited, or if racial/ethnic prejudices keep them out of certain jobs, they may turn to entrepreneurship rather than face unemployment or chronic underemployment, as evident particularly in Canada (Li 2001).

Immigrant communities are often characterized by extensive social networks, formed of strong ties between and among individual members that tend to be close and stable (Zhou 2004). Such strong ties can be critical when the immigrant is considering setting up a business. Ethnic social networks provide immigrant entrepreneurs with opportunities to interact with and obtain help from co-ethnics in establishing a business, in the subsequent growth of the enterprise, and in finding employment in co-ethnic businesses. However, weak ties, those that tend to be more superficial and lacking in emotional investment, may be equally critical in the continued success of a business enterprise (Coleman 1988), particularly if the entrepreneur wishes to expand the business and include larger numbers of non-ethnics within her clientele.

Geographers tend to stress the spatial context when assessing the rise of immigrant entrepreneurship (Hiebert 2002; Kaplan and Li 2006; Li 2009). Traditionally, immigrant entrepreneurs set up businesses in inner-city ethnic enclaves, which offer resources in the form of ethnic financial, social, and human capital. Concentrations of people belonging to a particular ethnic community could provide business owners with a ready and protected market, a steady stream of co-ethnic workers, and, just as important, a sense of familiarity and trust. Existing social networks within these communities could also be instrumental in garnering sufficient capital or credit to start a business.

Ethnic stores and services often attract other ethnic businesses, leading to agglomeration and the development of an ethnic space in which multiple ethnic businesses buttress one another, drawing in co-ethnic customers who may be able to meet all their retail and service needs in one locale, as demonstrated by Oberle in previous chapter. Ethnic enclaves (both residential and businesses) often develop in inner cities, particularly in places where new immigrants first settle. As the work of geographer Wei Li demonstrates, increasingly, newly arrived immigrants are choosing to settle in the suburbs rather than the urban cores of US cities, resulting in the formation of ethnoburbs, suburban ethnic clusters in large metropolitan areas that are multi-ethnic communities in which "one ethnic minority group has a significant concentration," but is not necessarily the majority (Li 2009, 1). The clustering of immigrant/ethnic businesses in suburban areas has spawned the research on *ethnoburbs*. In her case study, Li (2009) focuses on the Chinese ethnoburb of San Gabriel Valley, California (also see Yu, chapter 7, of this volume), but argues that similar community formations exist in other areas of North America such as the Vietnamese ethnoburb in the Washington suburbs of northern Virginia (Wood 1997), each with clusters of immigrant/ethnic businesses.

Ethiopians or Bolivians have not formed ethnoburbs in metropolitan Washington, as their businesses and residences are not especially concentrated; indeed, they represent a more heterolocal pattern. Typically, Ethiopian and Bolivian businesses are found in the suburbs amidst a diversity of other immigrant-run enterprises, which is a pattern noted by Spanish geographer Pau Serra (2012). Their businesses, however, support community maintenance through the creation of ethnic socio-commerscapes that immigrants use to obtain goods and services as well as to gather with co-ethnics (Chacko 2003). They also create culturally infused gathering places that individuals outside of the ethnic group seek out. Chief among these arenas is the cluster of Ethiopian restaurants in the Shaw neighbourhood in the District of Columbia that draw a diverse range of clientele, though being in an area where relatively few Ethiopians reside.

Finally, there is a tendency to view immigrant entrepreneurship strictly in terms of the agency of individual immigrants or the workings of an ethnic economy. While these factors are important, this study shows that institutional programs and policies created by local jurisdictions (say, a county or town ordinance) are also extremely important in encouraging or discouraging entrepreneurship. Inclusive policies can positively influence the manner in which immigrants are received by the communities in which they settle and the resources that are made available to them to assist them in setting up new businesses (Wang and Li 2007; Price and Chacko 2010; Walker and Leitner 2011). Similarly, exclusionary policies can drive out immigrant businesses or make them less visible and more informal (Varsanyi 2011; Singer et al. 2009). In general, not enough attention is paid to the role of local governance in supporting immigrant entrepreneurship.

## Methodology

This research relies upon a mixed methodology of quantitative and qualitative approaches to examine the social, spatial, and economic strategies of immigrant entrepreneurs in Metropolitan Washington. Census bureau data on the foreign-born and ancestry groups from Bolivia and Ethiopia in the metro area were analysed to demonstrate the size of the population, areas of immigrant settlement, and socio-economic indicators.

In this study, immigrant-owned businesses and self-employment were equated with entrepreneurship.

Qualitative methods such as focus groups and in-depth interviews were used to understand the motivations, experiences, and social networks of immigrant entrepreneurs. Separate focus groups (one each) were conducted with Bolivian (11 participants) and Ethiopian (6 participants) entrepreneurs, as well as in-depth interviews with eight to ten individual immigrants from each group who live in the Washington area. After a few Ethiopian and Bolivian entrepreneurs known to the researchers were contacted to see if they were willing to take part in the focus groups and interviews, a snowball method was used to recruit additional immigrant entrepreneurs. The entrepreneurs who participated in the study had established businesses such as restaurants, grocery stores, cafes, travel agencies, construction companies, house-cleaning services, a consulting company, and taxicab services. The rationale for the focus groups was to allow group members to discuss the issues, trajectories, and concerns of entrepreneurship among themselves, with guidance from the moderators. The focus groups were conducted in Spanish or English and were recorded and later translated and transcribed.

Individual entrepreneurs were asked open-ended questions on their businesses and business practices during in-depth interviews in order to gauge their perceptions and understandings of being an immigrant entrepreneur. This was in addition to the closed-ended questions that elicited personal information including length of stay, level of education, place of residence, and the kinds of jobs she or he held before and after moving to the United States. The relative roles of local and federal government policies and incentives in facilitating entrepreneurship were also assessed. Details about each person's experiences leading to settling in Washington and starting a business in the area were solicited, as were the reasons for establishing a particular kind of business, where and how the person obtained financial capital, and the population groups that the business catered to and drew upon for employees.

From focus group contacts, interviews, and local directories, a database of Bolivian and Ethiopian businesses was constructed. The location of these businesses was mapped and compared with the residential settlement patterns of these two groups, which contributed to a more detailed understanding of the spatial patterning of immigrants and their businesses.

## Group Characteristics of Ethiopians and Bolivians

Initial cohorts of Ethiopian and Bolivian immigrants to the Washington area were political and economic migrants with a high level of education. Civil war in Ethiopia and the chronic poverty in Bolivia were the primary push factors that impelled them to leave their countries in the 1970s and 1980s. The 35,000 Ethiopians in the Washington area form one of the oldest Black African immigrant communities, constituting one-quarter of the foreign-born population from sub-Saharan Africa living in the region (US Bureau of the Census 2009–11). Currently, Ethiopians in Washington account for a sizable percentage (22 per cent) of the 162,000 Ethiopian-borns living in the United States (ibid.). Most Ethiopians who left in the wake of the Marxist Revolution of 1974 were among the elite of their country. The Washington area was a prime destination for these immigrants who wished to settle in the nation's capital, which already had a small Ethiopian community. Over time, they were joined by compatriots who arrived as refugees and asylum seekers, on diversity visas, or through family reunification programs. This well-educated immigrant population found jobs largely as salaried workers (84 per cent), although 7 per cent are self-employed (ibid.).

Whereas Ethiopians are an important part of the new African migration to the United States, Bolivians make up a tiny fraction (0.3 per cent) of the 21 million immigrants from Latin America (US Bureau of the Census 2009–11). Early Bolivian arrivals came as labour migrants fleeing the economic turmoil of the early 1980s, when hyperinflation ravaged the Bolivian economy. They were drawn to Washington by the availability of jobs and, at the time, relatively little competition from other Latino immigrants. Today's Bolivian population, most of whom settled in Washington since the 1980s, are representative of Bolivia's multi-racial society of mixed Indian and European ancestry. The community grew via chain migration, by taking advantage of family reunification preferences, through diversity visas, and by developing an ethnic economy in which small Bolivian businesses in construction and home services grew through employing other co-ethnics and sponsoring other immigrants. This too is a well-educated community, in which 81 per cent are salaried workers and 10 per cent are self-employed (US Bureau of the Census 2009–11). They are part of a large and diverse Hispanic community in metropolitan Washington which totalled nearly 800,000 in 2010 (Pumar 2012).

Currently, Ethiopian and Bolivian immigrants in metropolitan Washington number 35,345 and 32,822 respectively (US Bureau of the Census 2009–11). Forty-four per cent of the Bolivian-born population in the United States resides in the Washington area, whereas just 22 per cent of the US Ethiopian-born population lives there. Persons with Bolivian and Ethiopian ancestry in the metropolitan area (immigrants and their descendants) are higher, at 43,878 Bolivians and 39,369 Ethiopians (ibid., S0201). This slightly larger cohort has quite high levels of socio-economic attainment. For example, 86.8 per cent of Ethiopians and 85.3 per cent of Bolivians have a high school degree or higher. Moreover, nearly one-third (30 per cent of Ethiopians and 27 per cent of Bolivians) have a bachelor's degree or higher. The median household income for Bolivians in metropolitan Washington was $69,083 in 2009–11, considerably greater than the median household income for the United States as a whole, at $51,484, but less than the median income for all of metropolitan Washington ($87,653). Similarly, Bolivians show higher rates of homeownership than other Hispanics, with 57 per cent living in owner-occupied housing, yet not as high as the overall US rate of 65 per cent (ibid.). In comparison, Ethiopians have a lower median household income ($51,797) and lower rates of homeownership, at 42 per cent. These lower figures may reflect the large number of refugees and asylum seekers who joined the existing well-off Ethiopian population in Washington after the passing of the Refugee Act in 1980. Between 1982 and 1990, at least 90 per cent of African refugees to the United States were from Ethiopia (US Department of Homeland Security 2008).

Table 11.1 contrasts the major occupational divisions for Ethiopians, Bolivians, and the general population in metropolitan Washington. Most Ethiopians and Bolivians work as salaried workers in the private sector, where over 80 per cent are employed. They are also more likely to be self-employed than the metropolitan area's population as a whole, with Bolivians being twice as likely to be self-employed (at 10 per cent) than the population as a whole, at 4.9 per cent. Each ethnic group is also over-represented in particular employment niches. Ethiopians are four times more likely than the general population to be employed in the transportation sector. In contrast, Bolivians are three times more likely to be employed in construction than the population as a whole; many work for Bolivian-owned construction firms. Both of these groups are well represented in the educational, health care, and management fields, as well as in the restaurant and food services fields, where many small businesses have emerged.

Table 11.1 Occupational divisions of Bolivians and Ethiopians in Metropolitan Washington, 2009–2011

| Major occupational divisions, 2009–2011 | % of Bolivian population in Metropolitan Washington | % of Ethiopian population in Metropolitan Washington | % of all Metropolitan Washington population |
|---|---|---|---|
| Construction | 18.3 | 0.3 | 6.3 |
| Professional, scientific, and management and administrative and waste management services | 15.8 | 12.9 | 20.4 |
| Transportation and warehousing, and utilities | 2.5 | 16.8 | 3.6 |
| Educational services, health care and social assistance | 12.7 | 17.7 | 19.1 |
| Retail trade | 8.6 | 15.8 | 8.1 |
| Arts, entertainment, recreation, accommodation, and food services | 15.5 | 12.8 | 8.2 |
| Private wage and salary workers | 81 | 83.7 | 70.5 |
| Self-employed workers | 10 | 7 | 4.9 |

Source: US Bureau of the Census 2009–11, Selected Population Profile (S0201).

## *Creating Ethnic Business Spaces*

The challenge that immigrants face in diverse and dispersed gateway cities such as Washington is to create spaces that can support communities, build ethnic economies, and over time grow to serve the general population. As was articulated in one Bolivian focus group, "All of us with businesses want to enter the American market. We may start working with Latinos or Bolivians, but always with the mindset to enter the world of the Anglo-Americans, the natives, because we know this will pay better."

Figures 11.1 and 11.2 map the distribution of immigrant Bolivians and Ethiopians by census tract, along with the locations of their ethnic businesses. In the case of the Bolivians (figure 11.1), there is considerable overlap between where businesses are located and pockets of residential concentration. Most of the businesses are in northern Virginia, and this is where most Bolivian immigrants live, which is the pattern that one would expect for an ethnic economy that relies heavily on providing services and goods to co-ethnic consumers. Such businesses as restaurants, markets, travel agencies, and money transfer companies are deeply embedded in the ethnic economy. Since Arlington County was the earliest important area of settlement for Bolivians, many of the oldest businesses started there, especially along Columbia Pike in South Arlington. As Bolivians settled in Falls Church and Fairfax County, more businesses followed them, especially on Route 50. Yet there is no commercial centre for Bolivians; many of their businesses share older retail spaces with other immigrant-owned businesses in the suburban areas of the metropolitan region, where immigrants tend to cluster. As for the scattered businesses in Montgomery County, Maryland, these are mostly restaurants or other services that cater to the Bolivians and other Hispanics in that county. Of the few Bolivian-owned businesses in the District of Columbia, most are restaurants that cater to pan-Latino consumers.

These businesses, however, also play an important role as gathering places for Bolivians eager to consume familiar tastes of home. Often they are busiest on the weekends, when scores of people gather to break bread and visit with friends. Interestingly, there is a popular Italian restaurant in Arlington owned by a Bolivian that serves Italian dishes. But on weekends a Bolivian menu is added to the offerings, which brings a loyal clientele from all parts of northern Virginia, who dine on *sopa de mani* (peanut soup) and *salteñas* (Bolivian empanadas), among other Bolivian dishes. This is an example of a crossover business strategy, serving a familiar food (Italian) that appeals to a general market, but also providing for the co-ethnic customer (in this case Bolivians) to enjoy the flavours of home. As table 11.1 shows, 12.8 per cent of the Bolivian population is employed in the "arts, entertainment, recreation, accommodation, and food services" areas, which is well above the average for the region.

Not all Bolivian-owned businesses are tied to the ethnic economy, nor are all Bolivian businesses shown on figure 11.1. For example, this map does not capture the range of small construction companies in the metro area that employ Bolivian men in a variety of areas, but especially for drywall installation, tile work, and fencing. Nor can it capture the various domestic services, such as childcare, eldercare, and cleaning services, that Bolivian women often engage in through both formal and informal channels. Most of these businesses are clearly targeting a general market that relies upon these services. Figure 11.1 does include a few high-tech companies developed by Bolivians that are also located in northern Virginia. Here the location is less about being near the immigrant community than being in an economically dynamic region that is supportive of high-tech

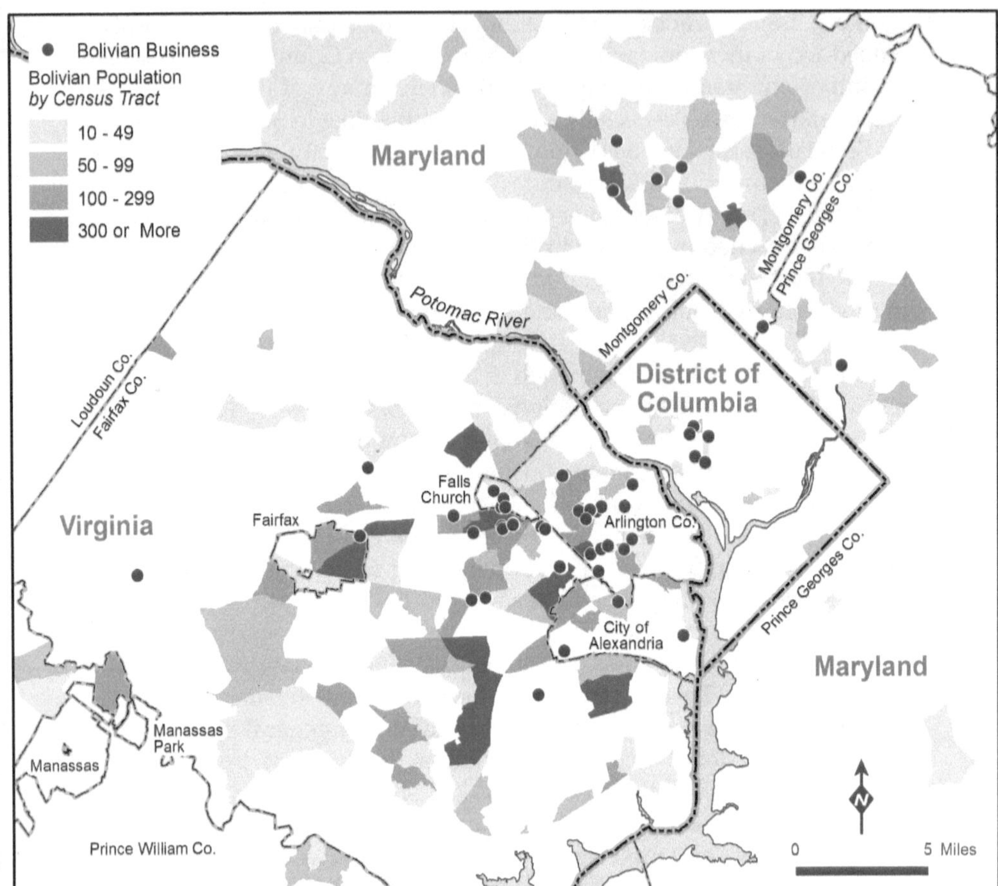


Figure 11.1 Distribution of Bolivian population and businesses in the Washington Metropolitan Area

Source: US Bureau of the Census 2009–11; Hispanic Yellow Pages and Bolivian Entrepreneurs

industries. One company based in Northern Virginia, Data Ventures, is owned by a longtime Bolivian-born US resident. The company develops software for financing and telecommunications. Its success in the United States has led it to form relations with other South American countries, namely, Argentina, Chile, and Bolivia. The owner has experimented with outsourcing some computer software development to Bolivia, because

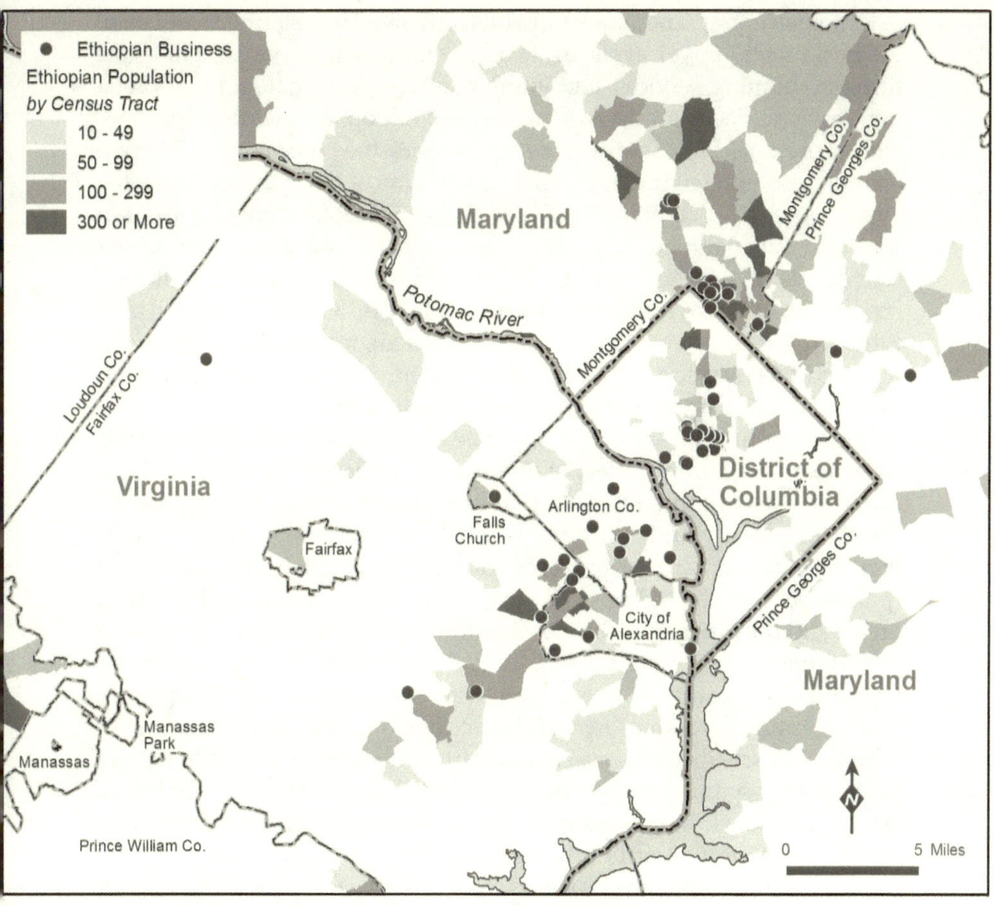


Figure 11.2 Distribution of Ethiopian population and businesses in the Washington Metropolitan Area.

Sources: US Bureau of the Census 2009–11 and Ethiopian Yellow Pages (2011).

there are professionals there who can do this work at lower costs. At the same time, the uncertain political and business climate in Bolivia makes this kind of investment difficult. These Bolivian "Argonauts" do not have the robust high-tech connections described by Saxenian (2006), but their work speaks of the role of highly skilled immigrants in forming local and transnational business ventures.

Ethiopian entrepreneurs are most often involved in restaurants and retail establishments, travel agencies, taxi and limousine services, real estate firms, accounting services, and immigration-related legal and interpretation services (Focus group and interviews). Ethiopians are disproportionately represented in a few of these endeavours: the percentage of Ethiopians engaged in retail trade (15.8 per cent) is almost twice the metropolitan average of 8.1 per cent; nearly 17 per cent of Ethiopians are employed in transportation, warehousing, and utilities, which only account for 3.6 per cent of overall employment in the area and is likely a reflection of their concentration in taxi and limousine services. Approximately 13 per cent of Ethiopians are employed in "arts, entertainment, recreation, accommodation, and food services," probably due to their involvement in the restaurant business, compared to 8.2 per cent overall (see table 11.1).

Ethiopian immigrants tend to locate their entrepreneurial ventures close to their communities. While the Ethiopian population is both found in Washington, DC, and scattered in suburbs in the neighbouring states of Maryland and Virginia, Ethiopian businesses trace a southwest to northeast arc, largely concentrated in the District of Columbia (particularly in the Adams Morgan, Columbia Heights, and U Street areas) and the inner suburbs of Arlington, Falls Church, and Alexandria in Virginia and the Silver Spring and Takoma Park areas of Maryland (figure 11.2), mostly following areas with concentrations of Ethiopians. The exceptions are areas in the District of Columbia, where Ethiopian immigrants set up stores and restaurants in neighbourhoods that were undergoing commercial gentrification, but in which they did not live in large numbers. Taking advantage of the relatively low rents and accessibility of DC neighbourhoods such as Adams Morgan and the U Street corridor, Ethiopian entrepreneurs established stores and restaurants there while they were still in transition. An Ethiopian entrepreneur who bought property in the U Street area for a restaurant in the early 1990s said, "We came here when nobody else wanted to. We took a chance. There were still shootings in this neighborhood when I bought the townhouse on 9th." Ethiopian enterprises in these neighbourhoods were indeed instrumental in their economic revival (Chacko 2003, 2008).

It should also be noted that the Ethiopian population concentrations in the suburbs also have significant numbers of people belonging to other ethnic and immigrant groups. The clusters of Ethiopian businesses may be considered sociocommerscapes, which provide services but also function as locales where members of the community can interact socially. The owner of a grocery store in Silver Spring noted that his Ethiopian clients were not "efficient shoppers with lists"; they liked to spend time talking to the owner and staff and with other Ethiopian customers. Even though his store was small, the owner placed a couple of small round tables and chairs so that shoppers could sit down and enjoy a beverage and snacks (sold at the store). Stores such as these also offer customers information on current events related to the diasporic community through bulletin boards and posters. The Ethiopian sociocommerscapes in Alexandria, Silver Spring, and the District of Columbia are among the oldest business clusters formed by the immigrant group. Restaurants, stores, and services there draw co-ethnics from the central city, suburbs, and even exurban areas. However, the large number of new Ethiopian eateries that are being established in the metropolitan area may have led to over-saturation in some neighbourhoods, while the intense competition between these restaurants has resulted in the failure of a few to thrive or even survive. At least two Ethiopian restaurants in the U Street area of the District of Columbia, with its distinctive ethnic imprint, have failed.

The locations of Ethiopian businesses in the Washington metropolitan area clearly reflect their target markets. Older Ethiopian restaurants in Washington, DC, located in the heart of the city offered native cuisine to the diaspora as well as to Washingtonians who are interested in sampling food from around the world. Suburban locations close to co-ethnic clusters, on the other hand, have ensured that community members would patronize these establishments and services, helping them stay in business particularly during the crucial initial stages. A large ethnic clientele can even help a business expand as a growing diasporic population accesses its goods and services. As with all enclave-market targeted services, however, the growth of these businesses will be limited unless they are able to expand their market beyond the ethnic community. While some Ethiopian immigrant businesses such as accounting, tax, legal, and interpretation services still largely cater to the co-ethnic population, it should be noted that several, particularly restaurants and taxi services, have a clientele from across the metropolitan area. Still others, such as grocery stores and travel agencies located as they are in heterolocal neighbourhoods, are likely to draw patrons from other ethnic and immigrant groups in addition to the co-ethnic population.

### *Community and Social Networks in Bolivian and Ethiopian Entrepreneurship*

The structure of the social networks that immigrant entrepreneurs are involved in influence their ability to mobilize resources to start or develop a business enterprise. Social networks are comprised of ties that people have with each other. Strong ties are often direct ties within the community and can be those of kinship, friendship, or a relation to ethnic community or place of origin. There is usually a strong degree of overlap of these various ties for those belonging to an ethnic immigrant community, as people are linked in multiple ways through multiple social networks (familial, religious, ethnic, hometown, etc.). Dense social networks are underlain by trust and obligation; co-ethnics may feel obligated to provide advice or funds, while trusting that the borrower will repay the loan in a timely manner.

Light, Bhachu, and Karageorgis (1990) identify three ways in which migration networks can enhance entrepreneurship: (1) by providing access to low-cost co-ethnic labour, (2) by supplying relevant economic and business information, and (3) by offering mutual aid and assistance beyond information. Some immigrant businesses such as stores and restaurants may rely on co-ethnic labour, which includes work by family members. This strategy keeps costs down as co-ethnics and family are often willing to work longer hours for less pay as their options may be limited or they have a vested interest in the success of the enterprise. In most Ethiopian restaurants in the Washington area, owners and co-ethnics are the hosts, waiters, and cashiers; however, those employed to cook and clean are Latinos. Similarly, there are Bolivian construction businesses and cleaning services that are entirely staffed by co-ethnics.

Immigrant entrepreneurs often suffer from lack of access to the capital needed to start or grow a business. Most commercial banks are reluctant to loan money to a new entrepreneur who may have little collateral and a limited credit history (see Li and Lo, next chapter). Foreign-born entrepreneurs may also be unfamiliar with or misunderstand local business regulations and practices due to a lack of proficiency in English, inexperience, and various cultural barriers. Established migrants thus become the newcomers' social resources in starting businesses. Through community social networks, Bolivian and Ethiopian immigrant entrepreneurs often tap into familial, hometown, collegial, school, religious, and other social networks when they seek to establish a business. For Ethiopian entrepreneurs, such networks serve as sources of capital through revolving credits (*ekub*), in which members of the group contribute a sum of money, but at each drawing a different person is the recipient of the group's savings, until everyone has benefited from the arrangement. For Bolivians, a similar practice exists, referred to as *pasanaco*. By pooling resources, the immigrant entrepreneurs are able to make relatively large sums of money available, which could provide the needed capital to start a business. Social networks are also useful for identifying business partners and even provide markets for products and services. In addition, such networks provide opportunities for potential entrepreneurs to learn from the successes and failures of co-ethnic immigrants with established businesses in the area.

The Washington area also has ethnic organizations that help fellow immigrants and co-ethnics set up businesses. The Ethiopian Business Association (EBA) provides financial advice and help to Ethiopians who wish to start business ventures. The Enterprise Development Group (EDG) is an affiliate of the well-known and established Ethiopian Community Development Council (ECDC), a non-profit community-based organization that was founded in 1983 to help resettle Ethiopian refugees and promote development of refugee and immigrant groups in the United States. The EDG seeks to provide affordable financial services to immigrants and refugees. Using money obtained from the Federal Office of Refugee Resettlement, the Small Business Administration, the Calvert Foundation, and even some commercial banks, EDG offers loans of up to $50,000. The organization analyses the business plans and credit histories of applicants before giving out loans, but also uses non-traditional collateral to secure a loan (Chacko 2009). Ethiopian immigrants and refugees who availed themselves of EDG's micro-financing have used the money to buy taxicabs or set up stores and other small business ventures.

In contrast, Bolivian focus group participants lamented the lack of a Bolivian organization that could help co-ethnics establish businesses. As one Bolivian noted, "There needs to be a service for the community to invest in itself." Organizations such as the Bolivian-American Chamber of Commerce (based in New York but with staff in Washington) are more focused on forging international investment and trade than on supporting local entrepreneurship. Much of the energy to support immigrant businesses in the Washington region is subsumed under a pan-Latino label that does not differentiate by country of origin.

Over the last three decades, the close social networks within the Bolivian and Ethiopian immigrant groups have led to the formation of a distinct type of ethnic economy – the ethnic niche, with a clustering of ethnic entrepreneurs in the same occupations or businesses. Taxi and limousine services for the Ethiopians and childcare services and construction for the Bolivians are examples of ethnic niche occupations for these groups. These are also the occupations into which new arrivals from Ethiopia and Bolivia are likely to be inducted. New immigrants may acquire skills by working in a co-ethnic business, but can also learn how to access lines of credit, customers, and business opportunities from compatriots who are already established in these enterprises.

### *Institutional Structures and Entrepreneurship*

In response to the difficulties that immigrant entrepreneurs face in obtaining credit (see also Li and Lo, next chapter) and navigating the bureaucracy needed to start a business, many immigrant-friendly cities have established units in institutions such as local chambers of commerce and economic development corporations that cater to the needs of immigrant entrepreneurs. Some cities also have initiatives such as free business classes in languages spoken by the major immigrant groups. For example, the Greater Washington Hispanic Chamber of Commerce assists businesses and entrepreneurs in the DC area develop the networks and tools necessary for successful businesses in a culturally familiar setting. In collaboration with Arlington County, a part of Washington's inner suburbs (see maps), the Hispanic Chamber of Commerce started BizLaunch, which provides direct outreach and educational services to the Hispanic business community in Spanish. Among the free services offered are information on how to start and operate a business, counselling and technical assistance, data, and assistance with analysis to identify clients, suppliers, and market trends.

The Latino Economic Development Center (LEDC) is another unit that assists Latino entrepreneurs in the metropolitan area. LEDC operates the Community Asset Fund for Entrepreneurs, a lending subsidiary that offers business loans to qualified startups and existing businesses that have difficulty in obtaining credit from mainstream financial institutions. The fund helps entrepreneurs obtain loans ranging from $5000 to $50,000. In 2012 alone, LEDC's lending program made eighty loans to businesses in the District, Maryland, and Virginia, totalling more than $1 million, supporting businesses that ranged from restaurants to cleaning services and hair salons (see ledcmetro.com). Within the District of Columbia, the Small Business Resources Center provides training, workshops, and computer software to help local entrepreneurs. Although this service does not focus solely on immigrants (it is available to all residents of Washington), it is useful to immigrant entrepreneurs (Latino Economic Development Center 2012).

Reflecting the diversity of Washington, DC, the government of the District of Columbia has established an Office on Asian and Pacific Islander Affairs, an Office on Latino Affairs, and, most recently, an Office on African Affairs. The DC Office on African Affairs (OAA) acts as a liaison office between the African community and the mayor of DC and DC government agencies. The OAA seeks to build the capacity of local entrepreneurs and support them. It regularly holds seminars and training sessions to assist new entrepreneurs by providing information on what they need to establish or grow a business in the District of Columbia, on possible sources of capital, and on how to conduct business with local and federal governments.

Across the Potomac River in Arlington County, Bolivian and other immigrant groups have worked closely with county officials and the school system to gain support and even political recognition. In 2000, one in four residents in Arlington County was foreign-born, and so county officials have created various programs to assist newcomers in job training and community building. In 2003 the county elected a Salvadoran immigrant, Walter Tejado, to its county board. Four years later, Tejado pushed through a county Statement of Inclusion that promoted immigrant inclusion and encouraged officials in other Virginia jurisdictions to do the same regardless of legal status. In 2008, Bolivian-born educator Emma Violand-Sanchez was elected to the Arlington school board, and in 2013 she became the chair of this body. This is the first time a Bolivian immigrant was elected to a local office, and her success underscores the advantages of building a demographic base as well as long-term relationships with officials in local government.

In the Bolivian case, Arlington County also supported the formation of Escuela Bolivia in 1998, a weekend Spanish-language and cultural program run by Bolivian volunteers but held in an Arlington public school. Recently, Escuela Bolivia rebranded itself as Edu-Futuro with a focus on leadership training for Latino youth as well as language training for youth and adults. Edu-Futuro's programs work to prepare Hispanic youth for college and jobs in the skilled workforce. Most of the founding members of this organization are Bolivian, yet the value of serving the larger immigrant and Spanish-speaking population in Arlington County could not be ignored. This strategy of nested identities, of being both Bolivian and Latino, allows for a relatively small group to build coalitions and increase their political and socio-economic clout.

Local governments have also responded to the problem of a lack of linguistic fluency in English faced by many new immigrants. All the close-in jurisdictions within metropolitan Washington (Fairfax, Montgomery, Prince George's, and Arlington counties, along with Alexandria, Falls Church, and the District of Columbia) provide government documents in multiple languages. These are especially important for immigrants with limited English skills to get the business applications and licensing that they need. For example, in 2004 the District of Columbia government passed the Language Access Act to ensure equal access to public services, programs, and activities for residents who have limited or no proficiency in English. Among the six languages in which the District of Columbia's government makes its forms available are Amharic, the primary language of Ethiopian immigrants in the area, and Spanish.

These institutional responses are examples of local jurisdictions constructing inclusive strategies that foster both social and economic integration. While it is hard to prove, intuitively it makes sense that localities that are supportive of immigrant groups and ethnic minorities, seeing them as part of a diversity advantage rather than as a liability, are more likely to nurture higher indices of social and economic attainment among these groups. A county official in Wheaton, in Montgomery County, Maryland, noted that the area was both "accommodating and capitalizing on the immigration trend" by promoting Wheaton's "funky, ethnic mix that makes it fell like a true urban environment without the urban problems" (Price and Singer 2008, 160). In contrast, jurisdictions (such as Prince William County and Herndon, VA) that insist upon more exclusionary measures such as English-only laws and imposition of 287g ordinances that deputize local police to arrest and detain undocumented migrants, have seen a decline in immigrant businesses in general, and Latino ones in particular.

## Challenges and Opportunities for Bolivian and Ethiopian Entrepreneurs

The nature of Bolivian and Ethiopian immigrants' human and cultural capital, as well as the socio-economic and cultural contexts of the receiving society, affects their job prospects. Both immigrant groups exhibit levels of college education that are higher than that of the general population. However, educational and occupational credentials obtained in Ethiopia or Bolivia are often not recognized in the United States, the immigrants may not be fluent in English, or their work experience in the sending country not valued. The structural barriers these immigrants face in the receiving society may force them to work in low-paying jobs in the ethnic sector, engage in retraining to obtain the required credentials, or look for gainful employment in new areas (as demonstrated by Lysenko and Wang in chapter 9, this volume). The *blocked mobility* of Bolivian and Ethiopian immigrants thus presents both a challenge and an opportunity. Locked out of the well-paying white-collar positions they may have initially preferred, educated immigrants may turn to entrepreneurship. By starting a business and working hard, they hope to achieve economic security and the American Dream of a successful and fulfilling life for themselves and their children. Many of our immigrant interviewees also believed that, given their situation, entrepreneurship was the fastest path to a markedly better financial and social status, with the rewards of success outweighing the risks (focus group and interviews).

Another challenge that immigrants face is anti-immigrant sentiment and local government policies that are not immigrant-friendly. The rapid rise in the number of unauthorized immigrants in the United States, estimated at around 11 million in 2010 (Passel and Cohn 2011), has resulted in a backlash that particularly targets Hispanic communities. US states such as Arizona and Alabama passed laws that many consider anti-immigrant in 2010 and 2011, respectively. These laws require police to determine the immigration status of anyone who is arrested or detained when there is "reasonable suspicion" that they are not in the United States legally. Alabama's bill also obliges public schools to determine the immigration status of all students, requires parents of foreign-born students to report the immigration status of their children, and effectively stops the enrolment of unauthorized immigrants in any public college. Although these laws purportedly target only unauthorized migrants, cities, counties, and states that are perceived as unfriendly to immigrants may inhibit even legal immigrants from setting up businesses within their jurisdictions, which may be associated with fear and negativity by the immigrant community at large. Moreover, immigrant businesses that draw on a clientele of co-ethnic and fellow immigrants are most likely to establish themselves in areas with large immigrant populations in places that welcome them. Local policies either assisting or targeting the undocumented in the United States are distinct from the Canadian case, where the percentage of the undocumented is much smaller.

The states, cities, and communities that form the Washington metropolitan area have different laws and responses in reaction to rising levels of unauthorized immigrants. In Prince William County, Virginia, police officers are required to question all criminal suspects about their immigration status once an arrest is made, while the town of Herndon, Virginia, entered into a controversial federal program, 287g, which deputized local police to check and enforce immigration status and detain undocumented individuals (Svajlenka 2010). By contrast, Washington, DC, Arlington County, and the city of Baltimore in Maryland are considered "sanctuary cities," where law enforcement is directed not to ask for evidence of immigration status. In the Washington metropolitan area, cities such as Falls Church, Virginia, Washington, and Silver Spring, Maryland, have diverse non-profit organizations that provide various kinds of assistance to immigrants.

Growing numbers of American cities, including Cleveland, Detroit, Dayton, Baltimore, and Chicago, are portraying themselves as immigrant-friendly through outreach to foreign-born populations. Outreach can take the form of city-run classes in the languages of the major immigrant groups, information made available in multiple languages in public libraries, English and citizenship classes, and training to start micro-enterprises. Avowedly immigrant-friendly cities are looking to immigrants to assist in reviving their local economies in a time of declining population and economic recession. They hope to draw high-skill immigrants to work in the IT, scientific, and other professional sectors, as well as working-class immigrants to set up small businesses. The mayor of Baltimore, Maryland, even held a town hall meeting in a public library to explain the city's new immigrant-friendly policies to an increasing population of foreign-born, whom she credits with stemming the city's decline in population and to whom she looks to revive the city's neighbourhoods and economy (Morello and Lazo 2012).

By providing culturally appropriate support, training, and activities that reflect the needs of immigrant entrepreneurs, cities can simultaneously improve their economic futures and those of immigrants living in them. Immigrants arrive with a variety of skill sets, which can result in the diversification of entrepreneurship in the area, while the operations they establish can be catalysts for economic revival and growth. The clustering of the Bolivian and Ethiopian populations and their businesses in cities and counties perceived as immigrant-friendly and which provide support services to the immigrant community is a testament to the importance of local policies in welcoming or driving away immigrants and immigrant entrepreneurs.

## Conclusions

The Washington metropolitan area is the prime destination in the United States for Bolivian and Ethiopian immigrants, groups that do not have a tradition of engaging in business enterprise. Prevented from partaking in mainstream white-collar jobs due to structural barriers, many of these immigrants have turned to entrepreneurship as a pathway to socio-economic advancement. Bolivian entrepreneurs engage largely in domestic services (housekeeping and childcare) and construction, while Ethiopian entrepreneurs are most often involved in the taxi and limousine, restaurant, real estate, and travel agency businesses. Both groups of immigrants are likely to locate their enterprises in areas with residential clusters of co-ethnics. The co-location of business and residences for each group is helpful, as business owners can more easily tap into the co-ethnic labour pool, while the concentrations of co-ethnics can provide a ready market for goods and services. However, even in areas where there is little overlap between Bolivian or Ethiopian residential and commercial clusters, ethnic sociocommerscapes draw members of the community from across the region. As immigrant entrepreneurs expand their enterprises beyond traditional ethnic businesses, they develop new ventures in locations that draw on a larger clientele that includes mainstream Americans and fellow immigrants from other ethnic groups. Both groups of immigrants also serve the wider Washington population through their food, taxi and limousine, domestic work, and construction services. Bolivians and Ethiopian immigrants help co-ethnic entrepreneurs through informal mutual aid and mentoring as well more formal ethnic institutions. A multitude of overlapping and intertwined social networks allow co-ethnics to draw on the collective knowledge, expertise, and experience of entrepreneurs from within each community. Numerous ethnic non-governmental organizations as well as those that cater to the larger immigrant community provide support through information, advice, loans, and assistance with bureaucratic paperwork. In addition, local governments can play a very important role in assisting immigrant entrepreneurs. The response of cities, counties, and states in the Washington metropolitan area to immigrants settling or doing business within them varies considerably. The District of Columbia and the inner suburbs where most of the immigrant enterprises are located are more immigrant-friendly, and actively court immigrants by offering services to help them establish and develop businesses. Evidence suggests that the rise of businesses owned or co-owned by immigrants can improve the quality of life in once-blighted neighbourhoods and help revive local economies. Hence, policymakers at local (city, county) and state levels would do well to support entrepreneurship as a critical element in building and promoting economic growth. Local governments can assist entrepreneurs by establishing venture-capital funds for new businesses in their jurisdictions, providing mentoring services to new business owners, keeping taxation rates at a level that is attractive for businesses, ensuring that transportation and other infrastructure used by entrepreneurs function well, and through the construction of local amenities such as incubator facilities (offices for small or new businesses that keep costs down through shared overhead expenses) whose goal is to assist new entrepreneurs. While this research reported on Bolivian and Ethiopian small business enterprises related to retail, the food industry, the service industry, and transportation in the Washington area, the involvement of immigrant entrepreneurs in establishing firms related information technology, particularly in the high-tech cluster in northern Virginia warrants further research.

## REFERENCES


Airriess, C. 2006. "Scaling Central Place of an Ethnic-Vietnamese Commercial Enclave in New Orleans, Louisiana." In *Landscapes of the Ethnic Economy*, ed. D. Kaplan and W. Li, 17–33. Lanham, MD: Rowman and Littlefield.

Aldrich, H., and R. Waldinger. 1990. "Ethnicity and Entrepreneurship." *Annual Review of Sociology* 16 (1): 111–35. http://dx.doi.org/10.1146/annurev.so.16.080190.000551.

Bonacich, E. 1973. "A Theory of Middleman Minorities." *American Sociological Review* 38 (5): 583–94. http://dx.doi.org/10.2307/2094409.

Borjas, G.J. 1986. *The Self-Employment Experiences of Immigrants*. NBER Working Paper Series, no. W1942. www.nber.org/papers/w1942.pdf.

Chacko, E. 2009. "Africans in Washington, D.C.: Ethiopian Ethnic Institutions and Immigrant Adjustment." In *The African Diaspora in the United States and Canada at the Dawn of the 21st Century*, ed. J.W. Frazier, J.T. Darden, and N.F. Henry, 243–56. Binghamton, NY: Global Academic Publishing.

Chacko, E. 2008. "Washington: From Bi-racial City to Multiethnic Gateway." In *Migrants to the Metropolis: The Rise of Immigrant Gateway Cities*, ed. M. Price and L. Benton-Short, 203–25. Syracuse, NY: Syracuse University Press.

Chacko, E. 2003. "Ethiopian Ethos and the Making of Ethnic Places in the Washington Metropolitan Area." *Journal of Cultural Geography* 20 (2): 21–42. http://dx.doi.org/10.1080/08873630309478274.

Coleman, J.S. 1988. "Social Capital in the Creation of Human Capital." *American Journal of Sociology* 94: S95–120. http://dx.doi.org/10.1086/228943.

Hiebert, D. 2002. "The Spatial Limits to Entrepreneurship: Immigrant Entrepreneurs in Canada." *Tijdschrift voor Economische en Sociale Geografie* 93 (2): 173–90. http://dx.doi.org/10.1111/1467-9663.00193.

Hou, F., and S. Wang. 2011. "Immigrants in Self-Employment." Component of Statistics Canada Catalogue no. 75-001-X. Ottawa, Canada.

Kaplan, D., and W. Li. 2006. *Landscapes of the Ethnic Economy*. Lanham, MD: Rowman and Littlefield.

Latino Economic Development Center. 2012. www.ledcmetro.org/

Li, P. 2001. "Immigrant's Propensity to Self-Employment: Evidence from Canada." *International Migration Review* 35 (4): 1106–28. http://dx.doi.org/10.1111/j.1747-7379.2001.tb00054.x.

Li, W. 2009. *Ethnoburb: The New Ethnic Community in Urban America*. Honolulu: University of Hawaii Press.

Li, W., G. Dymski, M.W.L. Chee, H-H. Ahn, C. Aldana, and Y. Zhou. 2006. "How Ethnic Banks Matter: Banking and Community/Economic Development in Los Angeles." In *Landscapes of the Ethnic Economy*, ed. D. Kaplan and W. Li, 113–33. Lanham, MD: Rowman and Littlefield.

Light, I., and E. Bonacich. 1991. *Immigrant Entrepreneurs: Koreans in Los Angeles, 1965–1982*. Berkeley, Los Angeles: University of California Press.

Light, I., P. Bhachu, and S. Karageorgis. 1990. *California Immigrants in World Perspective: The Conference Papers, April 1990*. Vol. 5, "Migration Networks and Immigrant Entrepreneurship." Los Angeles: Institute for Social Science Research, UC.

Lo, L. 2006. "Changing Geography of Toronto's Chinese Economy." In *Landscapes of the Ethnic Economy*, ed. D. Kaplan and W. Li, 83–96. Lanham, MD: Rowman and Littlefield.

Morello, C., and L. Lazo. 2012. "Baltimore Puts Out Welcome Mat for Immigrants, Hoping to Stop Population Decline." *Washington Post*, 24 July.

Oberle, A. 2006. "Latino Business Landscapes and the Hispanic Ethnic Economy." In *Landscapes of the Ethnic Economy*, ed. D. Kaplan and W. Li, 149–163. Lanham, MD: Rowman and Littlefield.

Passel, J.S., and D. Cohn. 2011. "Unauthorized Immigrant Population: National and State Trends, 2010." Press release from the Pew Hispanic Center, 1 February. www.pewhispanic.org/2011/02/01/unauthorized-immigrant-population-brnational-and-state-trends-2010/.

Price, M., and E. Chacko. 2010. "Immigrants as Entrepreneurs: How U.S. Cities Promote Immigrant Entrepreneurship." In *Inclusive Cities for All: Urban Policy and Practice for Immigrants*, ed. B. Colin and B. Kadioglu, 100–14. Paris: UNESCO.

Price, M., I. Cheung, S. Friedman, and A. Singer. 2005. "The World Settles In: Washington DC as an Immigrant Gateway." *Urban Geography* 26 (1): 61–83. http://dx.doi.org/10.2747/0272-3638.26.1.61.

Price, M., and A. Singer. 2008. "Edge Gateways: Immigrants, Suburbs and the Politics of Reception in Metropolitan Washington." In *Twenty-first Century Gateways: Immigrant Incorporation in Suburban America*, ed. A. Singer, S. Hardwick, and C. Brettell, 137–68. Washington, DC: Brookings Institution Press.

Pumar, E.S. 2012. *Hispanic Migration and Urban Development: Studies from Washington DC*. Bingley, UK: Emerald Press.

Saxenian, A.L. 2006. *The New Argonauts: Regional Advantage in a Global Economy*. Cambridge, MA, London: Harvard University Press.

Serra, P. 2012. "Global Businesses 'from Below': Ethnic Entrepreneurs in Metropolitan Areas." *Urbani Izziv* 23 (s2): s97–s106. http://dx.doi.org/10.5379/urbani-izziv-en-2012-23-supplement-2-008.

Singer, A., S. Hardwick, and C. Brettell. C. 2008. *Twenty-first Century Gateways: Immigrant Incorporation in Suburban America*. Washington, DC: The Brookings Institution.

Singer, A., J.H. Wilson, and B. DeRenzis. 2009. "Immigrants, Politics, and Local Response in Suburban Washington." In *Survey Series for the Metropolitan Policy Program*. Washington, DC: The Brookings Institution.

Svajlenka, N. 2010. "The Rescaling of US Immigration Policy: A Socio-spatial Analysis of Enforcement in Herndon, Virginia." Unpublished master's thesis, Department of Geography, George Washington University.

Teixeira, C. 2006. "Residential Segregation and Ethnic Economies in a Multicultural City: The Little Portugal of Toronto." In *The Landscapes of Ethnic Economy*, ed. D. Kaplan and W. Li, 49–65. Lanham, MD: Rowman and Littlefield.

Teixeira, C., W. Li, and A. Kobayashi. 2012. *Immigrant Geographies of North American Cities*. Don Mills, ON: Oxford University Press.

US Bureau of the Census. 2009–11. American Community Survey (ACS). Accessed through American FactFinder, Washington, DC.

US Department of Homeland Security. 2008. *Yearbook of Immigration Statistics: 2007*. Washington, DC: US Department of Homeland Security, Office of Immigration Statistics.

Varsanyi, M.W. 2011. "Neoliberalism and Nativism: Local Anti-Immigrant Policy Activism and an Emerging Politics of Scale." *International Journal of Urban and Regional Research* 35 (2): 295–311.

Walker, K., and H. Leitner. 2011. "The Variegated Landscape of Local Immigration Policies in the United States." *Urban Geography* 32 (2): 156–78. http://dx.doi.org/10.2747/0272-3638.32.2.156.

Wang, Q., and W. Li. 2007. "Entrepreneurship, Ethnicity and Local Contexts: Hispanic Entrepreneurs in Three U.S. Southern Metropolitan Areas." *GeoJournal* 68 (2–3): 167–83. http://dx.doi.org/10.1007/s10708-007-9081-0.

Wood, J. 1997. "Vietnamese-American Placemaking in Northern Virginia." *Geographical Review* 87 (1): 58–72. http://dx.doi.org/10.2307/215658.

Zelinsky, W., and B.A. Lee. 1998. "Heterolocalism: An Alternative Model of the Sociospatial Behaviour of Immigrant Ethnic Communities." *International Journal of Population Geography* 4 (4): 281–98. http://dx.doi.org/10.1002/(SICI)1099-1220(199812)4:4<281::AID-IJPG108>3.0.CO;2-O.

Zhou, M. 2004. "Revisiting Ethnic Entrepreneurship: Convergences, Controversies, and Conceptual Advancements." *International Migration Review* 38 (3): 1040–74. http://dx.doi.org/10.1111/j.1747-7379.2004.tb00228.x.

# 12 Financing Immigrant Small Businesses in Canada and the United States

WEI LI AND LUCIA LO

Immigrant entrepreneurs have been recognized as an important force for boosting economic productivity and countering the population aging trend. The percentage of immigrants owning and operating a business in Canada and the United States has been on the rise since the postwar years, and has been higher than that of the native-born population (as illustrated in the two previous chapters). A recent report indicates that in 2010, 18 per cent of all US small-business owners are immigrants, higher than their share in the total population (13%) and labour force (16%) (Fiscal Policy Institute 2012). Moreover, the largest share of such businesses consists of professional and business services (18% of total). Immigrant-owned or co-owned small businesses (with one to 99 employees) employed 4.7 million people and generated $776 billion in total receipts in 2007. A similar picture emerges in Canada. Self-employment since 1980 has grown faster among immigrants than among the native-born (Hou and Wang 2011, 3). In the late 2000s, 19 per cent of immigrant workers were self-employed, compared to 15 per cent of the Canadian-born population (Grant 2011). In contrast to the US situation, immigrant self-employment in Canada overall is highly concentrated in trade and transportation. In Census Metropolitan Areas (CMAs), immigrants' representation in business and professional services is lower than that of the native-born. However, their industrial distributions are similar outside the CMAs. A comparison of similarly ranked US and Canadian metropolitan areas finds that immigrant self-employment rates are higher in most Canadian cities (Lo and Li 2012, table 6.5).

Immigrant entrepreneurship operates across a continuum, from owners of part-time, home-based businesses and contingent worker contractors at the lower end, to self-employed professionals and small unincorporated owner- or family-operated businesses in the middle range, to owners of businesses that require significant capital investments at the upper end. It is widely held that most immigrant entrepreneurs operate in the more vulnerable small business sector, despite the booming of such businesses in many metropolitan areas (see, for instance, chapters 10 and 11 of this volume). For immigrants, the odds of starting and running a successful business are not very promising. In both Canada and the United States, the five-year survival rate for small- and medium-sized businesses is only slightly over 50 per cent (Fisher and Reuber 2010). While immigrants and non-immigrants face similar challenges, immigrants often encounter some additional barriers due to language issues, the unfamiliarity of the receiving country environment, and a difference in business culture between the origin and destination countries (Teixeira, Lo, and Truelove 2007; Workforce Planning Hamilton 2012). Such challenges also vary across national boundaries, in our case Canada and the United States.

Financing is one of the most commonly cited obstacles immigrants face when starting up or maintaining their businesses. This is evident in research everywhere (e.g., in Canada: Froschauer 2001; Teixeira 2001; Teixeira, Lo, and Truelove 2007; and Uneke 1994; in Europe: Barrett 1999; Boissevain et al. 1990; and Bruder, Neuberger, and Räthke-Döppner 2011; and in the United States: Bates 1989, 1997; Butler 1991; Feagin and Imani 1994; and Hodge and Feagin 1995). Since the financing environment in which firms operate impacts their performance (Kalckreuth and Murphy 2005; Levine 2005) and that environment affects smaller firms more than others (Beck et al. 2004; Ghosal and Loungani 1996), it is important to understand the financing impediments small firms face.

The primary focus of this chapter is on immigrant entrepreneurs' concerns about and needs for financial services. It seeks to (1) explore the financial needs of immigrant businesses, immigrant entrepreneurs' expectations of banks, and the barriers they face; and (2) gauge their opinion on different types of banks, their access to and dealings with them, and their assessment of the fairness and effectiveness of the financial services they receive or lack thereof.

After a brief review of the relevant literature on immigrant entrepreneurship and financial institutions, this chapter will report on an exploratory empirical study involving both experienced and potential immigrant entrepreneurs that was conducted in the San Francisco and

Vancouver areas, before concluding with some recommendations for breaking down some barriers to immigrant entrepreneurship.

## Immigrant Entrepreneurship and Financial Institutions

The literature on immigrant entrepreneurship is huge, with a fair portion focusing on why entrepreneurship is more common among some immigrant groups than others and how different groups deal with the challenges of entrepreneurship. Several explanations have been advanced through the years (see Lo 2009 and Light and Gold 2000 respectively for a short and lengthy review). The cultural thesis suggests that some groups are conditioned with certain characteristics, such as habits of saving, norms of sharing resources, and a propensity to take risk, that make them more inclined towards business development (Bonacich and Modell 1980). The structural approach argues that it is the roadblocks in the regular labour market such as racial discrimination or non-recognition of foreign credentials that push immigrants into small businesses (Light 1972). The third approach emphasizes the importance of resource mobilization in enhancing entrepreneurial capacity (Waldinger, Aldrich, and Ward 1990). Ethnic resources in the form of social ties and networks can be more readily mobilized within a larger group, with a long history of settlement and institutionally complete. Class resources are potential advantages conferred by higher education, official language proficiency, prior work/business experience, and accumulated wealth. Any ability to mobilize these resources increases the rate of entrepreneurship provided that the opportunity structure – manifested through the local social economic environment in terms of such things as business vacancies, degree of competition, and government policies regulating immigrant business ownership (ibid.), and the larger political institutional environment consisting of immigration and citizenship policies, taxation system, health and safety laws, and so on (Rath 2000) – is there. The latter, known as a mixed embeddedness approach, contextualizes the interaction of micro-level cultural forces and the broader social, economic, and political setting of the destination society. In this regard, financial institutions, in their own right as business entities and as keepers of private financing in the eyes of the public, play a role.

We posit that financial institutions utilize social capital or ethnic assets (Dymski and Li 2004) to tap into certain immigrant markets, and so possibly enhance the financial integration of these immigrant groups to the detriment of others. This raises the question of how social capital or ethnic assets are used by various types of financial institutions. Ethnic assets refer to the situation when banks possess a familiarity with an immigrant group's native language(s), cultural traditions, and business practices in their country of origin. They can be seen as both bonding and bridging social capital in the financial world. Bonding social capital refers to ties of people who share significant, often in-group, similarities, and bridging social capital is about ties to people who are significantly different and most often belonging to a different group (Putnam 2007, 143–4). Considerable differences exist in the type of ethnic assets banks possess and the way they are used. The ethnic financial sector, consisting of banks and branches owned by individuals belonging to particular immigrant groups or banking institutions from their country of origin (Li and Lo 2008; Li et al. 2002), is by definition most likely to possess ethnic assets or bonding social capital, but may choose not to use them. Mainstream banks, or banks with no ethnic or diasporaic affiliations, may not have ethnic assets to begin with, but may acquire bridging social capital over time by hiring or promoting immigrants from a particular group, or training existing employees to work more effectively with that group (Li, Oberle, and Dymski 2009). Bank customers, immigrant entrepreneurs included, can utilize social capital to receive information from or services provided by financial institutions that they feel more bonded or comfortable with.

Financial institutions can be classified as mainstream or ethnic (Li and Lo 2008; Li et al. 2002), with the former usually larger and having no affiliations to any ethnic-minority or immigrant groups as owners or targeted clientele. In any region or country, the magnitude of the ethnic financial sector, if any, is shaped by the regulatory regime governing financial institutions. For example, the different banking charters in Canada and the United States have given rise to two very different banking sectors (Clement 1977; Tickell 2000). Canada traditionally has stricter rules on the banking industry. The nation-wide branch-banking arrangement, and the forbidding of a single individual or group of associated individuals to hold more than 10 per cent of voting shares in a bank have led to a relatively small number of major banks (Freedman 1998), and thereby a more centralized and oligopolistic banking sector (Trichur 2007), with the big five – Bank of Montreal, Canadian Imperial Bank of Commerce, Royal Bank of Canada, Scotia Bank, and Toronto-Dominion Bank – accounting for over 85 per cent of the country's total bank assets (Li and Lo 2008). This also explains why, in Canada, there are no ethnic domestic banks, but rather ethnic credit unions, which are regulated primarily at the provincial level. In addition, it was only in 1980 when foreign banks were allowed to set up wholly owned subsidiaries. A minority of these foreign banks purposely pursued their diasporic communities and hence can be categorized into the ethnic financial sector.

In contrast, the US banking sector is subject to regulations at both the federal and state levels, with some states such as California pursuing more permissive policies than others. The forbidding of interstate banking until 1994 has dampened the domination of the sector by a few banks, as has been in the case of Canada. Meanwhile the Community Reinvestment Act has promoted the establishment of community banks, making it easier for immigrants or ethnic minority groups to establish small ethnic banks. In addition, like most of the foreign banks in Canada, foreign banks in the United States primarily act as global outposts managing their headquarters' cross-border financial flows (Hultman and McGee1990; ÓhUallacháin 1994), perhaps with the exception of the Chinese case in California (Dymski and Li 2004). These regulatory differences mean that the ethnic banking sector, consisting of banks and branches owned by individuals belonging to particular immigrant groups or banking institutions from their country of origin (Li and Lo 2008; Li et al. 2002), is comparatively much smaller and less conspicuous in Canada. We anticipate that mainstream and ethnic-based banks play different roles, and adopt different strategies towards immigrants. In the context of promoting or sustaining immigrant entrepreneurship and overcoming the challenges they face, we ask: to what extent do financial institutions use ethnic assets differently? How do immigrant entrepreneurs assess their relationship with the different types of financial institutions?

## Study Areas and Methods

This study focuses on two North American metropolitan areas: Vancouver in Canada and San Francisco in the United States. Both are immigrant gateways and financial sub-centres on the west coast, characterized by a rapid influx of immigrants – the majority from Asia – since the 1970s (calculation based on the Canadian and US censuses). Both are representative of the recent suburban transformation due to immigration, but differ in terms of financial dynamics due in part to a difference in their larger regulatory regimes (table 12.1). As such, these two

Table 12.1 Population and banks in the San Francisco Bay Area and Vancouver CMA

| 2007–2011 | San Francisco Bay Area | 2011 | Vancouver CMA |
|---|---|---|---|
| **Total population** | 5,805,052 | **Total population** | 2,228,700 |
| Total bank branches* | 1358 | Total bank branches* | 343 |
| Pop. per bank branch | 4275 | Pop. per bank branch | 6498 |
| **Total foreign-born pop.** | 1,882,709 | **Total immigrant pop.** | 913,310 |
| FB pop. per bank branch | 1386 | Immig. per bank branch | 2263 |
| **Foreign-born in total pop.** | 32.4% | **Immigrants in total pop.** | 41.0% |
| % Asian in FB pop. | 57.0 | % Asian in immig. pop. | 68.7 |
| % African in FB pop. | 1.7 | % African in immig. pop. | 3.1 |
| % Latin Amer. in FB pop. | 29.8 | % Latin Amer. in immig. pop. | 3.8 |
| % European in FB pop. | 8.8 | % European in immig. pop. | 19.0 |
| % Oceanian in FB pop. | 1.3 | % Oceanian in immig. pop. | 2.6 |

* Bank branch information was collected in 2009 from FDIC, bank websites, and the yellow pages.
Sources: Calculations based on 2011 Canadian National Household Survey and 2007–2011 American Community Survey 5-year estimates (United States).

cities are the most alike in a cross-national study of immigrant financial integration.

Consisting of twenty-one incorporated municipalities and an unincorporated area, and serving as the Canadian gateway to the Asia Pacific, the *Vancouver Census Metropolitan Area* is the second largest immigrant "port of entry" after Toronto. Immigration to Vancouver is fuelled in large part by contemporary globalization (Mitchell 2004). With immigrants constituting 41 per cent of its population, Vancouver has a high percentage of wealthy business-oriented immigrants (Edgington, Goldberg, and Hutton 2006; Ley 2003, 2010; Li 1992, 1997, 2005). The city is also home to one foreign global bank, HSBC (the seventh-largest overall bank in Canada), a total of twenty foreign banks, and the only ethnic-specific financial institution, One Filipino Cooperative of B.C. (or OneFilCoop). The *San Francisco Bay Area* has a long history of immigrant settlement, with immigrants engaged in various economic sectors ranging from high-tech production in Silicon Valley to serving dinners in Chinatown or cleaning houses in Los Gatos. San Francisco has both burgeoning foreign banking and an ethnic financial sector (Li, Oberle, and Dymski 2009; Li and Park 2006; Saxenian 2002; Wang 2010; Wong 1998, 2005).With the help of an immigrant-serving agency in each study area,[1] we had intended to conduct in-depth interviews with immigrant entrepreneurs in both metropolitan areas. However, the process proved to be very difficult in the San Francisco Bay Area. While we managed to conduct in-depth interviews with fifteen immigrant entrepreneurs in Vancouver in the summer of 2010, we could only do three in the San Francisco Bay Area. Subsequently, three focus groups with largely prospective immigrant entrepreneurs attending self-employment training sessions were conducted in the Bay Area in the summer of 2011. Such unintended and consequent methodological difference makes it difficult to directly compare and contrast the situations between the two study areas, since the needs and concerns of existing/experienced entrepreneurs can be very different from would-be/novice entrepreneurs. In this chapter, we only meant to explore what on-the-job and prospective immigrant entrepreneurs have to say about their needs for and expectation of financial services as well as their assessment of ethnic versus mainstream banks. It will be interesting to see if their views and opinions are similar or not. We are cautious of any claims of representation and generalization.

The total number of participants in the study is thirty-nine: eighteen immigrant entrepreneurs plus twenty-one prospective immigrant entrepreneurs, including fourteen Asians and one Latin American in Vancouver, and three Africans, five Asians, fifteen Latin Americans, and one European in the San Francisco Bay Area. There are twenty men and nineteen women. The businesses of the on-the-job entrepreneurs include florists, food distribution, graphic design, import/export trade, tile wholesale, wine making, and operating a daycare, a medical clinic, an online bookstore, and a spa. The interviews, varying in length between 40 and 90 minutes, and focus groups, varying in length between 90 and 120 minutes, were conducted in Chinese (Cantonese or Mandarin), English, or Spanish (with translation). They were audio-recorded whenever authorized, then transcribed verbatim or translated if they were not conducted in English.

Participants are diverse in terms of their immigration background and history, their possession of human and financial capital, and their preparation for doing business. They moved to Canada and the United States as investors, skilled workers, or family members. Mirroring the national average, our Canadian participants are mostly skilled or investor migrants. Nine out of fifteen have obtained at least a college degree, mostly completed in their country of origin. A high percentage of our US participants, by contrast, came for family reunification, with a lower percentage being college-educated, which is also somewhat reflective of the overall US immigrant profile. Most of the current entrepreneurs run small local mom-and-pop shops; only a few have transnational operations. Their roads to entrepreneurship also diverge. Some, including a Chinese woman in Vancouver and a Filipina in the Bay Area with twenty-five years' business experiences in the garment industry in the Philippines, have been successful business owners in their respective country of origin. Most others are first-time business owners and have not previously envisioned owning a business as their occupational choice. For quite a few, their businesses have little to do with what their educational or former employment backgrounds. For instance, two former architects now own a tile wholesale and a grain export business respectively; one journalism degree holder is the owner of a printing shop; a former education major owns an import/export trade business; a library-science degree holder now runs a daycare; and the Filipino former garment shop owner wants to go into the arts trade. Such phenomena are more prevalent in Vancouver, reflectiing the widely reported non-recognition or devaluation of home country academic credentials among skilled migrants in Canada (Alboim et al. 2005; Aydemir and Skuterud 2005; Ferrer and Riddell 2008).

## Empirical Findings

### *Financing Issues*

Despite diverse backgrounds and experiences, our immigrant participants face or anticipate some common challenges in owning and operating a business in Canada or the United States. The most frequently mentioned obstacles, as previously reported (Teixeira, Lo, and Truelove 2007; Workforce Planning Hamilton 2012; Chacko and Price, previous chapter), include non-familiarity with the regulatory, taxation, and other business-environment-related matters, lack of financial resources or credit history, and difficulties in developing markets.

On the financing side, not all immigrant-owned businesses require formal bank financing, and they do not need financing all the time. As it is well-documented in the literature (e.g., Light and Bonacich

1988), many immigrants start with their own savings, financial support from their family/friends, or credit from informal lending circles. Unlike what Chacko and Price (this volume) found among Ethiopian immigrants in Washington, DC, none of our interviewees in Vancouver used any micro-financing from community organizations. Nine out of the fifteen Vancouver entrepreneurs started off their businesses by pooling financial resources from family members or friends. Some stated this was sufficient during their startup phase. This observation corresponds to what Chavis, Klapper, and Love (2011) found in their analysis of small firms in over one hundred countries: a universal trend of younger firms relying less on bank financing and more on informal financing. However, we are not sure from our study whether this is the result of information asymmetry or not. What is clear is that many of our participants experienced difficulties getting a bank loan.

Discussions with current small-business owners show that, as expected, immigrant entrepreneurs' financial needs vary by the sector, size, and development stage of their business. Among the prospective entrepreneurs in the Bay Area, lack of startup money is considered a major problem. As a Spanish-speaking American participant said, "Basically all the obstacles are financial and a little bit (is) the language" (FG1 sm1). As is well documented in the literature (e.g., Light and Bonacich 1988), many would start with their own savings, financial support from their family/friends, or credit from informal lending circles. The prospective entrepreneurs are quite cautious of the risks of entrepreneurship, thereby preferring to start small. Many of our prospective entrepreneurs actually indicated that once they start their business, they would stay on their current jobs until the business runs smoothly.

The inability to obtain start-up funding is problematic, especially for recent immigrants who do not possess a credit history, collateral, and the business records that are often required for securing a business loan from banks in Canada or the United States. One established business owner in Vancouver commented, "[It is] difficult [for newcomers] to deal with banks because not everyone has a property [i.e., house] that can be used as collateral for bank loans ... They will have difficulty getting a loan if they do not have any form of collateral" (V12). Another Vancouverite said, "When you go to the bank, it doesn't matter how good your portfolio or your business plan is. They do not want to lend to a new business. They want something tangible or something to guarantee their payment" (V14). Lack of business experiences also prevents business newcomers from obtaining government-backed small business loans, which in the case of the United States "have a 2-year experience requirement, [and are] not for new businesses" (FG2 m2). Lack of a good credit history is definitely a major obstacle to starting a business, especially among those with lesser human capital and who have trouble making ends meet. This is the experience of current entrepreneurs as well as the perception of prospective ones. A prospective Hispanic entrepreneur in the Bay Area summed it up: "Most of us have a dream to start a business, but don't have the capital to do it, because we all work 95 percent and don't have the luxury to save up that much money ... A lot of us have the issue just with bad credit, and with bad credit we can't do anything" (FG3 em3).

A related problem is the mismatch between the needs for funding and its availability at different stages of business development. It is most difficult to obtain funding at the crucial start-up stage as most financial institutions do not want to take risks. But when the business grows and becomes profitable, the entrepreneur will be solicited by other banks. As one Vancouver entrepreneur put it, "When I started my business here, everything need[ed] money: machines, the store, and employee salary. *But at the beginning, the business was running under a deficit and the income is unstable. Banks wouldn't lend money to us*. So we had to borrow money ... *But after five years, they were 'Wow, we're gonna lend you the money right away!'* ... There were many other banks offering us loans now" (V2, emphasis added). This sentiment is shared by some other established entrepreneurs, also in Vancouver (V12, V14).

Some immigrant entrepreneurs also consider current bank loan programs too general and too rigid. They suggest that banks design different terms of references or requirements for different types of business. An entrepreneur who exports Canadian grains to a South American country provided an example he found absurd. He said, "When I approached the bank and asked for a letter of credit, they said they could provide a letter for 50,000 dollars maximum, with the condition that my client in (the South American country) needs to have the money in his bank account. My bank here will check that there are sufficient funds to pay the transaction and then will produce the letter. This is not viable for my business because clients don't pay the money right away. They request 60 to 90 days to pay the invoices" (V15).

*Banking with Different Institutions*

None of our participants, whether or not they were actually running a business at the time of being interviewed, is unbanked, that is, they all have a relationship with formal financial institutions and bank accounts of some kind. For those who have business accounts, they often separate their personal account from their business finance. Some even choose to bank with different financial institutions for different purposes.

Wherever available, our participants prefer non-profit financial institutions such as credit unions because credit unions offer better services than commercial banks although sometimes their rates may be higher. They also think the smaller banks offer better and more personalized services than the big ones. Given the choices, they prefer using the former to the latter. As stated by one Bay Area participant, "Credit unions would work better for us to build credit and start up [the business]; as for banks, if you have business, they work with you" (FG2 f4). However, several big banks such as TD Bank in Canada and Citibank in the United States are also mentioned by our participants as more business-friendly and providing the services they need.

With respect to bank loans, most complain that banks charge too high an interest rate and are not flexible at all. High interest rates add cost to their business operations. Many immigrants are very rate conscious; they are willing to put considerable efforts into seeking the best possible rate. As one Vancouverite noted, "It is really painful and time-consuming to deal with multiple accounts in multiple banks, not mentioning the fees of each account. It is just not worth it. So I think one bank is enough (for personal financing). In terms of the business loans, I wouldn't mind dealing with different banks and see which one provides lower rates" (V3). Another actually opted for a personal loan instead of a commercial loan because of its lower cost. He said, "The best commercial rate I found was HSBC at 2.5% plus prime, but I would still need to put the business up for collateral. That doesn't make sense to me. They also charge a monthly fee for loans and lines of credit even if you don't use it" (V14).

Some current as well as prospective business owners are very specific about what they think banks should offer small businesses. One said a one-year interest-free loan or lower rate for the first year will help incubate a business. Another said, "The banks can give us a bit more of a break for the loans too, up to a certain amount. It has to be easier to get loans ... maybe up to $25,000" (V10). Others want banks to do more for minority communities. As one potential Hispanic entrepreneur in San Francisco Bay Area put it:

> We are the minority. They [the banks] should have considered that the minority has to have the opportunity with the low interest to develop [into] a better community. If you really want to help the small communities to develop and be a good part of the community, they need to lower those interests that [are] killing people. High interest rates are demoralizing for us. (FG3 ef1)

As for government-backed loans, most Canadian participants are aware of them, but feel such loans are too little to make a difference or too cumbersome procedure-wise to bother trying, especially when financial needs are higher. However, sometimes local government policies make a difference. For example, the San Francisco City government offers special finance programs to redevelop formally deprived areas such as "South of Market [Street] Area" now housing companies like Twitter. Some of our participants are aware of this or they are direct beneficiaries of such programs. One Bay Area immigrant entrepreneur said, "San Francisco has a governmental program that helps small business owners by offering them loans, but not here. If the government could loan my business money, that would be great" (BA1). One recipient of such a loan, who at the time also applied for SBA funding, said, "The construction loan requires us to pay for some cost to apply. So I went to the city [San Francisco] and got the funding from the Urban Renewal Unit of the government" (BA2). This indicates that specially designed and spatially specific government programs can and do benefit small business owners.

Negative perceptions of both banking and government institutions get some participants rather frustrated. For example, one Canadian entrepreneur, who was unflattering towards both, blatantly said:

> The reason they offer you the loan is based on the government's perspective – how much you can help the government – not based on how to help you on your business. Thus, for [our] self-employed business, which couldn't hire many people, we already lose at the beginning … *From point A to Point B, no one knows how that's gonna work, but at least you need to try, and that requires money to input to try the things whose result is unknown*. Unfortunately, in Canada, this kind of funding is very slim. So it is very hard for an entrepreneur here in Canada … In my opinion, all financial institutions are

> seeking money from us. Based on our previous experience, all the financial products are for the benefit of the bank, not for us. If they want to lend you the money, there are usually traps they made for you. I don't think they are that nice people ... I think many North American banks provide services based on how much money you have in the account. (V4, emphasis added)

### *Banking on Ethnic Assets*

Immigrants often do not trust financial institutions, although they long to establish trust. Immigrants are more likely to want a personalized relationship with bankers than do the native-born population (Li et al. 2001, 2002). One Bay Area Chinese business owner proclaimed, "Chinese people don't like to press buttons when they consult the bank via phone. We like to directly talk with the real person, not the machine" (BA1). A Vancouver business owner enjoyed the fact that "when you know the manager, you don't need to go down to make a transaction, you can just call. You build up a relationship with the bankers" (V14).

Because of the perceived importance of relationship banking, our participants look at high bank-employee turnover negatively. Complained one Vancouver participant, "My account managers at Bank X have been changing all the time, which makes [it] very hard to build any trustable and sustainable business relationship with them ... A long-term and stable relationship with the financial account manager is also important to show customers the credibility of banks" (V4). In general, our participants with a personal working relationship with the bankers they deal with are more satisfied with the services they receive.

Banks that possess ethnic assets (Dymski and Li 2004) or are familiar with an immigrant group's native language(s), cultural traditions, and country-of-origin business practices often fare better in our immigrant entrepreneurs' assessment of their financial services and products because these banks understand their needs better and offer more personalized service. As noted by one Bay Area business owner, banks "still have some levels of disparities between their service to the Chinese and to other groups ... maybe because of a cultural difference or a language difference" (BA1).

In Canada, the regulatory regime governing the banking sector makes establishing small banks with immigrant ownership rather difficult, if not impossible, and sometimes foreign-owned banks fill the void by vigorously targeting their co-ethnic clientele. Immigrant entrepreneurs take notice of this in their search for the best banking services. One, for instance, noted, "I tried all the Big Five banks in Canada as well as other foreign banks and after comparing them, I chose Bank X [a foreign bank]" (V7). She was satisfied with the services and products provided by this bank. Another participant clearly stated that he decided to use a large foreign bank for his business account because of his familiarity and knowledge of the bank since his childhood in the country where that bank is one of the dominant banks. Although also approached by another bank from the same general region, he did not switch as he was unfamiliar with their practices and did not understand them. To him, good service and support is the most important, not a lower service charge (V12). Relationship banking is sometimes transferable. For example, a Korean business owner took over the business from its former Korean owner who had had a long relationship with a Korean banker in one of the Big Five banks in Canada. So even when he was new in the business, the banker had already known the business well and provided him with a loan to purchase the property. He has since worked with the same bank for eighteen years. Even so, he seriously considered switching to a local Korean credit union. He said, "The Korean one understands Korean people, so I think they are better for Korean people" (V10). All these examples show that if all other things are equal, the ethnic financial sector will have a comparative advantage over mainstream banks, as Gleisner, Hackethal, and Rauch (2010) have illustrated in their study of Turkish immigrants' bank nationality choice in Germany.

In the United States, the abundance of ethnic banks offers more potential for immigrant small business owners. In the Bay Area, the presence of many Chinese American banks makes their co-ethnic immigrants more attuned to banking with them. These Chinese American banks often locate inside Chinese residential or business areas and proactively target the Chinese (Ahn 2010; Li, Lo, and Oberle 2014). The recent economic crisis and its consequent financial landscape may have further benefitted certain Chinese American banks. For example, East-West Bank, an LA-based Chinese American bank, acquired the long-established but failing San Francisco–based United Commercial Bank at the peak of the financial crisis, and became the largest Chinese American bank in the United States. The acquisition further expands their presence and operations in the Bay Area. One Oakland Chinatown business owner had a thirty-six-year relationship with another Chinese-American bank, but made it clear he "now mainly banks with

East-West Bank because of a tight and trustable personal relationship with the manager there." He added, "Whenever I need loans, East-West Bank would lend to me" (BA3).

Other immigrant groups, however, do not have the luxury of having a large presence of "their own" banks (Dymski et al. 2010; Li and Lo 2008). For example, there used to be a single local Korean-American bank, but it was acquired by an LA-based Korean bank. There is also only one commercial branch of an LA-based Latino bank in the Bay Area, but it does not serve retail businesses.

Interestingly, despite ethnic banks' efforts to reach beyond their co-ethnic clientele (Li et al. 2002), many immigrants are not aware of the existence of such ethnic banks. For instance, quite a few prospective Asian or Latino entrepreneurs in the Bay Area have never heard of their most prominent co-ethnic banks. Immigrant groups not served by co-ethnic banks often bank with large mainstream banks, especially those actively hiring immigrant co-ethnics who speak the language and understand the culture. However, the dominance of such banks is also perceived as a problem among these prospective business owners.

> People are not aware of the other banking institutions. We are always going with Wells Fargo and Bank of America because they speak Spanish. Unfortunately, our people lack knowledge and are not aware of the other institutions that offer better opportunities for our community. Out of ten immigrants, five are banking with Wells Fargo and five are banking with Bank of America. Because some are unaware of their opportunities or some are afraid of asking because they are illegal. With respect to requisites asked by Wells Fargo and Bank of America, the most known banks, they need to be more flexible and we need to be better informed. Lots of people prefer to keep their money at home because they are never well informed. (FG3 SM3)

### *One Telling Story*

Perhaps the most telling story among our participants is that of a salon owner who was a successful businesswoman in her hometown in China. Her experiences testify to changing immigrant profiles, immigrant entrepreneurs' adjustment to a new business environment, relationships with financial institutions, and transnational financial flows. This woman came to Canada with her husband on an investor visa, and put $120,000 upfront to obtain her visa. Her Canadian business experiences and financial integration process, however, have been rather difficult. She said:

> In China I was a successful businesswoman; but my English was not good, so I was hoping to learn the language first, then get a job before going into business. My husband, son, and I arrived in Canada with several dozens of thousands Canadian dollars, but soon realized it's too expensive to live here. We didn't mean to earn much here but to stay for our son's education. I tried approaching all kinds of odd jobs, including working in a grocery store, a restaurant, and a beauty salon, but they all need English and/or Cantonese. *Had I got a job with $2k/month I'd settle with it*, but I was unable to land such a job. So I decided to start my business. My husband was a manager in a home decoration business back in China, but now doing manual job, and never getting a stable and good one.
>
> I have rather good business sense and ideas; but my English isn't good enough. Every penny I spent here in Canada was brought to or remitted from China. My family keeps sending money to support our daily expenses. So we largely rely on my sister's money to survive. More successful business people from China are back to China already, leaving us small business people here.
>
> I have this business but it does not make profits. I had a business plan to purchase some new and better equipment to upgrade my business and to attract different clientele instead of just walking-in customers seeking cheap service. *I submitted a loan application five times wanting to borrow $30,000, but was turned down all the times.* They require my business to have been open for at least 3 years and with profit. I told the bank managers I had a good business plan and had assets. I bought a BMW with $60,000 cash I brought with me from China; but I am now stuck, unable to get the $30,000 loan. I tried the bank I am with three times, other banks twice … I felt really helpless; the more I tried, the worse my credit became. (V6, emphasis added)

Her story points out that compared to the United States, Canadian immigration policy favours wealthy or well-to-do immigrants with lower investment thresholds and also immigrants with higher human capital.[2] These immigrants, however, do not necessarily want to operate a business as their first choice of occupation, but many do not have the official language skill to land a well-paid mainstream job.

Therefore, owning/operating a business is one of the few options they have, as demonstrated in the case of chapters 10 and 11. In a sense, consistent with the literature on immigrant entrepreneurship, they are being pushed into entrepreneurship. In many cases, their prior business experience and knowledge in their countries of origin are not directly transferrable to their new home. Some have a rocky relationship with banking institutions despite having already established a working relationship with them. The stringent lending requirements make it difficult, if not impossible, for new or not-yet-successful business owners to obtain the business loans necessary to make their business thrive. Repeated failures in dealing with banks further deteriorate their credit scores, leaving some people like this woman trapped in an asset-rich but cash-flow-poor situation (Ley 2003, 2010). Her financial dependence on her family in China and the recent trend of return migration among Chinese Canadians that she mentioned also indicate a pattern of settlement and integration that is different from that of previous generations of immigrants, something very interesting but beyond the scope of this paper.

## Conclusions

In this chapter, we examined the intermediary roles of financial institutions in the immigrant financial integration process from the perspective of current and prospective immigrant entrepreneurs in Vancouver and San Francisco. This study connects theories about social capital and ethnic assets with empirical evidence to add an ethnic layer to examining the financial needs and expectations of immigrant entrepreneurs, as well as their assessment of the services provided by different types of financial institutions.

Our different approaches (primarily focus group discussions with prospective entrepreneurs in the San Francisco Bay Area and semi-structured interviews with current business owners in Vancouver) and different sample sizes make it difficult to directly compare and contrast the situations between the two study areas. Yet despite the different institutional contexts (banking regulations, immigration policies, and immigrant profiles) between Canada and the United States, something similar emerges from discussions with established immigrant entrepreneurs in Vancouver and with prospective entrepreneurs in San Francisco with respect to immigrants' need for business financing and their expectation of banks. Irrespective of their immigrant status, their class origin, and their current/proposed business activities, they saw a financing need for their business at some point, they expected trust and help from banks, and they preferred dealing with bankers who understand their culture and language. Yet they saw banks, especially large mainstream banks, as being not entirely fair and effective in dealing with immigrants. They opined that financial institutions are not addressing the needs of new immigrant businesses and their integration into the financial mainstream. Bank credit is not readily available, especially at the startup stage. While the lack of credit history and business records explain the difficulty, some entrepreneurs viewed the financial sector's unfamiliarity with and perhaps scepticism towards the business practices of certain immigrant groups as key to the mismatch between immigrant businesses needing financing and bank financing being readily available. It is not surprising then their general perception of banks was rather negative. Perhaps the following quote best summarizes what immigrant entrepreneurs face and need: "It doesn't matter what type of bank you have in your home country or your education, when you come here you are like a newborn. You don't have connections; you don't know anything about the new country. I found that lots of people live on their savings ... There has to be an easier way to get a loan. That really is necessary" (V10).

Generally speaking, our immigrant entrepreneurs found smaller and ethnic-based banks a bit more accommodating. To them, relationship banking is important, suggesting that banks that utilize ethnic assets will more likely gain their trust. In reality, however, not all immigrants are aware of the players in the ethnic financial sector. This suggests more work to do for banks who want to capitalize on their ethnic assets.

Banks are profit-making entities. They do not consider themselves social service agencies with a mandate to help integrate immigrants; and they reject many loan applications from immigrants on the basis that their business proposals are terribly prepared (Teixeira, Lo, and Truelove 2007). However, we believe something can be done by banks and other governmental and non-governmental organizations to ease the agony of immigrant entrepreneurs.

First, there is a clear need for more training, educational, or bridging programs that provide prospective immigrant entrepreneurs with knowledge about doing business in Canada and the United States. They should be exposed to related regulations and procedures, sources of startup funding, business proposal writing, and mentoring possibilities.

A lack of preparation and understanding of Canadian/US business operation is detrimental to the successful establishment of a business. In some cases, misunderstandings may be as fundamental as not realizing that taxes will be owed on business income.

Second, the search for these support resources in the wider community can represent a major challenge for new or prospective immigrant entrepreneurs. Governmental and non-profit organizations should arrange for better information dissemination efforts. Strategies should be implemented to improve communication with potential immigrant entrepreneurs from certain groups for whom a lack of trust in governments and everything about dealing with them remain a major cultural barrier for new immigrants interested in starting a business.

Third, the financial sector and immigrant entrepreneurs can attempt to become partners in the creation and support of viable businesses. Instead of acting as filters preventing/guarding access to capital, financial institutions might rather be mentors or partners for new immigrant businesses. Teixeira, Lo, and Truelove (2007) suggest that financial institutions run pilot projects among those immigrants or immigrant groups whose loan applications would normally have been rejected. This is something that would give banks the opportunity to see how their businesses fare and thus enable them to devise future loan policies – government-backed or not – to nurture new immigrant-owned businesses.

Finally, while immigrants naturally feel comfortable interacting with those sharing common language and culture, they are often reluctant to go beyond their own community and connect with others in the broader community. Initiatives to help create inter-ethnic networks may mean better business and partnership opportunities.

## ACKNOWLEDGMENTS

The authors are grateful to the Canadian Social Sciences and Humanities Research Council (SRG Award #410-2010-2692), the US National Science Foundation (Award #0852424), and the Canadian embassy in the United States (Canadian Studies Research Grant) that enabled the fieldwork conducted for the research that lead to this article; to Tung Chan and Gary Dymski for their contributions and guidance in the research process; and to A New America in the Bay Area and S.U.C.C.E.S.S. in Vancouver for their help in recruiting participants for this study. We thank all interviewees and focus group participants for the time and insights they shared with us; and Brenda Garza and Wan Yu at Arizona State University and Julia Mais at York University for their invaluable research assistance.

## NOTES

1 We do not claim participants in our study are representative of all immigrant entrepreneurs, as they are mostly either then-current or previous entrepreneurship trainers or trainees of the two immigrant-serving organizations we collaborated with. However, they reflect the type of immigrants who are serious about entrepreneurship.

2 See Kobayashi, Li, and Teixeira 2012 for details.

## REFERENCES

Ahn, H. 2010. "Social Capital and Overseas Chinese Economy: A Comparison of Korean and Chinese Ethnobanks in California." *Journal of the Economic Geographical Society of Korea* 12 (4): 642–65.

Alboim, N., R. Finnie, and R. Meng. 2005. "The Discounting of Immigrants' Skills in Canada: Evidence and Policy Recommendations." *IRPP Choices* 11 (2): 1–26. http://archive.irpp.org.

Aydemir, A., and M. Skuterud. 2005. "Explaining the Deteriorating Entry Earnings of Canada's Immigrant Cohorts, 1966–2000." *Canadian Journal of Economics / Revue Canadienne d'Économique* 38 (2): 641–71. http://dx.doi.org/10.1111/j.0008-4085.2005.00297.x.

Barrett, G. 1999. "Overcoming the Obstacle? Access to Bank Finance for African-Caribbean Enterprise." *Journal of Ethnic and Migration Studies* 25 (2): 303–22. http://dx.doi.org/10.1080/1369183X.1999.9976687.

Bates, T. 1989. "The Changing Nature of Minority Business: A Comparative Analysis of Asian Nonminority, and Black-Owned Businesses." *Review of Black Political Economy* 18 (2): 25–42. http://dx.doi.org/10.1007/BF02895231.

Bates, T. 1997. *Race, Self-Employment, and Upward Mobility: An Illusive American Dream*. Baltimore: Johns Hopkins University Press.

Beck, T., A. Demirguc-Kunt, L. Laeven, and R. Levine. 2004. *Finance, Firm Size and Growth*. Working Paper no. 10983. NBER.

Boissevain, J., J. Blaschke, H. Grotenberg, I. Joseph, I. Light, M. Sway, R. Waldinger, and P. Werbner. 1990. "Ethnic Entrepreneurs and Ethnic

Strategies." In *Ethnic Entrepreneurs*, ed. R. Waldinger, H. Aldrich, and R. Ward, 131–56. Newbury Park, CA: Sage.

Bonacich, E., and J. Modell. 1980. *The Economic Basis of Ethnic Solidarity: Small Business in the Japanese American Community*. Berkeley: University of California Press.

Bruder, J., D. Neuberger, and S. Räthke-Döppner. 2011. "Financial Constraints of Ethnic Entrepreneurship: Evidence from Germany." *International Journal of Entrepreneurial Behaviour& Research* 17 (3): 296–313.

Butler, J.S. 1991. *Entrepreneurship and Self-help among Black Americans*. New York: State University of New York Press.

Chavis, L.W., L.F. Klapper, and I. Love. 2011. "The Impact of the Business Environment on Young Firm Financing." *World Bank Economic Review* 25 (3): 486–507. http://dx.doi.org/10.1093/wber/lhr045.

Clement, W. 1977. *Continental Corporate Power Economic Linkages between Canada and the United States*. Toronto: McClelland & Steward Ltd.

Dymski, G., and W. Li. 2004. "Financial Globalization and Cross-Border Co-Movements of Money and Population: Foreign Bank Offices in Los Angeles." *Environment and Planning A* 36 (2): 213–40. http://dx.doi.org/10.1068/a35286.

Dymski, G., W. Li, C. Aldana, and H. Ahn. 2010. "Ethnobanking in the USA: From Antidiscrimination Vehicles to Transnational Entities." *International Journal of Business and Globalisation* 4 (2): 163–91. http://dx.doi.org/10.1504/IJBG.2010.030667.

Edgington, D.W., M.A. Goldberg, and T.A. Hutton. 2006. "Hong Kong Business, Money andMigration in Vancouver, Canada." In *From Urban Enclave to Ethnic Suburb: New Asian Communities in Pacific Rim Countries*, ed. W. Li, 155–83. Honolulu: University of Hawaii Press.

Feagin, J.R., and N. Imani. 1994. "Racial Barriers to African American Entrepreneurship: An Exploratory Study." *Social Problems* 41 (4): 562–84. http://dx.doi.org/10.2307/3096989.

Ferrer, A., and W.C. Riddell. 2008. "Education, Credentials and Immigrant Earnings." *Canadian Journal of Economics / Revue Canadienne d'Économique* 41 (1): 186–216. http://dx.doi.org/10.1111/j.1365-2966.2008.00460.x.

Fiscal Policy Institute. 2012. "Immigrant Small Business Owners: A Significant and Growing Part of the Economy." http://www.fiscalpolicy.org/immigrant-small-business-owners-FPI-20120614.pdf.

Fisher, E., and R. Reuber. 2010. "The State of Entrepreneurship in Canada." Ottawa: Industry Canada. http://www.ic.gc.ca/eic/site/061.nsf/eng/rd02474.html.

Freedman, C. 1998. "The Canadian Banking System." Bank of Canada Technical Report no. 81. http://www.bankofcanada.ca/research/technical-reports.

Froschauer, K. 2001. "East Asian and European Entrepreneur Immigrants in British Columbia, Canada: Post-Migration Conduct and Pre-Migration Context." *Journal of Ethnic and Migration Studies* 27 (2): 225–40. http://dx.doi.org/10.1080/13691830020041589.

Ghosal, V., and P. Loungani. 1996. "Firm Size and the Impact of Profit-Margin Uncertainty on Investment: Do Financing Constraints Play a Role?" Board of Governors of the Federal Reserve System, International Finance Discussion Paper no. 557.

Gleisner, F., A. Hackethal, and C. Rauch. 2010. "Migration and the Retail Banking Industry: An Examination of Immigrants' Bank Nationality Choice in Germany." *European Journal of Finance* 16 (5): 459–80. http://dx.doi.org/10.1080/13518470903314410.

Grant, T. 2011. "Attracting the Entrepreneurial Immigrant." *Globe and Mail*, 31 July. http://www.theglobeandmail.com/news/national/time-to-lead/attracting-the-entrepreneurial-immigrant/article595877/.

Hodge, M., and J.R. Feagin. 1995. "African American Entrepreneurship and Racial Discrimination." In *The Bubbling Cauldron: Race, Ethnicity, and Urban Crisis*, ed. M. Smith and J. Feagin, 99–120. Minneapolis: University of Minnesota Press.

Hou, F., and S. Wang. 2011. *Immigrants in Self-employment*. Catalogue no. 75-001-X. Ottawa: Statistics Canada, Perspectives on Labour and Income.

Hultman, C.W., and L.R. McGee. 1990. "The Japanese Banking Presence in the United States and Its Regional Distribution." *Growth and Change* 21 (4): 69–79. http://dx.doi.org/10.1111/j.1468-2257.1990.tb00534.x.

Kobayashi, A., W. Li, and C. Teixeira. 2012. "Immigrant Geographies: Issues and Debates." In *Immigrant Geographies in North American Cities*, ed. C. Teixeira, W. Li, and A. Kobayashi. Toronto: Oxford University Press.

Kalckreuth, U., and E. Murphy. 2005. *Financial Constraints and Capacity Adjustment in the United Kingdom: Evidence from a Large Panel of Survey Data*. Bank of England, Working Paper no. 260.

Levine, R. 2005. "Finance and Growth: Theory and Evidence." In *Handbook of Economic Growth*, vol. 1, ed. P. Aghion and S. Durlauf, 864–934. Amsterdam: North Holland, Elsevier.

Ley, D. 2003. "Seeking Homoeconomicus: The Canadian State and the Strange Story of the Business Immigration Program." *Annals of the Association of American Geographers* 93 (2): 426–41. http://dx.doi.org/10.1111/1467-8306.9302010.

Ley, D. 2010. *Millionaire Migrants: Trans-Pacific Life Lines*. West Sussex, UK: Wiley-Blackwell.

Li, P.S. 1992. "Ethnic Enterprises in Transition: Chinese Business in Richmond, B.C., 1980–1990." *Canadian Ethnic Studies* 24: 120–38.

Li, P.S. 1997. "Asian Capital and Canadian Business: The Recruitment and Control of Investment Capital and Business Immigrants to Canada." In *Multiculturalism in North American and Europe: Comparative Perspectives on Interethnic Relations and Social Incorporation*, ed. W.W. Isajiw, 363–79. Toronto: Canadian Scholar's Press.

Li, P.S. 2005. "The Rise and Fall of Chinese Immigration to Canada: Newcomers from Hong Kong Special Administrative Region of China and Mainland China, 1980–2000." *International Migration* (Geneva, Switzerland) 43 (3): 9–34. http://dx.doi.org/10.1111/j.1468-2435.2005.00324.x.

Li, W., G. Dymski, Y. Zhou, M. Chee, and C. Aldana. 2002. "Chinese American Banking and Community Development in Los Angeles County." *Annals of the Association of American Geographers* 92 (4): 777–96. http://dx.doi.org/10.1111/1467-8306.00316.

Li, W., and L. Lo. 2008. "Ethnic Financial Sectors: The US and Canada Compared." *Migracijske i etnicke teme* 24 (4): 301–22.

Li, W., L. Lo, and A. Oberle. 2014. "Bank Branch Network and Service to Immigrants." *Canadian Geographer / Géographe Canadien* 58 (1): 48–62.

Li, W., A. Oberle, and G. Dymski. 2009. "Global Banking and Financial Services to Immigrants in Canada and the US." *Journal of International Migration and Integration* 10 (1): 1–29. http://dx.doi.org/10.1007/s12134-008-0089-1.

Li, W., and E. Park. 2006. "Asian Americans in Silicon Valley: High Technology Industry Development and Community Transformation." In *From Urban Enclave to Ethnic Suburb: New Asian Communities in Pacific Rim Countries*, ed. W. Li, 119–33. Honolulu: University of Hawaii Press.

Li, W., Y. Zhou, G. Dymski, and M. Chee. 2001. "Banking on Social Capital in the Era of Globalization – Chinese Ethnobanks in Los Angeles." *Environment and Planning A* 33 (11): 1923–48. http://dx.doi.org/10.1068/a342.

Light, I. 1972. *Ethnic Enterprise in America: Business and Welfare among Chinese, Japanese and Blacks*. Berkeley: University of California Press.

Light, I., and E. Bonacich. 1988. *Immigrant Entrepreneurs: Koreans in Los Angeles, 1965–1982*. Berkeley: University of California Press.

Light, I., and S.J. Gold. 2000. *Ethnic Economies*. New York: Academic Press.

Lo, L. 2009. "Ethnic Economies." In *International Encyclopedia of Human Geography*, vol. 3, ed. R. Kitchin and N. Thrift, 608–14. Oxford: Elsevier. http://dx.doi.org/10.1016/B978-008044910-4.00155-3.

Lo, L., and W. Li. 2012. "Economic Experiences of Immigrants." In *Immigrant Geographies in North American Cities*, ed. C. Teixeira, W. Li, and A. Kobayashi, 112–37. Toronto: Oxford University Press.

Mitchell, K. 2004. *Crossing the Neoliberal Line: Pacific Rim Migration and the Metropolis*. Philadelphia: Temple University Press.

ÓhUallacháin, B. 1994. "Foreign Banking in the American Urban System of Financial Organization." *Economic Geography* 70 (3): 206–28.

Putnam, R. 2007. "E Pluribus Unum: Diversity and Community in the Twenty-First Century. The 2006 Johan Skytte Prize Lecture." *Scandinavian Political Studies* 30 (2): 137–74. http://dx.doi.org/10.1111/j.1467-9477.2007.00176.x.

Rath, J., ed. 2000. *Immigrant Businesses: The Economic, Political and Social Environment*. Basingstoke, Hants: Macmillan. http://dx.doi.org/10.1057/9781403905338.

Saxenian, A.L. 2002. *Local and Global Networks of Immigrant Professionals in Silicon Valley*. San Francisco: Public Policy Institute of California.

Teixeira, C. 2001. "Community Resources and Opportunities in Ethnic Economies: A Case Study of Portuguese and Black Entrepreneurs in Toronto." *Urban Studies* (Edinburgh, Scotland) 38 (11): 2055–78. http://dx.doi.org/10.1080/00420980120080934.

Teixeira, C., L. Lo, and M. Truelove. 2007. "Immigrant Entrepreneurship, Institutional Discrimination, and Implications for Public Policy: A Case Study in Toronto." *Environment and Planning C: Government and Policy* 25 (2): 176–93. http://dx.doi.org/10.1068/c18r.

Tickell, A. 2000. "Global Rhetoric, National Politics: Pursuing Bank Mergers in Canada." *Antipode* 32 (2): 152–75. http://dx.doi.org/10.1111/1467-8330.00126.

Trichur, R. 2007. "Canada's Big Banks Like Monopoly." *Toronto Star*, 14 Aug. 2007. http://www.thestar.com/business/2007/08/14/canadas_big_banks_like_monopoly_paper.html.

Uneke, O.A. 1994. "Inter-group Differences in Self-employment: Blacks and Chinese in Toronto." PhD diss., University of Toronto.

Waldinger, R., H. Aldrich, and R. Ward. 1990. "Opportunities, Group Characteristics, and Strategies." In *Ethnic Entrepreneurs: Immigrant Business in Industrial Societies*, ed. R. Waldinger, H. Aldrich, and R. Ward, 13–48. Newbury Park, CA: Sage Publications.

Wang, Q. 2010. "How Does Geography Matter in Ethnic Labor Market Segmentation Process? A Case Study of Chinese Immigrants in the San Francisco CMSA." *Annals of the Association of American Geographers* 100 (1): 182–201. http://dx.doi.org/10.1080/00045600903379083.

Wong, B. 1998. *Ethnicity and Entrepreneurship: The New Chinese Immigrants in the San Francisco Bay Area*. Boston: Allyn and Bacon.
Wong, B. 2005. *Chinese in Silicon Valley: Globalization, Social Networks, and Ethnic Identity (Pacific Formations)*. Lanham, MD: Rowman & Littlefield Publishers.
Workforce Planning Hamilton. 2012. "Winning Strategies for Immigrant Entrepreneurship in Five Communities – Final Project Report." http://workforceplanninghamilton.ca/publications/226.

# Conclusion

# 13 Immigrant Experiences and Integration Trajectories in North American Cities: An Overview and Commentary on Themes and Concepts

JOHN W. FRAZIER

The two largest nations of North America have much in common, including democratic ideals and their relatively recent, more inclusive immigration policies. These policies have had monumental impacts and greatly contributed to the transformation of the fabrics of metropolitan regions, settlement structures, and the landscapes of both Canada and the United States. Large metropolitan regions, especially established gateways, have been the major ports of entry for global migrants seeking better lives for themselves and their families, and are now multicultural regions with distinct ethnic settlement patterns within their increasingly complex metropolitan spatial structures. Changes in these regions have occurred due to a host of processes and policies, including the preservation of ethnic identities and immigrant preferences, housing affordability, economic opportunity, class differences, segmented assimilation, secondary migrations, racialization, and other complex factors, all part of multiple processes that create and maintain fractured spatial and social patterns that constitute our current metropolitan regions of North America.

The challenges for any coherent volume concerned with immigrant integration in two nations and their respective communities, is the unravelling of some of these complexities and explaining the processes and adjustments that lead to specific types of immigrant trajectories. In this case, housing and economic trajectories that benefit, slow, or counter this integration process for immigrant groups become the focal points. An additional challenge is then to compare and contrast these processes and patterns between the two nations.

This effort is made more difficult when national differences are considered. Some of the important differences between Canada and the

United States are highlighted by the editors (Introduction). Fundamental among them are the differences in their current immigration policies. The family reunification emphasis of US policy versus the Canadian system's allocation of points based on human capital (educational attainment and language proficiency) yield substantially different immigrant origins and profiles, which are further complicated by differences in refugee policies and the unique US immigration policy elements, such as the diversity visa, which may likely influence immigrant integration experiences in both the US housing and the labour markets. Also, despite espoused similar political and social ideals, elements of the two federal nations' policies and attitudes stand in stark contrast, the "multiculturalism" and state-funded integration programs of Canada versus the assimilation through self-sufficiency of the United States.

Other differences between the two nations occur at the state/provincial and local levels, where states and communities of the United States have self-classified by either their public actions or announcements. A significant number of US jurisdictions have chosen either to impose controls to regulate immigrants or to welcome them. This, of course, is tied to the issue of the millions of undocumented immigrants who have entered the United States, another difference between the two countries. No discussion of similar Canadian illegal entries is provided in this text.

In addition to government response, financial institutions, public and private, in Canada and the United States appear to have some important structural differences that yield different opportunities and results. Li and Lo highlight some of these differences in chapter 12, indicating that regulatory patterns, valuation of ethnic assets, and the perceptions of value by local community policies of assistance to small businesses, including immigrant business, are different for the two countries.

Perhaps nowhere do greater differences exist between the two nations than in the Black slave and Mexican legacies of the United States that dramatically shaped the life experiences of original "immigrants," the enslaved Blacks and Spanish-Mexican settlers, and their descendants, in the area that became the Borderlands. The racialization of Hispanics, particularly Mexicans and Puerto Ricans, is well documented. The Mexican legacy has consisted of US policies that periodically embraced cheap Mexican migrant labour and a long, historically permeable border, which enhanced flows of undocumented immigrants, who now total more than eleven million. Hispanics, particularly those of Mexican ancestry, and their settlements, labelled with negative stereotypes, contributed to policies of removal termed repatriation that sent Mexicans and Mexican Americans out of the United States. These stereotypes, frustrations, and anger over undocumented immigrants have contributed to long-term racialization and anti-immigrant feelings towards Hispanics (Reisinger 2011).

The well-documented formal policies and informal discriminatory actions against Africans and their African-American descendants include the impacts of the earliest laws (the Virginia slave code that labelled Blacks as chattel) to present-day racial discrimination that perpetuates persistent inequalities ranging from unequal schools (see Logan, Oakley, and Stowell 2011, for a contemporary analysis) to persistent job discrimination, and from racial steering cases (e.g., National Fair Housing Alliance 2006) to racially based predatory lending (Kaplan and Sommers 2006).

Although these US racial issues are more visible, have a deeper and lasting history, and receive more press coverage than similar issues in Canada, there are nonetheless some strong commonalities, including among immigrant experiences and the historical patterns of the treatment of Blacks in the two nations. A number of US and Canadian authors have drawn attention to them (Frazier, Darden, and Henry 2010), such as the differential experiences of the highly diverse Black immigrant populations in both nations when seeking affordable and accessible housing, quality education, employment, and other resources. A number of examples of these matters as Canadian racial issues will be presented later in the essay. Nonetheless, the belief here is that the magnitude of racial discrimination in the United States remains far more frequent and visible, while racism in both nations remains persistent.

While racism and racial separation are evident in both countries, it is the sheer magnitude of the Black and Hispanic populations, which together at approximately ninety million people exceed the total population of Canada, and their cultural, economic, educational, and geographic histories that make national comparisons extremely difficult. Persistent racialization and continued disparities have spilled into the twenty-first century. This uniquely "American" (US) challenge is exacerbated by the recent US immigration of Blacks and Latinos of highly variable socio-economic classes. The high financial and social capital of recent immigrants has been touted, but the diversity visa and refugee policy brings immigrants of much lower socio-economic status (cab drivers versus management positions), which make understanding their respective human geographies, integration experiences, and trajectories complicated.

Despite these observations and concerns, especially the functional and historical differences between Canada and the United States, it is extremely worthwhile to compare their immigrant integration experiences for many reasons. However, we must consider the importance of our differences as we analyse and generalize the experiences and trajectories of visible minorities and immigrants in the two nations. In short, we can learn much from one another about the common barriers to integration in free societies and the common adjustments made by particular groups in particular place-based contexts that contribute to positive trajectories of immigrant integration. In doing so, we may suggest policy changes to achieve our similar goals and to make immigrant integration a reality.

As we proceed with our understanding of immigrant trajectories and integration into receiving societies, it is important to note that not all immigrants are equal in terms of social capital and that some immigrants enter their receiving country under conditions of racialization that span centuries and persist. The current inequalities among immigrant groups may be more pronounced in the United States due to the difference in historic and contemporary immigration policies. Further, social capital inequalities exist *within* the same ethnic immigrant group in both nations. We know that time in the respective nations can contribute to gains in socio-economic status and class differences among co-ethnics, contributing to within-group inequalities. In addition, Canada and the United States are class-based societies rooted in social capital differences. These, of course, influence the housing and economic trajectories that are so important to immigrant integration. Class differences also influence the nature and location of immigrant settlements. Racialization and ethnic differences add additional layers to the complexity of spatial form and structure that both influence trajectories and reflect difference. As noted above, centuries of exposure to the forces of class differences coupled with racialization in the United States has created somewhat unique visible underclass settlements and experiences that far exceed the experiences of Canadians. Despite this admission, the truths of social capital differences and racism do sometimes escape our Canadian colleagues because they are clearly present in smaller form and perhaps have less visible processes. Thus, national differences may have impacts on scope and complexity but do not exempt Canadian visible minorities and immigrants from discrimination and spatial isolation, processes that influence both disparities and immigrant integration trajectories, including the barriers that block positive experiences and trajectories in housing and socio-economic well-being that are essential to integration.

Immigrant adjustments often vary by culture, class, and location. Adjustments also are influenced by the colour line, a topic well documented in the United States. When exploring the housing experiences of "new African immigrants" in Toronto, Teixeira explained the differential success of Cape Verdeans in interacting with and being accepted by the residents of "Little Portugal," compared to Angolans and Mozambicans. One of his key explanations was that they "in general have lighter skin than" those other two groups, and 65–75 per cent of Cape Verdeans reside in and around "Little Portugal" as a result (Teixeira 2010). Thus, the colour line seems to influence the well-being of Blacks in both nations in particular contexts.

Clearly, immigrant risks are uneven in both nations and are influenced by a number of factors. National and local place contexts, cultural adjustments, social capital, and other forces lead to variable risk and different ethnic immigrant and visible minority experiences. It is important, then, to remind ourselves of the similarities and differences that influence our understandings of immigrant and visible minority experiences and trajectories in housing and economic standing in Canada and the United State and between and within immigrant and minority populations.

This text contains a wide variety of topics and places associated with immigrant integration in the two nations. The above commentary touches upon some of these consistent themes and concepts that help define the structure and purpose of this anthology and hopefully offer some insights into the collective effort of the editors' and authors' viewpoints and analyses. Below we offer a summary of some of the major themes and concepts that we believe define this text. We also offer some commentary as we proceed.

## The Expression of Themes and Concepts

In addition to the discussions of national similarities and differences, several repetitive themes and concepts appear throughout chapters in this volume and illuminate the nature of immigrant and visible minority experiences, adjustments to barriers, and trajectories towards integration in the US and Canadian societies. These are summarized below.

### *Immigrant Racial and Ethnic Inequalities/Disparities: Barriers and Risk*

These are readily apparent in both nations and influence immigrant integration, have impacts on a variety of immigrant groups, and also appear within immigrant groups. These disparities, pervasive across the nation or place-based, involve barriers that put immigrants at risk and detract from their integration through negative housing and economic experiences. In the history of both nations important historical events racialized the Black populations and framed the restrictions on freedoms that would last for centuries. The "Black codes" of the United States are well known, but perhaps less well known are the 1783 discriminatory measures against free Blacks in Nova Scotia that mirror the Black codes in the maintenance of white supremacy by restricting Black freedoms (Darden and Teixeira 2010). Legalized school racial segregation also was maintained in both countries until the 1950s. A number of chapters focus on various aspects of these disparities and can be summarized as follows.

Darden (chapter 2) rightly argues that, in the contemporary world, homeownership rates are an important measure of immigrant integration into the receiving society. His analysis using the concept of differential incorporation found an "inequality gap" that suggested a "European advantage" in both nations, although he also found different immigrant homeownership rates between the two nations (18% higher in Canada). His analysis of external forces on homeownership rates in the United States strongly suggested racial discrimination as a cause for lower homeownership among US minority populations. His case is strengthened by published reports of US racial steering lawsuits (e.g., National Fair Housing Alliance 2006) and recent settlements between the US Department of Justice and several large banks (e.g., Bank of America, Wells Fargo, and SunTrust), which agreed to pay hundreds of millions of dollars to settle charges of racial discrimination in lending (Savage 2011). The US attorney general, with state attorneys general, charged these banks with the racial steering of Blacks and Hispanics into sub-prime mortgages at higher fees than non-Hispanic Whites with the same credentials. Charges also included accusations of reverse redlining, or the targeting of minority neighbourhoods with predatory loans that resulted in high foreclosure rates (NBC News.com 2012 and *New York Times* reprints, 21 December 2011). The banks in question denied guilt, but agreed to pay the fines. There is no doubt that disparities are race- and ethnic-based and lead to risks for minorities, which likely include immigrants.

Haan and Yu (chapter 3) examined household formation and homeownership patterns in the two nations before the recent housing crisis in the United States by comparing the same arrival cohorts for foreign-born Blacks, Asian Indians, Mainland Chinese, and Filipinos, as measured against the native-born non-Hispanic White reference group and using longitudinal data. They reported that similarities exceeded differences in their analysis. They found that headship and homeownership rates varied substantially between "racial" groups. Blacks were more likely to be renter households, which explained lower homeowner rates.

Taken together, these two chapters, coupled with the race-based bank settlements reported above, indicate that race and ethnicity matter in homeownership patterns and that institutional racial bias persists in the US housing and lending markets, erecting barriers for the social and economic integration of minorities into the broader, White-controlled society.

Another measure of ethnic integration is the transferability of employment and educational attainment to suitable employment. Lysenko and Wang (chapter 9) addressed this challenge in the US context and found that underemployment rates vary by human capital (language proficiency, educational attainment) such that the foreign-born with relatively high social capital are more likely to be underemployed than those with less social capital. Simply stated, social capital does not translate into a good employment opportunity for US immigrants. This pattern of underemployment varies among the foreign-born on the basis of marital status, family size, and duration of US residency. The married foreign-born with larger families and longer stays in the United States have lower levels of underemployment. Clearly, class distinctions exist among the foreign-born and those who are underemployed likely earn less and experience a different housing trajectory than those who are not underemployed (not studied in this chapter). This is another important disparity between immigrants and native-born Americans. We will return to the notion of class distinctions in the next section.

On the Canadian side of housing and economic disparities, Pendakur and Pendakur (chapter 8) studied ethnic immigrant earnings disparities over a sixteen-year period, 1991–2006. Among their findings were that Canadian immigrant experiences reflect significant disparities that are getting worse. This is true for visible minorities, native, and foreign-born populations, and for both genders, as earnings disparities expanded over the sixteen-year study period. They noted that native-born Canadians did not experience the same trend. Disturbingly, this analysis concludes that Canada's visible minorities likely will not see an earnings convergence in their lifetimes. These earnings findings, of course, do not bode well for social and economic integration and have negative implications for future access to better housing trajectories.

### *Place-Making and Remaking: Culture, Class Dynamics, and Institutions*

Place-making refers to the construction or reconstruction of an area by an ethnic group that reflects the values and presence of that particular group, whether a wealthy white population or an ethnic group. This occurs at various scales, such as the transformation of the "New South" (Smith and Furuseth 2006) at the regional level, and more often at the micro-scale, where a particular group reflects its ethnic or cultural identity in a material sense (cultural or ethnic landscape). The place-making process is complex because it includes social, economic, political, class (social capital), and other influences (Frazier and Reisinger 2006). Social networks are important concepts related to place-making because they include both the connections and communications between the place-makers and the homeland. They both preserve ethnic identity and include the connections and influences (transnationalism) of the host society. Similarly, economic and local political influences have impacts on the place-making process, even through the reaches of the global economy at the local level. Local interests, financial and governmental, also often use federal policy and funding sources to create and recreate places through the process we term place-remaking.

As mentioned, both entrepreneurs and government policy can contribute to local place-making and place-remaking. Social capital, including that of both upwardly mobile and wealthy immigrants, for example, can influence the appearance and nature of the place being reconstructed. Entrepreneurs respond and plan for a particular market. Class distinctions and employment consequences among Black and Hispanic populations in the United States are well known. In a class-based and race-based society, institutional decisions create different benefits and barriers for different groups. The case of Africville in the next section supports this contention for the Canadian side. There are some similarities with that case in the United States, for example in Austin, Texas, where during the twentieth century a sequence of local and federal institutional decisions were used to camouflage blatant racism in the separation of Blacks from Whites by slum removal and revitalization. These included the 1928 Austin city plan, the 1956 US Highway Act, the 1970s Urban Revitalization Program, and the 1998 Smart Growth Initiative, all of which were used to reaffirm the "derelict nature" of Black East Austin for the purpose of removing Black neighbourhoods, in the name of the beautification and "reinvention" of urban space, to make way for a wealthy white elite (Skop 2010). The actions of class-based and race-based communities are often reflected in place-making and remaking.

Finally, culture and ethnicity also can influence not only the appearance and content of a newly constructed place (and the social places within that place), but also will help determine the nature of the ethnic institutions that accompany that group and help with their adjustment to a new society. Similarly, the receiving society can reshape an existing place and shape its future. Several examples in this text refer to making or remaking the places of immigrants in the United States and Canada.

Wan Yu (chapter 7) draws on the evolution of ethnoburbs in the San Gabriel Valley of California to illustrate the roles of globalization, social capital, the interaction between class and ethnic preference, and changing ethnic trajectories in housing experiences and economic integration in the United States. Despite the created conception of the model minority, recent Chinese Americans have long been an economically diverse population. For example, some remain trapped in ethnic economies, while others are drawn to elite suburbs, suburban Chinatowns, and ethnoburbs. Yu focuses on the changing housing experiences of the Chinese in suburban Los Angeles. Her findings underscore the importance of the global economy and also the impact of the length of stay by immigrants in a receiving society. These were contextualized in the dynamics associated with ethnic social capital and preferences for place-making and remaking in the L.A. suburbs, specifically Monterey Park and Rowland Heights.

The maturation of Chinese home seekers since their initial arrival in Monterey Park and the place changes of that community have led to a significant relocation to places like Rowland Heights. Yu's research subjects identified place changes that pushed them away from their former residences, including increasing prices, higher densities related to new home styles, and the loss of open spaces. Professional-class Chinese who could afford a better and more culturally pleasing environment were pulled to Rowland Heights. This research supports the notion that place matters and class matters because class differences allow certain members of an ethnic group to recreate places based on preferences.

An important theme in this book is that institutions create barriers or provide beneficial supports that influence immigrant integration trajectories in housing and the economy of the receiving nation. Government institutions create policies that not only determine immigrant entry, but also the well-being of immigrants on a class basis. When such institutional proposals are presented, they require scrutiny. For example, one of the most significant middle-class tax breaks in the United States involves the tax deduction of mortgage interest paid on the primary and secondary (non-rental) residences. This has led to billions of dollars of tax breaks for those who are able to own a primary home or a primary and vacation home in the United States. Homeownership, then, in a class-based society can result from institutional actions that create benefits for one group over others (renters). This type of policy also creates differential impacts on ethnic groups and immigrants. For example, foreign-born Blacks have a higher percentage of renter households than foreign-born Whites. Thus, this housing distinction also creates a "racial" and immigrant distinction. Such distinctions can influence place-making and remaking on a local basis.

The recent collapse of the US homeowner market has revealed a number of things about inequalities and policies designed to benefit particular groups by class. The first involved the importance of race and ethnicity in the continued discrimination related to racial steering and predatory lending, mentioned above. The second example involves the proposed legislative action to benefit one immigrant class over another. In October 2011, two US senators, Charles Schumer (D) and Mike Lee (R), presented the Visit-USA Act (5.1746), which would provide renewable "homeowner" visas to foreigners who invested a half-million dollars in a home purchase. The rationale was to assist the depressed US housing market with the injection of foreign investment. The bill won quick support from the US Chamber of Commerce and others with an economic interest in the business, housing, and travel markets. Foreign home sales and immigrant home purchases "totaled $82 billion in the 12-month period ending March 31, 2011, up from $66 billion the previous year" (*LA Times*, 20 October 2011). Opponents challenged the legislative proposal, not on assisting the wealthy but on the basis of US immigration policy and the foreign ownership of real property (*Wall Street Journal*, 20 October 2011). If enacted, it would certainly have guided foreign investment into particular places.

Ghosh (chapter 4) explores a topic that further complicates place-making on the basis of culture and class. This case study, operating at the opposite end of a continuum with the wealthy Chinese of Wan Yu's (chapter 7) ethnoburb, investigated the housing trajectory of one of Canada's poorest and disadvantaged populations, the refugee Tamils of Sri Lanka. They differ substantially from the typical Canadian immigrant because of poor social capital, troubled migration experiences, and settlements that reflect a place-making process different from that of other immigrants. The Tamils' place-making has taken on a time-based structure in two locations, first settling in inner-city Toronto in the 1980s and, in the 1990s, a new location in Toronto's suburbs. This pattern and structure is comparable to that of Liberian refugees of Minneapolis (Scott 2010), but the explanations are different. Ghosh's chapter provides strong evidence that culture matters in place-building. Ghosh argues that the Tamils' culture differs from that of other South Asian groups in important ways, including being shaped by their refugee status and flight experience and by their level of attachment to sociocultural institutions. Cultural differences and class led to their spatial separation from other groups residing in the same metropolitan region and to avoidance of public and private institutions that could have offered them assistance of various kinds. Tamil cultural values include a strong obligation to family, friends, and associates in and from their homeland. Despite the challenges of securing a foothold in Canadian society, they avoided the support of Canadian institutions in favour of reliance on family and friends for assistance, including housing. Despite some conflicts among family members and friends due to delayed or missed rent payments, Ghosh argues, this self-reliance and strong social networks have resulted in a place-making that is characterized by relatively high rates of homeownership in single-family detached houses in modest areas of Toronto's suburbs.

Other studies in the United States exhibit the importance of culture and place-making. Research in New York City illustrated the conflicting, or "contradictory," pressures faced by West Indian immigrants who realize that African Americans suffered a long history of discrimination by Whites, but who also wish to remain separate for reasons of cultural distinction (Foner 2001; Vickerman 2001) and, therefore, oftentimes distance themselves from African Americans. Despite these contradictory feelings, as New York City neighbourhoods were reshaped along racial, ethnic, and class relationships, a unique pattern of triple layering occurred, wherein Black immigrants settled (or were relegated to) areas of Black settlement (Waldinger 1996; Conway and Bigby 1992;

Crowder 1999; Foner 1998; Crowder and Tedrow 2001). West Indians were imbedded within the better sections of racialized American residential space. The immigrant ethnic enclaves were characterized by higher homeownership and households with higher social capital than the neighbouring African American areas. The third layer in this settlement structure was that of clusters of specific West Indian cultures. This pattern was later confirmed by Boswell and Jones (2006). More recently, this pattern was investigated, along with the potential of heterolocal settlements, for West Indians in Broward County, Florida. It confirmed the triple layering process existed in the Ft. Lauderdale area, but also determined that heterolocalism was present in the exclusive white suburbs of Broward County. West Indians, primarily Jamaican immigrants, while preferring residence in scattered suburban locations, maintained their ethnicity through place ties that depended on West Indian connections with various functions and places within a small area termed affectionately "Jam Hill" (Frazier 2011).

In this text, Murdie and Teixeira (chapter 5) focus on "Little Portugal" in Toronto as a place that has undergone change through a gentrification process common in North America. In their investigation of the positive and negative impacts of this place-remaking process, they identify several changes that typically occur in gentrified places, including a loss of ethnic identity associated with place, an increase in housing prices, a changing class structure, and the displacement of those of less means. Specifically, despite the benefits brought by gentrification, Murdie and Teixeira point to disrupted social networks, a loss of immigrant community due to residential and commercial displacements, and the loss of important modestly paying jobs in the declining ethnic economy typical of gentrified places. They employ a case study of Chicago neighbourhoods to argue for the disproportionate impact of gentrification on the foreign-born and native-born minorities of those gentrified neighbourhoods.

Portuguese settlements in Toronto, especially Little Portugal, have experienced changes in recent decades with the suburbanization of some Portuguese and their replacement by new immigrants and refugees. An argument can be made that gentrification has accelerated this process. The perceptions and experiences of Portuguese reported in Murdie and Teixeira's study help to clarify their beliefs about the impacts of gentrification. While some of their participants were pleased with the increasing housing values, some relationship existed between those positive reviews and the sale of their homes at a profit. Others saw this as a troubling scenario for the elderly and others on fixed incomes, and for those who could no longer find affordable housing. In particular, respondents were troubled by the changing housing styles brought about by gentrification, specifically the lack of smaller units at affordable prices. The typically positive perception of the stabilization of a declining neighbourhood also emerged among respondents. Yet, this was matched by concern at the loss of ethnic businesses, their products and services, and the jobs they provided at wages needed by the area's less skilled residents. Perhaps the best summary of the feelings of the Portuguese was a sense of the loss of power in their community, a place-based issue. This involved the loss of control due to the transplanting of gentrifiers' elitist attitudes that contribute to remaking the cultural/ethnic place, while simultaneously pushing out or displacing former community members. This reflects not only changing building facades, but also the remaking of a place with new values and class motivations.

There are certainly similar cases and outcomes to be found in the United States. Gentrification has occurred in many urban centres and, in large cities like New York, in multiple locations across a metropolitan region. These include Lower and Upper Manhattan, Chinatown, Harlem's 125th Street, and Long Island City, to name a few, and have captured the attention of scholars and activists alike. A recent documentary film, produced and directed by activist filmmaker Rachelle Gardner, *Harlem's Mart 125: The American Dream*, documents the history of the neighbourhood's experiences with urban renewal and Black displacement, the typical clash between the goals of economic development, and the forced movement of Black businesses and people. Another New York City example involves the forced relocations of twenty-three Bronx Terminal Market small ethnic businesses owned by Latino and Black entrepreneurs. Termed "eyesores" by Mayor Bloomberg and City planning officials, these businesses were moved under the threat of eviction notices in favour of a $300 million mall of one million square feet that was to enhance the quality of the neighbourhood adjacent to the then newly proposed Yankee Stadium in the Bronx. The outcomes were typical: the ethnic businesses that had long-served the product and service needs of ethnic families were eliminated and offered limited relocation assistance. A study of those business entrepreneurs found the City's relocation efforts to be "inadequate, misinformed, and misleading" (Ofori et al. 2010). Of the original twenty-three, only eight relocated within the Bronx.

The place-making of immigrants and the remaking of their places influence their housing and economic trajectories into the host society and are influenced by the class-based policies of national leadership, as well as by local economic development considerations.

### *Immigrant Experiences and Spatial, Commercial, and Institutional Structures*

Locations and geographic spaces express attraction, attachment, attitude towards community and the level of immigrant integration. Institutional structures vary at national levels, especially laws and regulations that determine immigrant entry and the nature of the institutional benefits available to immigrants.

Within national structures, volumes have been written on the significance of urban-metropolitan structures, such as gateway cities, and the reflexive influences between their characteristics and those of the immigrant populations residing there. Many of these are large metro areas, but in the United States of the late twentieth century also include emerging gateways. Also, similarities and differences within nations and between societal groups are often expressed on the basis of regional or local geography. These take the form of comparative area analyses, neighbourhood differences, spatial forms and spatial processes, and landscape types, where barriers block accessibility to resources, housing affordability varies, and settlement patterns and expressions of ethnicity influence life experiences and level of integration. Mensah and Firang (2010) have noted that Black neighbourhood concentrations in Montreal and the "longstanding overlap between race and class … but also because racism itself is often spatialized through residential segregation and other forms of spatial injustices in the location and allocation of urban resources" deplete the tax base needed to support education and other services (p. 45) and lead to neighbourhood deterioration, a process documented for US cities (Frazier, Anderson, and Hinojosa 2011).

There are historical and contemporary Canadian examples of local and regional experiences of immigrants and native-born Black descendants related to institutional and social systems. For example, Africville, an enclave in Halifax, was established in 1842 as a Black refugee settlement, similar to Black enclave settlements in the United States. Just as Blacks were placed in unwanted environments, or their locations became places for dumping or generating wastes, in the United States,

Africville evolved into a place with an infectious disease hospital, a waste dump, and a slaughterhouse andbecame a dirty, dangerous environment for its Black inhabitants, who had no sewers until the 1960s. Once perceived as a slum, Africville was levelled for a beautification project (Mensah and Firang 2010).

These concerns for the impacts of spatial, commercial, and institutional structures on Black immigrant experiences appear in a number of chapters. For example, Kataure and Walton-Roberts (chapter 6) address the different perceptions of ethnic enclave functions and their outcomes for immigrants in Canada and the United States. They maintain that ethnic clustering is viewed differently in the literature of the two nations, with Canadians having a positive view of ethnic enclaves as an immigrant choice and preference, and the US literature reflecting pessimism because it focuses on the consequences of a history of racial geography that produced negative outcomes. They see ethnic enclaves as leading to positive outcomes for visible minorities in Canada by providing opportunities for intercultural interactions and higher homeownership rates that will avoid underclass stigmatization. In this sense, geographic structure and geographic space can be seen as a tool for establishing equality and a positive trajectory in housing and economy in the host nation. The Kataure and Walton-Roberts case study, based in a Canadian suburb and focused on a South Asian population, support their contention that immigrants residing in a Canadian enclave hold positive images of the enclave. South Asians in this enclave, for example, achieved a relatively high rate of homeownership due to its affordability. Despite "some emerging concerns" that included racial stereotyping, lack of intercultural contact, and discrimination, there was a general positive feeling about life experiences and progress in this area of ethnic concentration.

It is useful to compare the experiences of South Asians in this Canadian example with those of Asian Indians in the United States, where experiences appear to vary by size of the Asian Indian settlement and its location. For example, Skop (2012) has argued that the small number (totally only a few thousand in 2000 and an increase in 2010, which although impressive, pales by comparison to that of other cities) of Asian Indians settling in locations like Phoenix, Arizona, avoid creating ethnic enclaves because they wish to remain invisible for a variety of reasons (this view is supported by Asian Indian scholars). In these cases, immigrant Asian Indian populations prefer scattered patterns of heterolocal settlements that support their invisibility and permit the retention of their ethnic identity. On the other hand, Asian Indian settlement structures in the borough of Queens, New York City, one of the nation's most diverse, yet ethnically separated counties, have enjoyed a forty-year settlement period and a population that has rapidly increased in recent decades. The total South Asian population of Queens, of which Asian Indians are the largest group, has been estimated at 186,000 persons. Research findings in Queens reveal a complex settlement structure for Asian Indians compared to the suburbs of Phoenix and Toronto (Brampton). In Queens, which has long been viewed as containing attractive New York City suburbs, Asian Indians have a small clustered settlement in Flushing, a larger presence in Richmond Hill, and a large and rapidly increasing settlement in eastern Bellerose and Queens Village neighbourhoods. In the latter, neighbourhoods are fast approaching ethic enclave status with 30–40 per cent shares of some neighbourhoods due to their relative affordability and suburban environments. There is a class-based and intra-ethnic character to these varied locations in Queens, but the South Asian groups residing there are far from invisible, economically, politically, religiously, commercially, or residentially (Frazier et al. 2011). Asian Indians, specifically, express their presence materially on the landscapes of Queens and reside in class-based neighbourhoods. Some of these enclaves have experienced overt ethnic tensions in the past few years, specifically with the host White population clashing with Asian Indian immigrants. In these cases, although the settlement choices have evolved from ethnic enclaves and sociocommercial landscapes, integration seems a distant ideal. Clearly, more comparative research is needed on the function and success of enclaves outside of the context of racialized geography imposed on Latinos and African Americans.

Another example in this text of the importance of spatial form and function is the study of landscapes by Oberle (chapter 10). His study of three Latino Midwestern landscapes in Chicago and Iowa explores their formation, size, and functions in their different historical contexts. In each case, he argues that these landscapes influence the immigrant experience through the ethnic economies that provide places for social interaction that enhance the community mainstream, and improve the trajectories towards integration with the receiving society. Oberle examined the "Capitol East" section of Des Moines, where nearly 12 per cent of the population is Latino, but the area has no long-term Latino enclave. As a result, the Capitol East commercial landscape is relatively small and compact. The increase in the Mexican population there and redevelopment efforts led to the creation of three Latino business clusters that, due to the small Latino market, serve multiple ethnicities.

By contrast, Chicago's Latino commercial landscapes are based, in part, on the city's long Hispanic legacy and growing Hispanic population, especially Mexicans. Pilsen, the core of Chicago's Latino and particularly native-born and foreign-born Mexican populations, contains a very large Latino commercial landscape. Thus, it differs from smaller, more recent Latino commercial landscapes in size, content, and function. It also differs from smaller commercial landscapes by an emphasis on ethnic cues, specifically on Mexican place name symbols, for example. It also has specialized products and services that draw Hispanics from a large trade area in the Chicago metropolitan region. The third Latino business landscape discussed by Oberle is Carpentersville, a small suburb on the commuting fringe of the metropolitan area. This historically working-class suburb is small and only recently experienced Hispanic population growth due to available affordable housing. Its small commercial landscape appears in three small clusters along two major thoroughfares.

Oberle argues that each of these three Midwestern Hispanic landscapes represents a composition that provides distinctive business and social functions based on the Hispanic population characteristics of its market. Capitol East serves a small but expanding Hispanic neighbourhood and other ethnics in the area, which explains its lack of a Mexican landscape expression. The Pilsen area of Chicago, by contrast, has a long Latino presence and large established Hispanic neighbourhood. It has numerous businesses that serve the wider metropolitan region and its multi-ethnic population. It also serves the local Latino market, especially those of Mexican and Mexican-American descent. Thus, its Mexican affinity shapes the commercial landscape. Carpentersville, with a small but growing Hispanic population over the past thirty years and a fringe location in the Chicago region, attracts a geographically scattered population from nearby suburbs. Its ethnic business types and numbers reflect this small, dispersed Hispanic market. Thus, Oberle demonstrates that an immigrant Latino commercial landscape morphology and content is indicative of the character, history, and size of the Latino population and that spatial structure is determined by function.

Chacko and Price (chapter 11) studied two immigrant groups in the more recent gateway of Washington, DC. There, Bolivians and Ethiopians constitute relatively equal populations, characterized as having relatively high social capital and high rates of entrepreneurship, distinct by their heterolocal settlement structures and "ethnic socio-commercescapes" that perform multiple functions. They create business opportunities that provide ethnic goods and services, ethnic employment, and, through social networks and place-based functions, the social interactions between co-ethnics who reside in scattered communities that preserve group affinity and ethnic identity. Chacko and Price also note the role of barriers, discussing the lack of transferability of immigrant credentials that create class distinctions and resulting "blocked mobility." The latter provides a motivation for self-employment in these two groups. These barriers, then, encourage the development of alternative pathways to improved socio-economic status that enhances the trajectory of immigrant integration. Lending behaviour distinctions, discussed previously, influence the integration of immigrants in Canada and the United States. In the case of Ethiopians and Bolivians, the availability of particular ethnic institutions that enhance small-business formation and functions, through loans and business-plan advice, varies. Bolivians lack access to the financial support of formal institutions and must rely on ethnic networks for loans from the pooling of co-ethnic resources, or investments by co-ethnic potential partners. Ethiopians, by contrast, in addition to access to ethnic social networks that provide capital with repayment, also rely on a variety of ethnic economic options that include loans, advice, and other services from ethnic institutions. These include the Ethiopian Business Association and the Enterprise Development Group, which is associated with the Ethiopian Community Development Council. All these activities and functions, whether providing alternative pathways or direct economic assistance through social and economic networks, foster immigrant integration.

Finally, Li and Lo (chapter 12) discuss the importance of the financing of small ethnic businesses within the context of institutional influences, positive and negative, for immigrant trajectories in integration within the receiving nations. After studying the needs, expectations, accessibility, and satisfaction of immigrant entrepreneurs, they argue that cultural conditioning relates to cultural differences, such as risk-taking behaviour, and to barriers that block access to the financial resources needed by ethnic small businesses. The structural barriers imposed by the receiving society and its institutions "push" some immigrants to entrepreneurship and towards the social networks and ethnic resources that contribute to financial support essential to integration. Despite the advantages, many immigrants simply fail to work through the institutional controls of the lenders. The knowledge available from ethnic leaders (ethnic assets) is unevenly used by immigrant entrepreneurs and leads to favouritism. Specifically, limitations of culture and language and the different institutional structures of banks in Canada and the United States create issues of unused assets or how assets are weighed in loan decisions. Lack of immigrant knowledge and credit history, coupled with lack of collateral, creates barriers to the financing of immigrant entrepreneurs. Li and Lo suggest that Canadian immigrant entrepreneurs perceive that government involvement in the financial system of Canada has resulted in unnecessary bureaucracy and loans too small to be helpful to small-business entrepreneurs. By contrast, these researchers noted, in the United States local government participation (San Francisco) seemed to be viewed in a positive light, as helpful to ethnic businesses. There also was satisfaction expressed by ethnic entrepreneurs with the presence and utility of ethnic banks. Again, differences related to ethnic banks were noted. Many more ethnic banks exist in the United States than in Canada due to the structural differences of their respective financial systems. The US ethnic banks also are located in the same vicinity as ethnic businesses and residential communities. The positive experiences of ethnic entrepreneurs with particular banking types and their functions, such as ethnic assets and good locations, provide the basis of positive trajectories into the host nation. Similarly, the unmet needs of immigrant entrepreneurs and their negative experiences, whether due to structural institutional barriers or the individual challenges created by language and culture, interfere with immigrant integration into a receiving society.

Themes and concepts here, helpful in summarizing the trends of immigrant integration into two nations, share important similarities, but also have substantial differences. General themes and concepts bring order to any investigation and, in the case of this book, provide a structure for interpreting theories and research findings.

## REFERENCES

Boswell, T.D., and T.A. Jones. 2006. "The Distribution of SES of West Indians Living in the United States." In *Race, Ethnicity, and Place in a Changing America*, vol. 1, ed. J.W. Frazier and E.L. Tettey-Fio, 155–82. Albany: SUNY Press.

Conway, D., and U. Bigby. 1992. "Where Caribbean Peoples Live in New York City." In *Caribbean Life in New York City: Socio-cultural Dimensions*, ed.

C.R. Sutton and E.N. Chaney, 70–8. Staten Island, NY: Center for Migration Studies.

Crowder, K.D. 1999. "Residential Segregation of West Indians in the New York/ New Jersey Metropolitan Area: The Roles of Race and Ethnicity." *International Migration Review* 33 (1): 79–113. http://dx.doi.org/10.2307/2547323.

Crowder, K.D., and L.M. Tedrow. 2001. "West Indians and the Residential Landscape of New York." In *Island in the City: West Indian Migration to New York*, ed. N. Foner, 81–114. Berkeley: University of California Press.

Darden, J.T., and C. Teixeira. 2010. "The African Diaspora in Canada." In *The African Diaspora in the United States and Canada at the Dawn of the 21st Century*, ed. J.W. Frazier, J.T. Darden, and N.F. Henry, 13–34. Albany: SUNY Press.

Foner, N. 1998. "Towards a Comparative Perspective on Caribbean Migration." In *Caribbean Migration: Globalised Identities*, ed. M. Chamberlain, 47–60. New York: Routledge.

Foner, N. 2001. "West Indian Migration to New York: An Overview." In *Islands in the City: West Indian Migration to New York*, ed. N. Foner, 1–22. Berkeley: University of California Press.

Frazier, J.W. 2011. "West Indian Patterns in Broward County, Florida." In *Race, Ethnicity, and Place in a Changing America*, vol. 2, ed. J.W. Frazier, E.L. Tettey-Fio, and N.F. Henry, 149–71. Albany: SUNY Press.

Frazier, J.W., R. Anderson, J. Hinojosa. 2011. "People on the Move in the U.S.: Black Movements and Settlement Structures." In *Race, Ethnicity and Place in a Changing America*, vol.2, ed. J.W. Frazier, E.L. Tettey-Fio, and N.F. Henry, 89–105. Albany: SUNY Press.

Frazier, J.W., J.T. Darden, N.F. Henry. 2010. *The African Diaspora in the United States and Canada at the Dawn of the 21st Century*. Albany: SUNY Press.

Frazier, J.W., B. McGovern, and N.F. Henry. 2011. "Asian 'Indian-ness' and Place Visibility: Landscapes in Queens, New York." In *Race, Ethnicity, and Place in a Changing America*, vol. 2, ed. J.W. Frazier, E.L. Tettey-Fio, and N.F. Henry, 339–62. Albany: SUNY Press.

Frazier, J.W., and M.E. Reisinger. 2006. "The New South in Perspective: Observations and Commentary." In *Latinos in the New South: Transformations of Place*, ed. H.A. Smith and O.J. Furuseth, 257–84. Burlington, VT, London: Ashgate.

Kaplan, D.H., and G. Sommers. 2006. "Lending and Race in Two Cities." In *Race, Ethnicity, and Place in a Changing America*, vol. 1, ed. J.W. Frazier and E.L. Tettey-Fio, 97–110. Albany: SUNY Press.

Logan, J.R., D. Oakley, and J. Stowell. 2011. "Public Policy Impacts on School Desegregation, 1970–2000." In *Race, Ethnicity, and Place in a Changing*

*America*, vol. 2, ed. J.W. Frazier, E.L. Tettey-Fio, and N.F. Henry, 33–44. Albany: SUNY Press.

Mensah, J., and D. Firang. 2010. "The African Diaspora in Montreal and Halifax. A Comparative Overview of the Entangled Burdens of Race, Class, and Space." In *The African Diaspora in the United States and Canada at the Dawn of the 21st Century*, ed. J.W. Frazier, J.T. Darden, and N.F. Henry, 35–48. Albany: SUNY Press.

National Fair Housing Alliance. 2006. "Housing Segregation Background Report: Brooklyn, New York." Washington, DC: NFHA. 10 October.

NBC News.com. 2012. "Wells Fargo Pays $175 Million to Settle Race Discrimination Probe." 30 May.

Ofori, E., et al. 2010. "Ethnic Small Business Relocations: A Case Study in the Bronx, New York, 2007." In *The African Diaspora in the United States and Canada at the Dawn of the 21st Century*, ed. J.W. Frazier, J.T. Darden, and N.F. Henry, 307–28. Albany: SUNY Press.

Reisinger, M.E. 2011. "Hispanics/Latinos in the United States." In *Race, Ethnicity, and Place in a Changing America*, vol. 2, ed. J.W. Frazier, E.L. Tettey-Fio, and N.F. Henry, 175–96. Albany: SUNY Press.

Savage, C. 2011. "Countrywide Will Settle a Bias Suit." *New York Times Reprint*, 21 December.

Scott, E.P. 2010. "Liberians and African Americans: Settlements and Ethnic Separation in the Minneapolis–St. Paul Metropolitan Area." In *The African Diaspora in the United States and Canada at the Dawn of the 21st Century*, ed. J.W. Frazier, J.T. Darden, and N.F. Henry, 271–86.

Skop, E. 2010. "Austin: A City Divided." In *The African Diaspora in the United States and Canada at the Dawn of the 21st Century*, ed. J.W. Frazier, J.T. Darden, and N.F. Henry, 109–22.

Skop, E. 2012. *The Immigration and Settlement of Asian Indians in Phoenix, Arizona 1965–2011*. Lewiston: Edwin Mellon Press.

Smith, H.A., and O.J. Furuseth, eds. 2006. *Latinos in the New South: Transformations of Place*. Burlington, London: Ashgate.

Teixeira, C. 2010. "Housing Experiences of New African Immigrants and Refugees in Toronto." In *The African Diaspora in the United States and Canada at the Dawn of the 21st Century*, ed. J.W. Frazier, J.T. Darden, and N.F. Henry, 61–80. Albany: SUNY Press.

Vickerman, M. 2001. "Jamaicans: Balancing Race and Ethnicity." In *New Immigrants in New York*, ed. N. Foner, 201–28. New York: Columbia University Press.

Waldinger, R. 1996. *Still the Promised City? African Amereicans and New Immigrants in Postindustrial New York*. Cambridge, MA: Harvard University Press.

# Contributors

**Elizabeth Chacko** is associate professor of geography and international affairs at the George Washington University, where she is currently the chair of the Department of Geography. She has conducted research in the United States, India, Ethiopia, and Singapore, focusing on the effects of immigration on cities, and issues related to immigrant identity. As a Fulbright Scholar Elizabeth spent six months in Singapore studying the integration of the Asian Indian immigrants in that country.

**Joe T. Darden** is professor of geography at Michigan State University. His research interests are urban social geography, residential segregation, immigration, and socio-economic neighbourhood inequality in multiracial societies.

**John W. Frazier** is a SUNY Distinguished Professor, Binghamton Campus, and has produced six books and authored numerous articles, including many dealing with the applied aspects of geography and racial/ethnic studies. He has served as a consultant to the federal HUD office, both the Planning and Fair Housing and Equal Opportunity divisions. Awards received include the James R. Anderson Medal of Applied Geography (1996), the highest honour bestowed by the Association of American Geographers for applied geography, the Binghamton University Equal Opportunity Outstanding Faculty Service Award (2006), and the AAG's Career Diversity Award (2009), which recognizes career contributions in research, teaching, and service related to diversity.

**Sutama Ghosh** is an associate professor in the Department of Geography at Ryerson University, in Toronto. Her research interests include migration and settlement geographies and transnationalism, focusing specifically on the challenges and triumphs of South Asians in Canada.

**Michael Haan** is an associate professor and Canada Research Chair in Population and Social Policy at the University of New Brunswick. His research interests intersect the areas of demography, immigrant settlement, labour market integration, and residential segregation.

**Virpal Kataure** graduated with a master of arts in geography from Wilfrid Laurier University, Waterloo, Canada, and a bachelor's degree in urban and regional planning from Ryerson University, Toronto. Her research interests include ethnic enclaves, suburban development and housing, second-generation South Asians, and urban settlement patterns. She currently practises as a planner in the Greater Toronto Area. Her latest publication, co-authored with Dr Margaret Walton-Roberts, is entitled "The Housing Preferences and Location Choices of Second-Generation South Asians Living in Ethnic Enclaves in South Asian Diaspora."

**Audrey Kobayashi** is professor and Queen's Research Chair at Queen's University, Kingston. She has researched and published widely in the area of immigration, as well as on various aspects of social justice, including antiracism. She is past president of the Association of American Geographers.

**Wei Li** is professor of Asian Pacific American studies and geography at Arizona State University. Author or co-editor of four scholarly books and two journal theme issues, with ninety other academic publications, she is the recipient of the 2009 Book Award in Social Sciences by the Association for Asian American Studies, the Distinguished Ethnic Geography CAREER Award by the Ethnic Geography Specialty Group of the Association of American Geographers (AAG), and the 2014 AAG Ronald F. Abler Distinguished Service Honors.

**Lucia Lo** is professor of geography at York University, Toronto. Trained as an economic geographer in the spatial science tradition, Lucia has brought innovative perspectives to examining immigrant integration and settlement issues. Her current research focuses on the role of immigrants in economic development and urban transformation. She is author of numerous academic publications and her co-authored book, *Social Infrastructure and Vulnerability in the Suburbs,* is in press.

**Tetiana Lysenko** is a doctoral student at the Department of Geography and Earth Sciences, University of North Carolina at Charlotte. Her current research projects examine issues related to women and immigrant and ethnic minorities in STEM disciplines and industries.

**John Miron** is professor of city studies, geography, and planning at the University of Toronto. He has published extensively in the area of housing demand, including two books with McGill-Queen's University Press. He is currently working on a book that explores how thinking about the city as a set of markets helps us to understand the economic organization of cities.

**Robert A. Murdie** is a professor emeritus in the Department of Geography, York University, Toronto. Professor Murdie has undertaken several studies concerning immigrant experiences in Toronto's housing market and has collaborated with other researchers investigating the housing situation and needs of newcomers in Montreal, Toronto, and Vancouver. Currently, he is co-investigator of a multi-year study on neighbourhood inequality, diversity, and change in six Canadian metropolitan areas.

**Alex Oberle** is an associate professor in the Department of Geography at the University of Northern Iowa. His areas of interest include Latino urban settlement in the United States, and his research has been published in the *Geographical Review* as well as in edited volumes from presses such as the University of Texas Press and the Brookings Institution Press.

**Krishna Pendakur**, Professor of Economics at Simon Fraser University, has spent the last twenty years studying statistical and econometric issues relating to the measurement of economic discrimination, inequality, and poverty. His paper "The Colour of Money: Earnings Differentials Among Ethnic Groups in Canada" (with Ravi Pendakur) published in the *Canadian Journal of Economics* in 1998 established that visible minorities face labour market disparity in Canada. He subsequently published widely on this issue and others in a variety of outlets including the *American Economic Review, The Economic Journal,* and the *Journal of Econometrics.*

**Ravi Pendakur** holds a PhD in sociology from Carleton University, Ottawa, an MA in environmental studies from York University, Toronto, and a BA in sociology and geography from the University of British Columbia. Before joining the Graduate School of Public and International Affairs at the University of Ottawa, he spent eighteen years as a researcher in a number of federal government departments, including Multiculturalism and Citizenship, Canadian Heritage, and Human Resources and Social Development. His research focuses primarily on diversity, with the goal of assessing the socio-economic characteristics of language and immigrant and ethnic groups in Canada and other settler societies.

**Marie Price** is a professor of geography and international affairs at George Washington University. A Latin American and migration specialist, her studies have explored human migration's impact on development and social change. She is a non-resident fellow of the Migration Policy Institute, a non-partisan think tank that focuses on immigration, and the vice-president of the American Geographical Society. Her current research is on the spatial dynamics of immigrant inclusion and exclusion.

**Carlos Teixeira** is a professor in the Department of Geography at the University of British Columbia–Okanagan. His current research focuses on the changing social geography of Canadian cities and the housing experiences of immigrants in Canadian cities. In 2005 he received the Order of "Infante Dom Henrique" from the Portuguese government – one of the highest awards a Portuguese citizen residing overseas can receive for work in service of the Portuguese diaspora.

**Margaret Walton-Roberts** is an associate professor and associate director of the International Migration Research Centre at Wilfrid Laurier University and currently associate dean of the School of International Policy and Governance at the Balsillie School of International Affairs, Waterloo, Canada. She is a human geographer with research interests in gender and migration, transnational networks, and immigrant settlement in Canada. Current research focuses on gender, care, and international migration in the context of India. Her forthcoming co-edited book *The Human Right to Citizenship: A Slippery Concept* will be published with the University of Pennsylvania Press.

**Qingfang Wang** is an associate professor of geography at the University of North Carolina at Charlotte. She is interested in place, as both worksite and residential location, as it interacts with race, ethnicity, immigration status, and gender in shaping individual labour market experiences and other aspects of socio-economic well-being. Funded by the National Science Foundation, US Department of Housing and Urban Development, the Kauffman Foundation, and other institutions, she has published mainly on immigrant labour market segmentation and ethnic entrepreneurship.

**Wan Yu** is a PhD candidate in the School of Geographical Sciences and Urban Planning of Arizona State University. She received her MA degree from Miami University of Ohio and her BS degree from Peking University, Beijing.

**Zhou Yu** is an Associate Professor of Community Studies at the University of Utah. His recent work is on the intersection between housing and immigration, with a particular focus on racial/ethnic differences in homeownership attainment and household formation. His publications have appeared in journals such as *Habitat International, International Migration Review, Real Estate Economics*, and *Urban Studies*. His research has been supported by institutions such as the US Department of Housing and Urban Development and the Russell Sage Foundation. He received his PhD in Planning from the University of Southern California.

# Index


Adams Morgan, District of Columbia, 314
affordable housing, 6
Africa, 25, 44
African American, 14, 60–1, 150, 158, 357, 365–6; blocked mobility thesis, 304. *See also* Black
African immigrants, 129, 130
African immigrant homeowners, 52
agency, 31, 110, 115, 207, 208–9, 212, 220–2, 306, 323, 334
Alexandria, Virginia, 314–15, 320
allocation, 156, 209, 216, 356, 368
American Community Survey, 77
Angola, 129–3, 133, 141
Arab and West Asian, 13, 247, 251
Arlington County, Virginia, 311, 318–19, 322
Asia, 25, 44
Asian, 11, 14–15, 28, 44, 54, 59–62, 64, 78, 99, 159, 194, 230, 244, 246, 256, 263, 268, 270, 272, 319, 333–4, 342
Asian Indian, 28, 64, 71–2, 75–6, 81–3, 86, 91–2, 361, 369–70
assimilation, 151
Atlanta, 24
Australia, 44

Baltimore, 322
banks. *See* mainstream banks
Black, 13–15, 45, 61–2, 64, 71–2, 75–6, 78, 85–6, 88, 91, 99, 121, 126, 147–8, 154–5, 158–9, 227–8, 244, 246–8, 250, 250–1, 256, 266, 270–2, 308, 356–7, 359, 367–9
Bolivian–American Chamber of Commerce, 317
Bolivian immigrants, 308–10
Boston, 50
brain waste, 13, 261
Brampton, Ontario, 33–4, 159
Brazil, 129, 141
business landscape, 15–16, 281–2, 293, 296
business plan, 317, 336, 343, 372

Calgary CMA, 50
Canada, 3, 5, 23, 28, 32, 43–4
Canadian Census Metropolitan Areas, 44
Canadian Human Rights Commission, 64
Cape Verde Islands, 129–3, 133
Caribbean and Bermuda, 44, 51

Census Metropolitan Areas (CMAs), 50–1, 113, 228, 241, 251, 254, 328
Central America, 44
Chicago, xv, 15, 50, 55–6, 121, 126–7, 147–8, 151, 281–4, 286, 288, 291–7, 299, 322, 366, 370–1
Chinese, 8–9, 28, 33–5, 62, 64, 71–2, 75–6, 81–3, 85–6, 88, 91–2, 99, 117, 133, 147, 149, 153, 155, 158–9, 176–7, 179–201, 215, 227, 244, 247–51, 255, 274, 276, 305, 332–5, 340–1, 344, 361, 363, 365
Chinese American banks, 341
Chinese ethnoburbs, 176
co-ethnic, 9, 11, 34, 155, 162, 164, 178–9, 264, 284, 303–6, 308, 310–11, 315–18, 321, 323, 340–2, 358, 372
cohorts, 72
collateral, 316–17, 336, 338, 373
coping, 6, 26, 29, 134, 178, 214, 216, 218, 221
core need, 24
cost, 3, 25, 27, 29–32, 62, 70, 72, 77, 90, 103, 108–12, 114–15, 129, 133–4, 136, 138, 141, 198, 209, 212–14, 216, 218, 221, 313, 316, 324, 338–9
credit history, 316, 335–7, 345, 373
credit unions, 332, 338

Dallas, 50
Decennial Census (US), 77
devaluation, 12–13, 263, 273, 335, 362
differential incorporation, 45
discrimination, 5–6, 12, 24, 26–7, 43, 45–6, 48–9, 57, 60–5, 98, 151, 154–5, 158, 162, 166, 168, 178, 209, 214, 219, 230–1, 252, 263–4, 276, 303, 330, 357–8, 360, 364–5, 369
District of Columbia, 306, 311, 314–15, 318–20, 323
diversity, 8, 14, 23, 25, 32–4, 44, 98, 122, 138–40, 146, 162, 168, 197, 208, 214, 252, 264, 266, 296, 308, 319–20, 356–7. *See also* variation

economic experience, 3–5, 10–12, 15–17, 98, 207, 209–15, 221, 281, 298–9, 360, 364
economic status, xiv

Edmonton CMA, 50
educational attainment, xiii, 12, 43, 45, 48, 58–9, 62, 85, 87, 150, 157, 164, 186–7, 215, 264–6, 269–70, 274–6, 356, 361, 365
earnings, 12–13, 31, 66, 69, 99, 102, 114, 157, 185, 208, 214–15, 218–19, 221, 227–33, 237–43, 247–55, 263, 361–2. *See also* wage
*ekub*, 316
entrepreneur, xv, 16–17, 131, 137, 187–8, 197–200, 207, 209, 219–21, 282–4, 298, 302–7, 312, 314–24, 328–32, 334–7, 339–40, 324, 344–7, 362, 367, 371–3
entrepreneurship, 16, 207, 219, 221, 282–4, 298, 302–7, 315–18, 321–4, 328–30, 332, 335–6, 344, 347, 371–7
Ethiopian Business Association, 317
Ethiopian Community Development Council, 317
Ethiopian immigrants, 314, 317, 320–3
ethnic assets, 330–2, 334, 344–5, 356, 373
ethnic banks, 332, 341–2, 373
ethnic economy, 137, 149, 154–5, 191, 284, 306, 308, 310–11, 317, 366
ethnic enclave, 9, 23, 34, 123, 126, 140, 142, 146, 148, 150–2, 154–9,

161–8, 178–80, 217, 220, 305, 366, 369–70
ethnic financial sector, 331–2, 334, 341, 345
ethnic niche, 264, 317
Europe, 25, 44
European, 6, 11, 17, 27–8, 43–6, 48–52, 55, 57–60, 62, 64–5, 99, 123, 135, 146–7, 151, 156, 168, 213, 228–9, 242, 244, 246–52, 255, 257, 263–4, 308, 333–4, 360
European advantage, 53
external forces, 43, 45

Family Class Migrants, 99
family ties, 106
Fairfax County, Virginia, 311
Falls Church, Virginia, 311, 314, 320, 322
family reunification, 10–11, 153, 308, 335–6
Filipino, 28
financial dynamics, 16–17, 332
financial institutions, 16, 63, 220–1, 318, 329–33, 337–40, 342, 344–6, 356
financial integration, 330, 333, 343–4
financial needs, 16–17, 329, 339, 344
financial services, 16, 317, 329, 334, 340
financing, 77, 298, 312, 317, 329–30, 335–6, 338, 344–5, 372–3
foreclosure, 61, 63
foreign bank, 332–4, 341
foreign-born, 44, 71
foreign-born Black, 28

gateway cities, 26, 50
gentrification, 7, 25, 29, 31, 121–3, 126–7, 130–3, 135–42
globalization, 9
Great Recession, the, 61, 63
Greenland, 44

headship rates, 78
heterolocal, 16, 303, 306, 366, 369, 372
hidden homelessness, 7, 29
Hispanic, 14–15, 27, 44–5, 60–2, 69, 72, 74, 77–8, 86–8, 93–4, 158, 220, 256, 265–6, 269–72, 275, 281–99, 302, 308, 311–12, 318–19, 321, 337, 339, 356–7, 360–2 *See also* Latino
Hispanic Chamber of Commerce, 318
homelessness, 25, 30
Home Mortgage Act, 63
homeownership, 5, 7, 24–5, 155
homeownership attainment, 70, 72
homeownership benefits, 47
homeownership rates, 43; inequality gap in, 55
household, 6, 26–31, 48, 51–2, 58–61, 63–4, 69–75, 77–80, 83–6, 88–94, 100–7, 110–17, 122, 129, 143, 155, 161–2, 164–5, 185–6, 189–90, 194, 201, 209, 212, 215–16, 218, 221, 232, 236, 253, 262, 265, 270, 276, 285, 309, 333, 361, 364, 366
household formation, 70, 72
housing, xiii–xv, 3–10, 17, 23–36, 43, 49–50, 52, 60–4, 69–78, 83, 85, 89–94, 98–103, 105, 107–16, 122, 129–30, 132–3, 132–43, 146–50, 152–5, 158, 162–4, 166–8, 176–85, 188–94, 196–201, 211, 218, 221, 265, 269–70, 289, 309–10, 339, 355–70
housing attainment, 90
housing affordability, 24–5, 30
housing careers, 7, 24–5, 28–9
housing experiences, 3, 24, 155, 167
housing inequalities, 6, 25–6
housing market, 25

housing preferences, 146, 162
housing trajectories, 6, 162
HSBC, 333, 338
Houston, 50
human capital, xiii, 10–13, 16–17, 71, 156, 209, 213, 215, 221, 262–4, 268, 272, 275–6, 305, 337, 343, 356, 361

identity, 126, 130, 142, 149, 153, 157, 168, 208, 214, 285, 293, 296, 298, 362, 366, 370, 372
immigrant business, 15–16, 220–1, 303–4, 306, 315, 317, 321, 329–30, 345–6, 356
immigrant entrepreneur, 16, 207, 220–1, 282–4, 302–7, 315–16, 318, 322–4, 328–32, 334, 336–7, 339–40, 342, 344–7, 372–3
immigrant gateway, 15–16, 50, 302, 332
immigrant housing trajectories, 196–9
immigrant housing types, 191–6
immigrant incorporation, 45
immigrant integration, xiii, xiv, xv, 3, 25
immigrant-owned, 16–17, 284, 298, 307, 311, 328, 335, 346
immigrant settlement, 23–4
immigrant suburbanization, 151
immigration, xii–xiv, 3–5, 10–11, 15–19, 25, 27, 32, 36–41, 44, 66–7, 70, 72, 75–6, 80, 95–7, 103, 106–7, 118, 120, 127, 141–3, 146, 149, 152–3, 157–9, 166–7, 169, 171–80, 191, 202–4, 207, 210–19, 258, 277, 280–4, 299, 314, 320–2, 326–7, 330, 333–4, 343–4, 349–50, 355–7, 364, 375, 377–8, 380–1
immigration policies, xv, 27, 44, 70, 146, 153, 158, 227, 344, 355–6, 358
immigration reform, xiii
income, 156
Indians, 28
inequalities, 6, 25–7, 357–8, 360, 364
inequality gap in homeownership rates, 55
institutional structures, 318, 368–9, 373
institutional ties, 107
integration, xiii–xv, 3–7, 10, 16, 23–5, 28, 30, 34, 36, 46, 69, 92, 115, 136, 146, 149, 151–2, 156–8, 163–4, 166–7, 176–7, 179–80, 201, 217, 220, 229, 252, 263, 303, 320, 330, 333, 343–5, 355–64, 368, 370, 372–3
intergenerational transfer, 47
internal characteristics, 43, 45
investor, 10–11, 191, 217–18, 232, 334–5, 342

labour market, xii, xv, 3, 71, 78, 99, 102, 113, 148, 151–3, 207, 209, 211, 213–18, 221, 227–9, 231–2, 240, 242, 251, 261, 263–4, 266, 269, 273–7, 298, 303, 330, 356
Language Access Act, 320
Latin America, 11, 16, 24–5, 69, 244, 246, 256, 263, 266–7, 269, 281–2, 290, 298, 308, 334
Latino, 15–16, 61, 70, 126, 220, 230, 281–99, 303, 308, 311, 316–20, 342, 357, 367, 370–1. *See also* Hispanic
Latino Economic Development Center, 318
lending circle, 17, 336
Little Portugal, 8, 26, 31–2, 123, 128, 135, 137–42

loan, 47, 60, 63, 178, 197, 216–17, 220, 316–18, 323, 336–40, 342–6, 372–3
Los Angeles, 8, 26, 30, 32, 35, 50

mainstream banks, 331, 334, 341–2, 345
Melanesia, 44
metropolitan areas, 77, 94
Mexican American, 220, 281, 283, 286, 288, 292–3, 296, 298, 371
Mexico, 44
Miami, 32, 50
Micronesia, 44
middleman minority thesis, 304
minority, 6, 8–9, 13, 24, 26–8, 30, 34, 45, 60, 69–76, 81, 83–8, 90–1, 93, 100, 116, 121, 126–7, 151, 153–7, 159–60, 167, 177–80, 191, 214, 227–42, 247, 250–7, 266, 304–5, 331–2, 339, 359–60, 363
Montgomery County, Maryland, 311, 320
Montreal, 13, 32, 50, 52–3, 98, 103, 111, 122, 139, 147, 149, 159, 230, 233–4, 238, 240–2, 331, 368
mortgage interest deduction, 47
mortgage lending, 25
motivation, 199, 207, 210–11, 219, 307, 367, 372
Mozambique, 129–3, 133, 141
multiculturalism, xiv

national origins quotas, 44
neoliberalism, xiv
network, 208–9, 211–12, 214–17, 220–1, 303, 305, 307, 315–17, 323
New York City, 24, 26, 32, 50, 126
New Zealand, 44
non-European immigrants, 43
non-Hispanic Whites, 45
non-pecuniary, 209, 213
non-White, xiv, 45, 60–3, 93, 113, 230, 232
normative values, xv
North America, 3–4, 6–8, 10–11, 17, 23–7, 30–3, 44, 146, 151–2, 156, 159, 166–8, 179, 200, 285, 290, 304–5, 332, 340, 355, 366

occupation, 156
occupational division, 309–10
Oceania, 44
official language, 231–2, 236, 253, 330, 343
Ottawa CMA, 50

paired tests method, 61
*pasanaco*, 316
place of birth, 46
place making and remaking, 362–4
policy implications, 32
Polynesia, 44
Portuguese, 8, 121, 123–5, 127–42
predatory lending, 63
Prince William County, Virginia, 320, 322

racialization, 148, 153, 355–8
racial discrimination in housing, 63
racialized minority, 71, 72
racially restrictive immigration policies, 44
racial minority, 24, 356–8
racism, 164
ranking of immigrant groups, expected and observed, 58
ratio of equality, 54
refugee, 7, 24, 26–7, 30, 33, 35, 47, 65, 98–101, 105–6, 112–13, 115–17, 129,

152, 217, 302, 308–9, 317, 356–7, 365–6, 368
relationship banking, 340–1, 345
remittance, 216, 221, 262
residential assimilation, 78
residential displacement, 121, 126, 132, 136
residential integration, 6
residential overcrowding, 113
residential segregation, 26–7
resources, 62, 113–14, 116, 134–5, 148, 154, 158, 208–10, 213, 221, 305–6, 315–16, 318, 330, 335–6, 346, 357, 368, 372
return migration, 178, 210, 344
returns, 209, 212–14, 218
risk, 24, 29–30, 61, 85–6, 89–90, 92, 162, 178, 209–12, 214–16, 217–21, 359–61, 372
rotating credit association, 304
rules, 28, 91, 208, 221, 331

St Pierre and Miquelon, 44
sanctuary cities, 332
San Diego, 50
San Francisco, 17, 26, 32, 50, 56, 329, 332–4, 339, 344, 373
saving, 63, 209, 212, 217–19, 316, 330, 336, 345
Scarborough, 32
segregation, 154
self-employed, 219, 273, 284, 308–10, 328–9, 339
self-employment, 10, 14, 61, 219, 261, 303–4, 307, 328, 334, 372
settlement process, 6
Silver Spring, Maryland, 314–15, 322
small business, 12, 219, 289, 304, 309, 317–18, 322, 324, 328–30, 336–9, 341, 343, 356, 372–3
Small Business Resources Center, 318
social capital, xiii, 99, 148, 159, 264, 276, 304, 330–1, 344, 357–9, 361–3, 365–6, 371
social housing, 27
social networks, 7–8, 25, 29, 157
social ties, 107
sociocommerscape, 303, 306, 314–15
South America, 44
South Asians, 7, 8 , 30, 34, 159, 162
South Parkdale (Toronto), 122–3
Sri Lankan Tamil, 30, 99
startup, 318, 336, 345
Statistics Canada, 65
strategy, 29, 101, 134, 207–9, 210, 212, 216, 218–19, 221, 302, 304, 311, 316, 319
structure, 208–9, 210, 217, 221
suburbs, 8, 30, 146, 152–3
suburban enclaves, 34
suburbanization, 34
suburbanization of immigrants, 8–9
suburban settlements, 25, 32

Takoma Park, Maryland, 314
Tamils, 7, 26, 30–2
Toronto, 7, 8, 13, 26, 30–4, 50, 52–4, 70, 98–100, 102–16, 121–5, 127–31, 135, 138, 140–2, 147, 149, 159–60, 230, 233–5, 239–42, 254, 285, 293, 304, 331, 333, 359, 365–6, 370 287g, 320, 322

unauthorized immigrants, 24, 321–2
uncertainty, 214, 217. *See also* risk
underemployment, 12–13, 209, 261–77, 303–4, 361
United States, 3, 5, 23, 29, 32, 43–4
United States, amendments to 1965 Immigration Act, 44

United States Department of Housing and Urban Development, 61, 63
United States Fair Housing Act of 1968, 63
U Street, 314–15

vacancy rates, 26
Vancouver, 13, 17, 26, 32, 50, 52–3, 94, 98, 113, 122, 148–9, 159, 230, 232, 234, 236, 239–42, 254, 330, 332–8, 340, 344, 346
variation, 35, 52, 55, 72–3, 77, 84, 90, 93, 208–9, 215, 217, 240, 253, 263, 285, 291
visible minority, 8, 13, 30, 34, 45, 69, 75, 86–8, 100, 116, 151, 153–4, 159–60, 167, 214, 227–37, 240–2, 247, 250–2, 254, 257, 359

wage, xiii, 11–12, 198, 207, 211, 215, 230–3, 247, 310. *See also* earnings
Washington, DC, 16, 50, 55–6, 94, 284–5, 299, 302, 314–15, 322, 336, 371
Washington Metropolitan Area, 312–13, 315, 322–3
West Central Toronto, 121, 123, 127–30, 138, 141
white, 13–14, 17, 28–9, 45, 49–50, 60–2, 64, 69, 71–2, 74–8, 81–8, 90–1, 93, 126, 139, 157, 165, 168, 191, 214–15, 227–8, 230–4, 237–42, 245–6, 251, 257, 265–7, 269–72, 274–6, 285, 360–6, 370

www.ingramcontent.com/pod-product-compliance
Lightning Source LLC
LaVergne TN
LVHW090801070826
844660LV00022B/1049

*9781442628380*